Research Lab. Copy

Green Radio Communication Networks

The importance of reducing energy costs, reducing CO_2 emissions, and protecting the environment are leading to an increased focus on green, energy-efficient approaches to the design of next-generation wireless networks. Presenting state-of-the-art research on green radio communications and networking technology by leaders in the field, this book is invaluable for researchers and professionals working in wireless communication.

Summarizing existing and ongoing research, the book explores communication architectures and models, physical communications techniques, base station power-management techniques, wireless access techniques for green radio networks, and green radio test-bed, experimental results, and standardization activities. Throughout, theoretical results are blended with practical insights and coverage of deployment issues. It serves as a one-stop reference for key concepts and design techniques for energy-efficient communications and networking, and provides information essential for the design of future-generation cellular wireless systems.

Ekram Hossain is a Professor in the Department of Electrical and Computer Engineering at the University of Manitoba, Canada, where his current research interests lie in the design, analysis and optimization of wireless/mobile communications networks, smart grid communications, cognitive, and green radio systems. He has received several awards including the University of Manitoba Merit Award in 2010 (for Research and Scholarly Activities) and the 2011 IEEE Communications Society Fred Ellersick Prize Paper Award.

Vijay K. Bhargava is a Professor in the Department of Electrical and Computer Engineering at the University of British Columbia, Canada. He has served on the Board of Governors of the IEEE Information Theory Society and the IEEE Communications Society, and was President of the IEEE Information Theory Society. He is currently serving a two year term as the President of the IEEE Communications Society.

Gerhard P. Fettweis is the Vodafone Chair Professor at Technische Universität Dresden, with 20 companies from around the world currently sponsoring his research on wireless transmission and chip design. An IEEE Fellow, he runs the world's largest cellular research test beds, coordinated the EASY-C project, and has received numerous awards. He began his career at IBM Research and has since developed nine start-up companies.

Green Radio Communication Networks

Edited by

EKRAM HOSSAIN
University of Manitoba, Canada

VIJAY K. BHARGAVA
University of British Columbia, Canada

GERHARD P. FETTWEIS
Technische Universität Dresden, Germany

CAMBRIDGE UNIVERSITY PRESS
Cambridge, New York, Melbourne, Madrid, Cape Town,
Singapore, São Paulo, Delhi, Mexico City

Cambridge University Press
The Edinburgh Building, Cambridge CB2 8RU, UK

Published in the United States of America by Cambridge University Press, New York

www.cambridge.org
Information on this title: www.cambridge.org/9781107017542

© Cambridge University Press 2012

This publication is in copyright. Subject to statutory exception
and to the provisions of relevant collective licensing agreements,
no reproduction of any part may take place without the written
permission of Cambridge University Press.

First published 2012

Printed in the United Kingdom at the University Press, Cambridge

A catalogue record for this publication is available from the British Library

Library of Congress Cataloguing in Publication data
Hossain, Ekram, 1971–
    Green radio communication networks / Ekram Hossain, Vijay K. Bhargava, Gerhard P. Fettweis.
        p.   cm.
    Includes bibliographical references and index.
    ISBN 978-1-107-01754-2 (hardback)
    1. Wireless communication systems – Environmental aspects.   2. Wireless communication
systems – Energy consumption.   I. Bhargava, Vijay K., 1948–   II. Fettweis, Gerhard P.   III. Title.
    TK5103.2.H675 2012
    621.384028'6–dc23   2012009106

ISBN 978-1-107-01754-2 Hardback

Cambridge University Press has no responsibility for the persistence or
accuracy of URLs for external or third-party internet websites referred to
in this publication, and does not guarantee that any content on such websites is,
or will remain, accurate or appropriate.

Vijay Bhargava
July 2012

For
our families

Contents

List of contributors		*page*	xvi
Preface			xxi

Part I Communication architectures and models for green radio networks — 1

1 Fundamental trade-offs on the design of green radio networks — 3
Yan Chen, Shunqing Zhang, and Shugong Xu

1.1	Introduction	3
1.2	Insight from Shannon's capacity formula	5
	1.2.1 SE–EE trade-off	6
	1.2.2 BW–PW trade-off	7
	1.2.3 DL–PW trade-off	8
	1.2.4 DE–EE trade-off	10
	1.2.5 Summary	10
1.3	Impact of practical constraints	12
1.4	Latest research and future directions	14
	1.4.1 SE–EE trade-off	14
	1.4.2 BW–PW trade-off	16
	1.4.3 DL–PW trade-off	17
	1.4.4 DE–EE trade-off	18
1.5	Conclusion	20

2 Algorithms for energy-harvesting wireless networks — 24
Vinod Sharma, Utpal Mukherji, and Vinay Joseph

2.1	Introduction	24
2.2	Energy-harvesting technologies	26
2.3	Point-to-point channel	28
	2.3.1 Model and notation	28
	2.3.2 Stability	29
	2.3.3 Delay optimal policies	30
	2.3.4 Generalizations	31
	2.3.5 Simulations	32

		2.3.6	Model with sleep option	34
		2.3.7	Fundamental limits of transmission	37
	2.4	MAC policies		39
		2.4.1	Orthogonal channels	40
		2.4.2	Opportunistic scheduling for fading channels: orthogonal channels	40
		2.4.3	Opportunistic scheduling for fading channels: CSMA	42
		2.4.4	Simulations for MAC protocols	42
	2.5	Multi-hop networks		44
		2.5.1	Problem formulation	45
		2.5.2	Simulations	48
	2.6	Conclusion		50

3 PHY and MAC layer optimization for energy-harvesting wireless networks   53

Neelesh B. Mehta and Chandra R. Murthy

	3.1	Introduction		53
	3.2	Physical layer design		55
		3.2.1	No CSI at transmitter and retransmissions	55
		3.2.2	Power management with channel state information	61
		3.2.3	Simulation results	66
	3.3	Cross-layer implications in a multi-node network		67
		3.3.1	Multiple access selection algorithms	69
		3.3.2	Energy harvesting, storage, and usage model	70
		3.3.3	Energy neutrality implication	70
		3.3.4	Performance analysis	70
		3.3.5	Numerical results	73
	3.4	Conclusion		74

4 Mechanical relaying techniques in cellular wireless networks   78

Panayiotis Kolios, Vasilis Friderikos, and Katerina Papadaki

	4.1	Introduction		78
	4.2	Background		79
		4.2.1	DTN architecture	79
		4.2.2	Routing in DTNs	80
	4.3	Mechanical relaying		81
		4.3.1	Mobile internet traffic mix	83
		4.3.2	Mechanical relaying strategies	86
	4.4	Real-world measurements		89
	4.5	Related standardization efforts		92
	4.6	Conclusion		93

Part II Physical communications techniques for green radio networks   97

5   Green modulation and coding schemes in energy-constrained wireless networks   99
Jamshid Abouei, Konstantinos N. Plataniotis, and Subbarayan Pasupathy

5.1	Introduction	99
5.2	System model and assumptions	101
	5.2.1 Performance metric	101
	5.2.2 Channel model	102
5.3	Energy consumption of uncoded scheme	103
	5.3.1 M-ary FSK	103
	5.3.2 M-ary QAM	106
	5.3.3 Offset-QPSK	108
	5.3.4 Numerical evaluations	110
5.4	Energy-consumption analysis of LT coded modulation	113
	5.4.1 Energy efficiency of coded system	114
	5.4.2 Energy optimality of LT codes	116
5.5	Numerical results	118
	5.5.1 Experimental setup	118
	5.5.2 Optimal configuration	119
5.6	Conclusion	122

6   Cooperative techniques for energy-efficient wireless communications   125
Osama Amin, Sara Bavarian, and Lutz Lampe

6.1	Introduction	125
6.2	Energy-efficiency metrics for wireless networks	126
	6.2.1 Instantaneous EE metrics	128
	6.2.2 Average EE metrics	129
6.3	Energy-efficient cooperative networks	130
	6.3.1 Single relay cooperative network	131
	6.3.2 Multi-relay cooperative network	136
	6.3.3 Multi-hop cooperative network	137
6.4	Optimizing the EE performance of cooperative networks	139
	6.4.1 Modulation constellation size	139
	6.4.2 Power allocation	141
6.5	Energy efficiency in cooperative base stations	143
6.6	Conclusion	146

7   Effect of cooperation and network coding on energy efficiency of wireless transmissions   150
Nof Abuzainab and Anthony Ephremides

7.1	Introduction	150
7.2	Relay cooperation in single link wireless transmissions	152

		7.2.1	System model	152
	7.3		User cooperation in wireless multicast transmissions	155

		7.2.1	System model	152
		7.2.2	Cooperation protocols	153
	7.3	User cooperation in wireless multicast transmissions		155
		7.3.1	System model	155
		7.3.2	Cooperation protocols	156
	7.4	Energy-cost minimization		158
	7.5	Stable throughput computation		158
	7.6	Performance evaluation		159
		7.6.1	Relay cooperation	159
		7.6.2	User cooperation	161
	7.7	Conclusion		162

Part III Base station power-management techniques for green radio networks   165

8   Opportunistic spectrum and load management for green radio networks   167
Oliver Holland, Christian Facchini, A. Hamid Aghvami, Orlando Cabral, and Fernando Velez

	8.1	Introduction		167
	8.2	Opportunistic spectrum and load management concepts		169
		8.2.1	Opportunistic load management to power down radio network equipment	169
		8.2.2	Opportunistic spectrum management to improve propagation characteristics	171
		8.2.3	Power saving by channel bandwidth increase or better bandwidth balancing	173
	8.3	Assessment of power-saving potential		174
		8.3.1	Example reflecting GSM networks	174
		8.3.2	Example reflecting LTE networks	180
		8.3.3	Example reflecting HSDPA networks	185
		8.3.4	Power saving by channel bandwidth increase or better bandwidth balancing	187
	8.4	Conclusion		188

9   Energy-saving techniques in cellular wireless base stations   190
Tao Chen, Honggang Zhang, Yang Yang, and Kari Horneman

	9.1	Introduction		190
	9.2	Energy-consumption model of RBS		191
	9.3	EE metric		192
	9.4	RBS energy-saving methods		194
		9.4.1	Time-domain approaches	195
		9.4.2	Frequency-domain approaches	196
		9.4.3	Spatial-domain approaches	197
		9.4.4	Performance comparison	198

	9.5	Layered structure for energy saving	199
		9.5.1 System model and assumptions	199
		9.5.2 Energy-consumption model of RBS	200
		9.5.3 Energy-aware handover mechanism	201
		9.5.4 Simulation study	203
	9.6	Conclusion	206

10 Power management for base stations in a smart grid environment — 209
Xiao Lu, Dusit Niyato, and Ping Wang

10.1	Introduction	209
10.2	Power management for wireless base station	210
	10.2.1 Green communications in centralized wireless networks	210
	10.2.2 Approaches for power management in a base station	212
	10.2.3 Open research issues	215
10.3	Power-consumption model for a base station	216
	10.3.1 Components of a base station	216
	10.3.2 Assumptions and power-consumption model for a macro base station	218
	10.3.3 Assumptions and power-consumption model for a micro base station	218
10.4	Optimization of power management in a smart grid environment	220
	10.4.1 System model	220
	10.4.2 Demand-response for base station in smart grid	222
	10.4.3 Optimization formulation for power management	223
	10.4.4 Performance evaluation	226
10.5	Conclusion	230

11 Cooperative multicell processing techniques for energy-efficient cellular wireless communications — 236
Mohammad Reza Nakhai, Tuan Anh Le, Auon Muhammad Akhtar, and Oliver Holland

11.1	Introduction	236
11.2	Cell splitting	238
11.3	A multicell processing model	239
	11.3.1 Transmission and channel model	239
	11.3.2 User-position-aware multicell processing	242
11.4	Multicell beamforming strategies	243
	11.4.1 MBF using instantaneous CSIT	243
	11.4.2 MBF using second-order statistical CSIT	245
	11.4.3 An iterative MBF using second-order statistical CSIT	246
11.5	Coordinated beamforming	248
11.6	Backhaul protocol	250
	11.6.1 A protocol for information circulation in the backhaul	250

		11.6.2	Power calculation for the ring protocol	251
		11.6.3	An effective sum-rate	252
	11.7	Performance evaluation		252
		11.7.1	Performance evaluation under ideal backhaul	252
		11.7.2	Performance evaluation under limited backhaul	254
	11.8	Cooperative routing		255
		11.8.1	Power-aware cooperative routing algorithm	256
	11.9	Conclusion		258

Part IV Wireless access techniques for green radio networks — 261

12 Cross-layer design of adaptive packet scheduling for green radio networks — 263
Ashok Karmokar, Alagan Anpalagan, and Ekram Hossain

	12.1	Introduction	263
	12.2	Related work	264
	12.3	Importance of cross-layer optimized design	265
	12.4	Why cross-layer adaptation is important for green radio networks	266
	12.5	Cross-layer interactions, models, and actions	267
	12.6	Cross-layer vs. single-layer adaptation techniques	271
	12.7	How to solve the cross-layer design problem	273
	12.8	Power savings in the cross-layer optimized system	276
	12.9	Other literature on energy-efficient cross-layer techniques	278
	12.10	Challenges and future directions	281
	12.11	Conclusion	282

13 Energy-efficient relaying for cooperative cellular wireless networks — 286
Yifei Wei, Mei Song, and F. Richard Yu

	13.1	Introduction		286
	13.2	Energy saving in cellular wireless networks		288
		13.2.1	Energy-saving techniques	288
		13.2.2	Energy-efficiency criteria	289
	13.3	Energy-efficient cooperative communication based on selective relay		290
		13.3.1	Relay selection schemes	291
	13.4	System model for the relay selection problem		293
		13.4.1	S2R channel	294
		13.4.2	R2D channel	294
		13.4.3	Energy model	295
		13.4.4	Objectives	296
	13.5	Problem formulation		296
		13.5.1	Relay states	296
		13.5.2	System reward	297
		13.5.3	Solution to the restless bandit problem	298

13.6	Distributed relay selection scheme		300
	13.6.1 Available relay candidates		300
	13.6.2 Relay selection process		301
	13.6.3 Cost evaluation		302
13.7	Simulation results and discussions		302
	13.7.1 System reward		303
	13.7.2 Error propagation mitigation		303
	13.7.3 Spectral efficiency improvement		305
	13.7.4 Network lifetime		305
13.8	Conclusion		306

14 Energy performance in TDD-CDMA multi-hop cellular networks — 309
Hoang Thanh Long, Xue Jun Li, and Peter Han Joo Chong

14.1	Introduction	309
14.2	Structure of relay stations and power consumption	309
	14.2.1 Random relay station (RRS) structure	311
14.3	Time-slot allocation schemes	312
	14.3.1 Fixed time-slot allocation (FTSA)	313
	14.3.2 Dynamic time-slot allocation (DTSA)	313
	14.3.3 Multi-link fixed time-slot allocation (ML-FTSA)	314
	14.3.4 Multi-link dynamic time-slot allocation (ML-DTSA)	315
14.4	System model	315
14.5	Simulation results and discussions	317
	14.5.1 Blocking and dropping probabilities for high and low data rate traffic	320
	14.5.2 Energy consumption for single-hop and multi-hop transmission using FRS	322
	14.5.3 Energy consumption for RRS structure	325
14.6	Conclusion	328

15 Resource allocation for green communication in relay-based cellular networks — 331
Umesh Phuyal, Satish C. Jha, and Vijay K. Bhargava

15.1	Introduction	331
15.2	Enabling green communication in cellular wireless networks	332
	15.2.1 Component level	332
	15.2.2 Equipment level	332
	15.2.3 Network level	332
	15.2.4 Computational complexity versus transmit-power-saving	333
15.3	Relay-based green CCN	333
	15.3.1 Implementation issues and challenges	334
	15.3.2 Advantages of fixed relay-based CCN	336
	15.3.3 Green performance metrics for resource allocation	337

15.4	Resource-allocation schemes for CCN: a brief survey		337
	15.4.1	Throughput maximization schemes	338
	15.4.2	QoS-aware transmit power minimization schemes	338
	15.4.3	Energy-aware green schemes	339
15.5	Design of a green power allocation scheme		339
	15.5.1	System model	340
	15.5.2	Green power allocation scheme	342
	15.5.3	Performance analysis of GPA scheme	344
	15.5.4	Adaptive interrupted transmission	345
	15.5.5	Simulation results	345
15.6	Green performance versus system capacity		351
	15.6.1	Performance analysis	352
15.7	Conclusion		354

Part V Green radio test-bed, experimental results, and standardization activities   357

16 How much energy is needed to run a wireless network?   359

Gunther Auer, Vito Giannini, István Gódor, Oliver Blume, Albrecht Fehske, Jose Alonso Rubio, Pål Frenger, Magnus Olsson, Dario Sabella, Manuel J. Gonzalez, Muhammad Ali Imran, and Claude Desset

16.1	Introduction		359
16.2	Energy-efficiency evaluation framework (E^3F)		360
	16.2.1	Small-scale, short-term system-level evaluations	361
	16.2.2	Global E^3F	361
16.3	Power model		363
	16.3.1	Base station power-consumption breakdown	363
	16.3.2	BS power consumption at variable load	366
16.4	Traffic model		367
	16.4.1	Deployment areas of Europe	367
	16.4.2	Long-term large-scale traffic models	368
	16.4.3	Statistical short-term traffic models	372
16.5	Green metrics		372
	16.5.1	Efficiency metrics vs. consumption metrics	373
	16.5.2	Energy-consumption metrics in cellular networks	374
16.6	Case study: energy efficiency of LTE		375
	16.6.1	Assessment methodology	375
	16.6.2	Small-scale short-term evaluations	376
	16.6.3	Large-scale long-term evaluations	377
16.7	LTE technology potential in real deployments		377
	16.7.1	Global radio access networks	378
	16.7.2	LTE system evaluation	380
	16.7.3	Evolution of LTE energy-efficiency over time	380
16.8	Fundamental challenges and future potential		381
16.9	Conclusion		382

17	**Standardization, fora, and joint industrial projects on green radio networks**		**385**
	Alberto Conte, Hakon Helmers, and Philippe Sehier		
	17.1	Introduction	385
	17.2	Standardization fora	386
		17.2.1 ETSI	387
		17.2.2 3GPP	389
		17.2.3 TIA and 3GPP2	394
		17.2.4 ATIS	395
		17.2.5 IETF/EMAN	395
		17.2.6 CCSA	396
	17.3	Consortium and joint projects	396
		17.3.1 NGMN alliance	396
		17.3.2 FP7 EARTH project	398
		17.3.3 GreenTouch initiative	400
	17.4	Synthesis and classification of energy-saving solutions for wireless networks	403
		17.4.1 Technology and component level	403
		17.4.2 Base station adaptation to traffic load	404
		17.4.3 Network architecture	405
		17.4.4 Heterogeneous networks	405
		17.4.5 Air interface	406
		17.4.6 Dynamic NW adaptation to traffic load	407
	17.5	Conclusion	407
	Index		409

Contributors

Nof Abuzainab
University of Maryland, USA

A. Hamid Aghvami
King's College London, UK

Auon Muhammad Akhtar
King's College London, UK

Osama Amin
University of British Columbia, Canada

Alagan Anpalagan
Ryerson University, Canada

Gunther Auer
DOCOMO Euro-Labs, Germany

Sara Bavarian
University of British Columbia, Canada

Vijay K. Bhargava
University of British Columbia, Canada

Oliver Blume
Alcatel-Lucent Bell Labs, Germany

Orlando Cabral
King's College London, UK

Tao Chen
VTT Technical Research Center, Finland

Yan Chen
Huawei Technologies, China

Peter Han Joo Chong
Nanyang Technological University, Singapore

Alberto Conte
Alcatel-Lucent Bell Labs, France

List of contributors

Claude Desset
IMEC, Belgium

Anthony Ephremides
University of Maryland, USA

Christian Facchini
King's College London, UK

Albrecht Fehske
Technische Universität Dresden, Germany

Pal Frenger
Ericsson Research, Sweden

Vasilis Friderikos
King's College London, UK

Vito Giannini
IMEC, Belgium

Istvan Godor
Ericsson Research, Hungary

Manuel J. Gonzalez
Technologies of Telecommunication and Information (TTI), Spain

Hako Helmers
Alcatel-Lucent Bell Labs, France

Oliver Holland
King's College London, UK

Kari Horneman
Nokia Siemens Networks, Finland

Ekram Hossain
University of Manitoba, Canada

Muhammad Ali Imran
University of Surrey, UK

Satish Kumar Jha
University of British Columbia, Canada

Ashok Karmakar
Ryerson University, Canada

Panayiotis Kolios
King's College London, UK

Lutz Lampe
University of British Columbia, Canada

Tuan Anh Le
King's College London, UK

Xue Jun Li
Nanyang Technological University, Singapore

Hoang Thanh Long
Nanyang Technological University, Singapore

Xiao Lu
Nanyang Technological University, Singapore

Neelesh B. Mehta
Indian Institute of Science, India

Utpal Mukherji
Indian Institute of Science, India

Chandra R. Murthy
Indian Institute of Science, India

Mohammad Reza Nakhai
King's College London, UK

Dusit Niyato
Nanyang Technological University, Singapore

Magnus Olsson
Ericsson Research, Sweden

Katerina Papadaki
King's College London, UK

Umesh Phuyal
University of British Columbia, Canada

Jose Alonso Rubio
Ericsson Research, Sweden

Dario Sabella
Telecom Italia, Italy

Vinod Sharma
Indian Institute of Science, India

Philippe Sehier
Alcatel-Lucent Bell Labs, France

Mei Song
Carleton University, Canada

Fernando Velez
King's College London, UK

Ping Wang
Nanyang Technological University, Singapore

Yifei Wei
Carleton University, Canada

Shugong Xu
Huawei Technologies, China

Yang Yang
Shanghai Research Center for Wireless Communications, China

F. Richard Yu
Carleton University, Canada

Honggang Zhang
Zhejiang University, China

Shunqing Zhang
Huawei Technologies, China

Preface

A brief journey through "Green Radio Communication Networks"

Currently, the information and communications technology (ICT) industry sector accounts for about 2–6% of the energy consumption worldwide, and a significant portion of this is contributed by the wireless and mobile communications industry. With the proliferation of wireless data applications, wireless technology continues to increase worldwide at an unprecedented growth rate. This has resulted in an increased number of installed base stations and higher demand on power grids and device power usage, causing an increased carbon footprint worldwide. Current wireless industry therefore needs to embrace eco-friendly green communication technologies at different levels – from components, circuits, and devices to protocols, systems, and networks. Since the rate of improvement in power efficiency of hardware devices lags data traffic growth in both the radio access and core networks, network scaling will be increasingly tied to energy consumption in future wireless protocols, systems, and networks. Hence, it is crucial to develop green technologies for wireless systems and networks to improve energy efficiency and reduce CO_2 emissions. Again, from the perspective of network operators, energy is a significant portion of their OPEX (Operational Expenses). Therefore, green radio technologies will help to reduce the operating costs of wireless networks.

Green ICT has become a critical agenda item around the world. In this context, many organizations and standard bodies throughout the world including the European Commission (EC), US Environmental Protection Agency, US Department of Energy, ISO, IEC, ITU-T, ETSI, ATIS, and the IEEE are working towards the vision of green communication networks. In particular, the EC is developing a comprehensive code of conduct on the energy consumption of broadband equipment. The IEEE is developing energy-efficient protocols for Ethernet (i.e. IEEE P802.3az protocol). There are many ongoing projects on green communication networks. For example, EU FP7 projects EARTH (Energy-Aware Radio and Network Technologies) and C2POWER (Cognitive Radio and Cooperative Strategies for Power Saving) focus on developing energy-efficient mobile communications systems. The Mobile VCE Green Radio project aims at developing new green radio architectures and radio techniques to reduce the overall energy consumption. GreenTouch, which is a consortium of ICT industry, academia, and non-governmental research experts, has an ambitious goal of improving the energy efficiency of the ICT industry by three orders of magnitude by 2015 compared to that in 2010. Japan's Green-IT project aims to develop energy consumption metrics and energy efficiency standards for networking equipments. Some mobile network operators have already set targets to reduce their carbon emissions significantly within the next ten years.

This book provides a comprehensive treatment of the state of the art of existing and on-going research on energy efficient wireless/mobile communications and networking techniques with an emphasis on cellular wireless networks. It consists of articles covering different aspects of green cellular radio communications and networking issues that include the following: architecture issues and performance models for green radio networks including energy-harvesting wireless networks; physical communication techniques for green radio, including novel modulation and coding techniques and joint physical (PHY) and medium access control (MAC) optimized techniques; dynamic power-management/energy-conservation techniques for base stations in cellular wireless networks; relaying and user cooperation techniques and energy-cognizant wireless protocols (e.g. for scheduling, dynamic power management) for green radio communications; standardization initiatives, test-beds, prototypes, practical systems and case studies.

This book contains 17 chapters which are organized into 5 parts. A brief account of each chapter in each of these parts is given below.

Part I: Communication architectures and models for green radio networks

From the perspective of green wireless networks, it is necessary to develop a clear understanding of energy consumption in current networks and the network elements, base sites, and mobiles, and to determine the best backhaul strategy for a given architecture. Different trade-offs involved in the design of green cellular systems need to be understood considering practical system aspects. It is important to determine what is the optimum deployment scenario for a wide-area network given a clearly defined energy-efficiency metric. An emerging paradigm for green wireless networks is the concept of energy harvesting. Analysis and modeling of green wireless networks based on energy harvesting is therefore becoming increasingly important.

In *Chapter 1*, Chen, Zhang, and Xu focus on a fundamental framework for green radio research and propose four fundamental trade-offs to construct this framework. These trade-offs are: (i) spectrum efficiency–energy efficiency (SE–EE) trade-off, (ii) bandwidth–power (BW–PW) trade-off, (iii) delay–power (DL–PW) trade-off, and (iv) deployment efficiency–energy efficiency (DE–EE) trade-off. The authors illustrate these trade-offs for point-to-point communications predicted by the Shannons capacity formula, which gives a set of monotonically decreasing curves for each of the fundamental trade-offs. In practical systems, network deployment and operation cost as well as the non-linear efficiency of the power amplifier and the processing power and circuit power need to be considered. With considerations of these issues, the trade-off relations usually deviate from the simple monotonic curves derived from Shannon's formula, which bring a new design philosophy for green radio networks. The authors review the current state of the investigation on these trade-offs and also outline a number of open research issues.

In *Chapter 2*, Sharma, Mukherji, and Joseph focus on modeling and analysis of an energy-harvesting green wireless network. First, the authors consider a point-to-point channel in an energy-harvesting communication system. The harvested energy is stored in a battery (energy queue) and the data to be transmitted is stored in a data buffer. The necessary condition for the stability of the data queue is obtained and a throughput optimal transmission policy is proposed when the energy is spent only in transmission. Also, a delay-optimal transmission policy is proposed that minimizes the average delay. Next, a more realistic case is considered with channel fading when the energy is also spent in processing and other activities and there may be leakage in the battery storing the energy. Also, the transmission policies are modeled considering the sleep and wake-up mode of an energy-harvesting node. Subsequently, the Shannon capacity of a point-to-point additive white Gaussian channel (AWGN) is obtained for an energy-harvesting transmitter. Second, the authors develop the transmission policies for a multiple access scenario. Third, the authors model and analyze the problem of jointly optimizing power control, routing, and scheduling policies for a multi-hop network with energy-harvesting nodes.

In *Chapter 3*, Mehta and Murthy study the implications of energy harvesting on the design and optimization of the physical (PHY) and medium access control (MAC) layers. In particular, the authors focus on the transmission power control at the physical layer for a single-hop communication scenario, and the interactions among multiple energy-harvesting relay nodes in a two-hop communications scenario. The primary design focus of PHY and MAC layers is to judiciously utilize all the harvested energy and ensure that energy is available for consumption when required. Other design objectives are energy-conservation and spectral-efficiency maximization. The authors investigate the effects of several important factors such as the energy-harvesting profile, availability or unavailability of channel state information, and energy-storage capability on the design of both single-hop, and relay-based two-hop cooperative communications.

In *Chapter 4*, Kolios, Friderikos, and Papadaki describe the concept of mechanical relaying and outline its benefits in cellular wireless networks. In mechanical relaying (MR) mobile terminals are entitled to store and carry the information messages while in transit and forward the data to the base station only when at favorable locations within the cell coverage area. Due to this store-carry-and-forward operation, significant gains in energy consumption can be attained by utilizing the elasticity of a plethora of different Internet applications (such as adaptive progressive video download, file transfers, software/firmware updates over the air (OTA), and RSS feeds). While intrinsically a delay-tolerant networking scheme, mechanical relaying can in fact boost the cellular system performance at no expense to the perceived user experience. The authors outline the deployment challenges of mechanical relaying in current and emerging mobile networks, open-ended research problems, and future avenues of research in this area.

Part II: Physical communications techniques for green radio networks

Future green radio networks will need to support multimedia data services at two or three orders of magnitude lower transmission power than currently used. This will of course require energy-efficient transmission and modulation techniques. More importantly, a holistic and system-wide design of the system that exploits the cross-layer interactions will be required.

In *Chapter 5*, Abouei, Plataniotis, and Pasupathy study the energy efficiency of some popular modulation schemes for energy-constrained wireless networks in fading channels. The authors demonstrate that the non-coherent M-ary frequency-shift keying (NC-MFSK) provides superior energy-efficiency performance in short-range wireless networks when compared with other sinusoidal carrier-based modulations such as M-ary quadrature amplitude modulation (MQAM), differential offset quadrature phase-shift keying (OQPSK), and coherent MFSK. Also, the authors analyze the energy efficiency of Luby transform (LT)-coded MFSK modulation when compared to classical BCH and convolutional-coded modulation as well as uncoded modulation. The LT-coded MFSK scheme provides higher energy efficiency over other uncoded and coded schemes due to the flexibility to adjust its rate according to the channel condition. The authors conclude that LT-coded MFSK modulation is a candidate green modulation and coding scheme for energy-constrained wireless networks.

In *Chapter 6*, Amin, Bavarian, and Lampe focus on the cooperative communications techniques for energy efficiency in cellular wireless networks. The authors first introduce the instantaneous and average energy-efficiency metrics that consider both the transmission energy and the transceiver system (consisting of analog and digital circuits) energy along with the data rate of transmission. The average energy efficiency of a single-relay cooperative communication system is evaluated considering selective decode-and-forward, incremental decode-and-forward, amplify-and-forward, and incremental amplify-and-forward-based relaying strategies. The authors also demonstrate how the gain in energy saving in a single-relay network can be improved through optimizing the modulation constellation size and the power allocation at the source and the relay under an average error rate constraint. For a multi-relay system, the authors also investigate the effect of relay selection and also the number of hops (in a multi-hop cooperative network) on the energy-efficiency performance. To this end, the authors discuss the base station cooperation technique, namely, the coordinated multipoint (CoMP) technique to improve the system-wide energy efficiency in cellular wireless systems.

In *Chapter 7*, Abuzainab and Ephremides focus on the energy efficiency of different physical and network layer cooperative techniques for two wireless transmission models in fading channels. The first model considers that a relay is used to assist the source node to deliver its data to the destination node. The second model considers multicast transmissions from the source node to two destination nodes and in this case user cooperation is utilized. That is, the destination node that first receives the data successfully can assist the source in transmitting the data to the other destination. Alamouti coding is used in the physical layer, while random network coding is used in the network layer. For both the transmission models, the energy cost is defined as the expected energy spent per successfully delivered packet. Simulation results show that with proper selection of the coding parameter, random network coding-based cooperative transmission technique achieves better performance than automatic-repeat request (ARQ)-based cooperative technique even when it is enhanced with Alamouti coding. Also, further improvements in the performance are achieved when random network coding is used combined with Alamouti coding. The results also show that the performance of user cooperation depends on the channel quality between the different nodes in the network.

Part III: Base station power-management techniques for green radio networks

For green radio communication networks, it is essential to develop techniques to achieve significant improvements in the overall efficiency for base stations, which is measured as radio frequency (RF) power out to total input power, and techniques that will reduce the required RF output power required from the base station while still maintaining the required quality-of-service (QoS). When a base station's energy supply is derived from renewable energy sources in a smart power grid, it is important to determine how this would be best used for communications. It will be necessary to develop sleep mechanisms that deliver substantial reductions in power consumption for base stations with no loads and techniques that allow power consumption to scale with load. Also, multi-cell processing techniques based on the cooperation among base stations can reduce the energy consumption at the base stations.

In *Chapter 8*, Holland et al. investigate the concepts of opportunistic spectrum and load management across multiple frequency bands (owned by an operator or a group of operators) to reduce the power consumption of base stations while satisfying the QoS requirements in the network. In particular, the authors focus on concepts such as powering down radio network equipments (i.e. base stations) using particular frequency bands by reallocating traffic loads to other bands at times of low load, and opportunistic spectrum usage to exploit the propagation characteristics of spectrum bands and reduce necessary transmission power. Using simulations of GSM, HSDPA, and LTE networks, the authors demonstrate the power savings achievable through these concepts. However, there is a tradeoff between the power saving and network capacity improvement.

In *Chapter 9*, Lu, Niyato, and Wang consider the problem of power management for base stations with renewable power sources in a smart grid environment. With the demand-response (DR) and demand-side management (DSM) features in smart grids, base stations powered by the smart grids can reduce the cost of power consumption by using an adaptive power-management method. The authors provide an overview of the existing approaches of power management for wireless base stations, which include base station power control through beamforming, base station assignment based on the dynamic connectivity patterns between mobile units and base stations, smart mode switching, and cooperative relaying. The authors propose an adaptive power-management method, which dynamically controls the power consumption from the electrical grid and from renewable power sources given the varied price and the amount of renewable power generation. A stochastic optimization problem is formulated and solved to obtain the best decision on power consumption in an uncertain environment, so that the power cost for the base stations can be minimized while satisfying the traffic demand in the network.

In *Chapter 10*, Chen et al. propose an energy-saving technique for the base stations in 3rd Generation Partnership Project (3GPP) Long-Term Evolution (LTE) systems where femtocells are overlaid with macrocells. The authors also provide an overview of the different energy-saving techniques, divided into time, space, and frequency domains, for the LTE base stations. The main idea here is to off-load the downlink traffic of macrocells to femtocells. Simulation results for a scenario with one base station (i.e. eNodeB) and multiple femtocells (i.e. HeNodeBs) show that, with the proposed method, the total RF power in the system can be reduced when the number of HeNodeBs is relatively small.

In *Chapter 11*, Nakhai et al. develop cluster-based multicell processing strategies to improve energy efficiency in cellular wireless communications. In two of the proposed strategies, user signals are globally shared by the coordinating base stations, using both instantaneous and second-order statistics of channel state information. The third one is an iterative solution using statistical channel state information. These three schemes are referred to as the multicell beamforming (MBF) strategies where the base stations share users' data via backhaul links and possess full global channel state information. The objective of the MBF strategies is to find a set of beamforming vectors for a number of simultaneously active users such that the overall transmit power in a virtual cell (i.e. a cluster of three base stations) is minimized, while a prescribed signal-to-interference-plus-noise ratio (SINR) target is maintained for each user. The last one is a coordinated beamforming (CBF) strategy based on a standard semidefinite programming formulation, where user signals are not shared among the coordinating base stations. In this case, the user terminals are served by their local base stations only and a number of base stations coordinate at the beamforming level to minimize their mutual intercell interferences. With CBF, the backhaul overhead is lighter when compared to MBF. The performance evaluation results show that MBF is more power efficient than CBF even when the backhaul signaling is considered.

Part IV: Wireless access techniques for green radio networks

In addition to using the power-saving protocols at the base stations, energy-efficient radio resource (e.g. transmission power, time-slot, frequency band) management and channel access techniques will need to be used to reduce the power consumption in green wireless networks. In this context, cross-layer design and optimization of wireless access techniques would be crucial to improve the energy efficiency at the system level.

In *Chapter 12*, Karmokar, Anpalagan, and Hossain present a cross-layer (physical and MAC) optimization technique for energy-efficient packet scheduling in wireless networks while maintaining the QoS metrics, such as bit error rate, packet delay, and loss rate within the required limit. The cross-layer technique considers the channel gains as well as the buffer occupancy and the traffic characteristics. The authors first consider the case when the channel is fully observable, and then they discuss the cases when it is either partially observable or delayed, as often occurs in real networks. Results are presented to show the advantage of cross-layer optimization to conserve energy in wireless data communication networks.

In *Chapter 13*, Wei, Song, and Yu focus on the energy-saving performance of cellular wireless networks with cooperative relaying among the users. The objective is to minimize energy consumption while satisfying certain QoS performance criteria for the users. The authors present an energy-efficient distributed relaying method based on the selection of a single relay. The relay selection problem is modeled as a stochastic restless multi-armed bandit problem and the solution is obtained by linear programming relaxation and primal-dual index heuristic algorithm. The problem formulation considers finite-state Markov channels, adaptive modulation and coding, and residual energy at the wireless nodes. The method can be implemented based on an RTS/CTS-based handshaking mechanism. Performance of the proposed method is compared with a memoryless relay selection method and a random relay selection method in terms of system reward, which is calculated as a function of the bit error rate of source-to-relay link, spectral efficiency of relay-to-destination link, energy consumption of delivering the data packets from source to destination, as well as residual energy.

In *Chapter 14*, Phuyal, Jha, and Bhargava focus on the energy-efficient resource allocation strategies in a cellular system using a relay-based dual-hop transmission approach. The benefits and implementation challenges of this approach are discussed. For downlink transmission under this approach, the authors propose a green power allocation (GPA) scheme between the base station and the relay station (where the total transmit power is constrained), which minimizes the required transmit power per unit achievable throughput (i.e. [J/bit]) and at the same time it guarantees a minimum end-to-end data rate required by a user. Performance of this scheme is compared with three other power allocation schemes, namely, throughput maximization power allocation (TMPA) scheme, uniform power allocation (UPA) scheme, and GPA with no QoS provisioning (GPANQ) scheme. It is observed that the minimization of J/bit generally degrades the achievable capacity of the network. To this end, the authors extend the optimization model for GPA to a multi-objective optimization model where the objective function accounts for both power consumption and network capacity.

In *Chapter 15*, Long, Li, and Chong investigate four different time slot allocation schemes for energy-efficient communication in cellular single-hop and two-hop TDD-CDMA systems. These are fixed time slot allocation (FTSA) and dynamic time slot allocation (DTSA) schemes for single-hop systems, and multi-link fixed time slot allocation (ML-FTSA) and multi-link dynamic time slot allocation (ML-DTSA) for two-hop relay-based cellular systems. The authors consider four cases of relay station architectures: one fixed and three random relay station structures. With fixed relay station (FRS) structures, the locations of the RSs are determined in advance based on a certain algorithm, while with random relay station structures (RRS) the RSs can be randomly placed around the BS. The total energy consumption is considered to be the summation of transmission energy and hardware energy. Simulation results show that with two-hop transmission in the optimal FRS structure, the blocking and dropping probabilities as well as the total energy consumption can be decreased significantly.

Part V: Green radio test-bed, experimental results, and standardization activities

The research on green communication technologies has started to take shape within and between industry and academia. Internationally there are many ongoing green projects that aim to reduce the carbon footprint through energy savings.

In *Chapter 16*, Auer et al. focus on the assessment of the overall energy efficiency of a 3GPP LTE network over an average European country based on the EARTH E^3F framework. For this assessment the authors consider realistic power consumption at the base stations and traffic models in 3GPP networks. Two energy-consumption metrics are considered: power per unit area, measured in [W/m^2], and energy per bit, measured in [J/bit]. Based on the simulation results, the authors conclude that there is a huge potential for energy savings at the base stations when the network is not fully loaded.

In *Chapter 17*, Conte, Helmers, and Sehier describe the energy efficiency-related activities conducted in important standardization bodies and fora, as well as by the relevant industrial and academic joint projects and consortiums. In particular, the authors focus on the activities conducted by ETSI (European Telecommunication Standard Institute) and its partners, 3GPP (Third Generation Partnership Project), IETF (Internet Engineering Task Force), and the China Communication Standard Association (CCSA). In order to assist and influence these standardization fora, several other groups/projects/consortiums have been created, including the NGMN (Next Generation Mobile Networks) alliance, the GreenTouch consortium, and the EARTH (Energy-Aware Radio and neTwork tecHnologies) project within the European Commission (EC) 7th Framework Programme for Research and Technological Development (FP7). The EARTH project proposes technical solutions to improve the power efficiency of wireless mobile networks at component level, link level, and network level. It proposes a tool called the energy-efficiency evaluation framework (E^3F) to analyze the energy efficiency of network solutions. GreenTouch is a non-profit research consortium founded by experts from industry, academia, government, and research institutions around the world which aims to define new, clean-slate technologies that will be at the heart of sustainable communication networks.

Part I

Communication architectures and models for green radio networks

1 Fundamental trade-offs on the design of green radio networks

Yan Chen, Shunqing Zhang, and Shugong Xu

1.1 Introduction

There is currently a global concern about the rise in the emission of pollutants and energy consumption. The carbon dioxide (CO_2) footprint of the information and communications technologies (ICT) industry, as pointed out by [1], is 25% of the 2007 carbon footprint for cars worldwide, which is similar to that of the whole aviation industry. Within the ICT industry, the mobile network is recognized as being among the biggest energy users. The exponentially growing data traffic in mobile networks has made the issue an even grander challenge in the future. In a data forecast report provided by Cisco [2], it has been pointed out that the global mobile data traffic will increase 26-fold between 2010 and 2015. In particular, unexpectedly strong growth in 2010 has been observed mainly due to the accelerated adoption of smartphones. For instance, China Unicom's 3G traffic increased 62% in a single quarter from Q1 to Q2 of 2010, while AT&T reported a 30-fold traffic growth from Q3 2009 to Q3 2010. The unprecedented expansion of wireless networks will result in a tremendous increase in energy consumption, which will further leave a significant environmental footprint. Therefore, it is now a practical issue and demanding challenge for mobile operators to maintain sustainable capacity growth and, at the same time, to limit the electricity bill. For instance, Vodafone Group has announced the goal of reducing its CO_2 emissions by 50% against its 2007 baseline of 1.23 million tonnes, by the year of 2020 [3]. Figure 1.1 gives examples of the green demand from mobile operators worldwide.

As has been pointed out in [4], the radio access part of the wireless network accounts for up to more than 70% of the total energy bill for a number of mobile operators. Therefore, developing energy-efficient wireless architectures and technologies is crucial to meet this challenge. Research actions have been taken worldwide. It is now an important trend for the wireless designers to take energy consumption and energy efficiency into their design frameworks. Vodafone, for example, has predicted that energy-efficiency improvement will be one of the most important areas that demand innovation for wireless standards beyond LTE [5].

Green radio research is a large and comprehensive area that covers all layers in the design of efficient wireless access networks. There have been efforts devoted to traditional energy-saving ways, such as designing ultra-efficient power amplifiers, reducing feeder losses, and introducing passive cooling. However, these efforts are isolated and thus cannot make a global vision of what we can achieve in five or ten years for energy

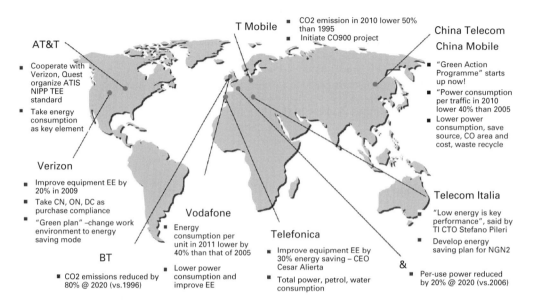

Figure 1.1 Global operators' demand on green communications.

saving as a whole. Innovative solutions based on top-down architecture and joint design across all system levels and protocol stacks are needed, which cannot be achieved via isolated efforts.

Green research projects with holistic approaches and joint efforts from the industry and the academia have sprung up all over the world during recent years. For instance, the EARTH (Energy Aware Radio and neTwork tecHnologies) project [6]–[7] under the European Framework Program 7, started to develop green technologies at the beginning of 2010. In the UK, *GreenRadio* [8] is one of the Core 5 Programs in Mobile VCE that has been set up since 2009. Most recently, the *GreenTouch* Consortium sets its 5-year research goal to deliver the architecture, specification, and roadmap needed to reduce the end-to-end energy-consumption per bit by a factor of 1000 from the current level by the year 2020. In addition, there are also active discussions in standardization organizations, such as ETSI, ATIS, and 3GPP, on energy-efficiency metrics and measurement, as well as studies for base station level or network level savings.

Instead of a survey that reaches every aspect of the matter, or a report elaborating one specific green research point, this chapter focuses on the fundamental framework for green radio research and strings together the currently scattered research points using a logical "rope." In this chapter we propose four fundamental trade-offs to construct such a framework. These were first introduced in [9]. As depicted in Figure 1.2, they are

- *Spectrum efficiency–energy efficiency (SE–EE) trade-off*: given the bandwidth available, to balance the achievable rate and the energy cost;
- *Bandwidth–power (BW–PW) trade-off*: given the target transmission rate, to balance the bandwidth utilized and the power needed;

1.2 Insight from Shannon's capacity formula

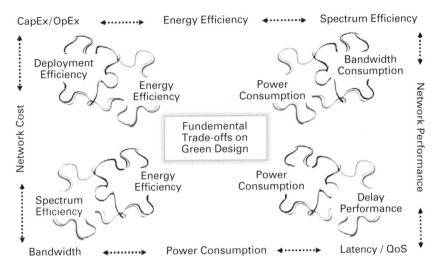

Figure 1.2 Four fundamental trade-offs form the core of green research.

- *Delay–power (DL–PW) trade-off*: to balance the average end-to-end service delay and the average power consumed in the transmission;
- *Deployment efficiency–energy efficiency (DE–EE) trade-off*: given the network traffic requirement, to balance the deployment cost, throughput, and energy consumption, in the network as a whole.

By means of the four trade-offs, key network performance/cost indicators are all strung together. In the rest of the chapter, we will elaborate in detail the definitions, justifications, practical concerns, as well as research directions for each of the trade-off studies. In particular, we shall show that in practical systems, the trade-off relations usually deviate from the simple monotonic curves derived from Shannon's formula, which brings a new design philosophy.

1.2 Insight from Shannon's capacity formula

Shannon's capacity formula [10] establishes a bridge between the maximum achievable transmission rate R and the received power $P^{(r)}$ for the point-to-point additive white Gaussian noise (AWGN) channel, i.e.

$$R = W \log_2 \left(1 + \frac{P^{(r)}}{W N_0}\right), \qquad (1.1)$$

where N_0 is the noise power density at the receiver and W is the system bandwidth. Though Shannon's ground-breaking formula has been known for more than half a century, people mainly look at it from the channel capacity point of view. However, as we will show later in this section, the formula actually gives us a fundamental insight into the energy-related trade-offs in the wireless point-to-point link transmission. In this section, we shall formally introduce the definitions of the trade-offs and sketch their behavior predicted by Shannon's capacity formula.

The following are the equivalent transformations of the above formula, which will be used in the characterization of the different trade-offs.

$$\frac{R}{W} = \log_2\left(1 + \frac{R}{W}\frac{E_b^{(r)}}{N_0}\right). \tag{1.2}$$

$$\frac{1}{T_b} = W\log_2\left(1 + \frac{1}{T_b}\frac{E_b^{(r)}}{WN_0}\right). \tag{1.3}$$

In the equations above, $E_b^{(r)}$ stands for the average energy per bit and T_b denotes the average transmission time per bit. They are introduced through the relations $E_b^{(r)} = P^{(r)}/R$ and $T_b = 1/R$. Further, considering a constant attenuation on the transmitted signal, denoted as a simple function of the transmit power $P^{(t)}$, namely $f(P^{(t)}) = \kappa_0 P^{(t)}/d^\alpha$, where κ_0 and α are the attenuation coefficient and exponent, respectively, we have

$$\frac{R}{W} = \log_2\left(1 + \frac{R}{W}\frac{E_b^{(t)}}{N_0}\frac{\kappa_0}{d^\alpha}\right). \tag{1.4}$$

1.2.1 SE–EE trade-off

Spectrum efficiency (SE), defined as the system throughput for unit bandwidth, i.e. bits/sec/Hz, is a widely accepted criterion for wireless network optimization. The peak value of SE is always among the key performance indicators of standardization evolution such as 3GPP. For instance, the target downlink SE of 3GPP increases from 0.05 bps/Hz to 5 bps/Hz as the system evolves from GSM to LTE. On the contrary, energy efficiency (EE), defined as the data rate achievable per unit of transmitted power, i.e. bits/sec/Watt, namely bits/Joule, was previously ignored by most of the research efforts and has not been considered by 3GPP as an important performance indicator until very recently.

Shannon's groundbreaking work on reliable communication over noisy channels showed that there is a fundamental trade-off between SE and received/transmitted EE. Informally speaking, a lower transmission rate leads to a lower transmitted power, for the same system bandwidth. Given the definitions above, SE can be expressed as $\eta_{SE} = R/W$ and the received EE as $\eta_{EE}^{(r)} = 1/E_b^{(r)}$. From (1.2), the SE–EE trade-off can be characterized by

$$\eta_{EE}^{(r)} = \frac{\eta_{SE}}{(2^{\eta_{SE}} - 1)N_0}, \tag{1.5}$$

which is depicted on the left-hand side (LHS) of Figure 1.3, where $N_0 = -174$ dBm. Seen from both the mathematical relation and the figure, η_{EE} converges to a constant, $1/(N_0 \ln 2)$, when η_{SE} approaches zero. On the contrary, η_{EE} approaches zero when η_{SE}

1.2 Insight from Shannon's capacity formula

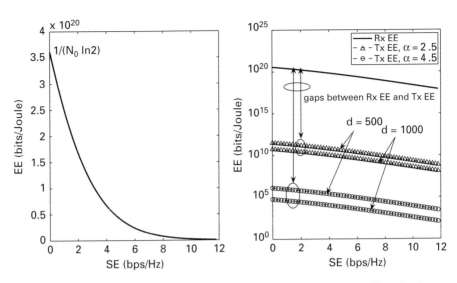

Figure 1.3 Illustration of the SE–EE trade-off. On the LHS, the figure shows the trade-off relation between SE and received EE from Shannon formula, while on the RHS, the figure depicts the transmit EE as function of the path-loss exponent α at different distance d.

tends to infinity. Similarly, considering the relation in (1.4), the transmit EE-SE trade-off can be expressed as

$$\eta_{EE}^t = \frac{\eta_{SE}}{(2^{\eta_{SE}} - 1)N_0} \cdot \frac{\kappa_0}{d^\alpha}, \quad (1.6)$$

as shown on the right-hand side (RHS) of Figure 1.3. The gaps between the received EE and the transmit EE depend heavily on the transmission channel degradation, i.e. the path-loss exponent α and the transmission distance d.

1.2.2 BW–PW trade-off

Bandwidth (BW) and power (PW) are both fundamental but limited resources in wireless communications. From the Shannon's capacity formula in (1.1) and (1.4), the relation between the transmit power, P^t, and the transmission bandwidth, W, for a given transmission rate, R, can be expressed as

$$P^t = W N_0 (2^{\frac{R}{W}} - 1) \cdot \frac{\kappa_0}{d^\alpha}. \quad (1.7)$$

The expression above exhibits a monotonic relation between PW and BW, as sketched in the LHS of Figure 1.4. The fundamental BW–PW trade-off shows that, to transmit at a given data rate, the expansion of the transmission bandwidth is preferred in order to reduce transmit power and thus achieve better energy efficiency. From (1.7), in the extreme case, the minimum power consumption is as small as $N_0 R \ln 2$ if there is no bandwidth limit.

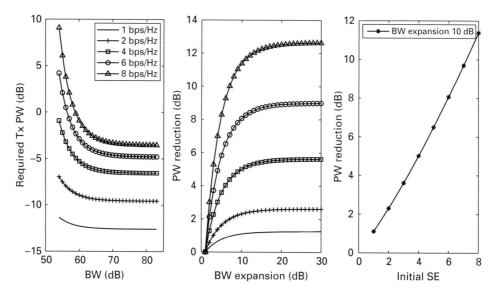

Figure 1.4 Illustration of the BW–PW trade-off derived from Shannon formula. The difference between the set of curves is the initial SE values before any BW expansion. The left figure gives the absolute value of the required transmit power while the middle one shows the PW reduction gain. The right figure depicts the PW reduction gain at 10 dB BW expansion at different initial SE values. $\kappa_0/d^\alpha = -140$ dB.

Figure 1.4 depicts the BW–PW trade-off from three different angles. Firstly, the leftmost figure shows the relation between the required transmit power and the system bandwidth, the trend of which behaves exactly as equation (1.7) predicts. The middle figure shows the PW reduction as function of the BW expansion. From (1.4), the reduction in the transmit power is the same as that in the received power. It can be observed from the figure that increasing the BW by ten (10 dB) brings considerable gain in PW reduction, no matter what the initial SE of the system is. Larger than 10 dB BW expansion, however, only adds marginal gain. Moreover, the higher the initial SE, the larger the PW reduction gain. It can be found from the right-most figure that expanding the BW 10 times brings less than 3 dB PW reduction gain to a system with the initial SE at 2 bps/Hz, but offers a larger than 10 dB gain to the system with the initial SE larger than 8 bps/Hz.

1.2.3 DL–PW trade-off

The metrics such as EE, SE, and BW, as described in the two trade-offs above, are important system performance criteria but cannot be directly observed by end users. Delay (DL) is different to these metrics and is usually taken as a measure of quality of service (QoS) and user experience. According to the scope of the definition, there are different types of delay. Two major ones are the physical (PHY) delay, defined as the time spent during the physical layer transmission, and the medium-access-control (MAC) delay, defined as the sum of both waiting time in the MAC layer data queue and transmission time in the PHY layer.

1.2 Insight from Shannon's capacity formula

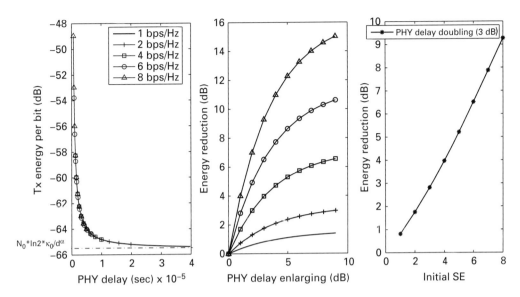

Figure 1.5 Illustration of the PHY DL–PW trade-off derived from Shannon's formula. The middle figure shows the gain of energy reduction as a function of the PHY delay increasing. The right figure shows the energy reduction gain provided by doubling the PHY delay at different initial SE values. $\kappa_0/d^\alpha = -140$ dB and $W = 200$ kHz.

Let us start with the simpler one, the PHY delay, for which the Shannon's capacity formula reveals most of the characteristics. For point-to-point transmission over AWGN channels, formulas (1.3) and (1.4) tell us the average energy per bit required to transmit a data bit in time T_b can be calculated as

$$E_b^t = N_0 T_b W \left(2^{\frac{1}{T_b W}} - 1 \right) \cdot \frac{\kappa_0}{d^\alpha}. \tag{1.8}$$

The above expression shows a monotonically decreasing relation between received energy per bit and PHY delay, as sketched on the left of Figure 1.5. The middle figure of Figure 1.5 shows that the higher the initial SE, the more energy reduction gain can be obtained from enlarging the PHY delay. For instance, doubling the PHY delay reduces the average transmit energy per bit by less than 2 dB for the initial SE of 2 bps/Hz but more than 6 dB for that of 6 bps/Hz. This is true for single symbol transmission or continuous symbol transmission (full buffer). However, the relation may change when we consider bursty data blocks, as will be shown later in Section 1.3.

The MAC delay, on the other hand, is closely related to the upper layer traffic arrivals and statistics. By Little's law [11], the average delay has a direct relation with the average queue length in the data queue. As a result, the design of transmission schemes shall cope with both channel uncertainties, traffic variations, and queue dynamics, which makes the characterization of DL–PW trade-off more complicated. Shannon theory alone is not enough to characterize the DL–PW in these scenarios. Other theoretical analysis tools are needed, such as queueing theory [11] and control theory [12]. Moreover, as technologies evolve, the types of future wireless services become diverse enough to have heterogeneous delay requirements. Therefore, in order to build a green radio, it is important to know when and how to trade tolerable delay for low power.

1.2.4 DE–EE trade-off

Deployment efficiency (DE), a measure of network throughput per unit of deployment cost, namely bits/$ or Mbits/$, is an important network performance indicator for mobile operators. The deployment cost consists of both capital expenditure (CapEx) and operational expenditure (OpEx). For radio access networks, the CapEx mainly includes infrastructure costs, such as base station equipment, backhaul transmission equipment, site installation, etc., while the key drivers for the OpEx are electricity bill, site and backhaul lease, and operation and maintenance costs. The scope of the EE definition in the previous trade-offs can either be for a single base station or for a network; the EE concept involved in the DE–EE trade-off is a metric for the whole network, namely a measure of network throughput per unit of network energy consumption, i.e. bits/Joule.

The two different metrics often lead to opposite design criteria for network planning. For example, to save the expenditure on site rental, base station equipment, and maintenance, network planning engineers tend to "stretch" the cell coverage as much as possible. However, the path loss between the base station and mobile users will degrade by 12 dB whenever the cell radius doubles if the path-loss exponent is four, which induces a 12 dB increase in the transmit power to guarantee the same signal strength for those users at the cell edges. Some simple calculations give the result that to provide cellular coverage for a given area, increasing the number of base stations will save the total network transmit power by the same factor.

Table 1.1 helps to understand the inner logic. Assume the reference cell radius is d_0, β and γ are two coefficients associated with the cell size shrinking scenario where $0 \leq \beta, \gamma \leq 1$, inter-cell interference is not considered, and the transmit power for all users is kept the same, derived from the SE requirement η_{SE} of the cell-edge user. Figure 1.6 further depicts the DE and EE performance at different β. The DE and EE values in the figure are normalized by that of the reference scenario. An implicit assumption is that the total traffic served by different scenarios on the given area A is the same. The leftmost figure shows that the improvement in EE via cell size shrinking depends heavily on the wireless channel environment, e.g. the path-loss exponent α. The larger the α is (faster degradation of the transmitted energy), the more benefit small cells could bring. As shown in the middle figure, the value of γ impacts the DE performance. Here, $1 - \gamma$ can be interpreted as the average cost reduction ratio per base station. Note that the increase in the number of cells adds extra cost in the backhaul and site maintenance. The constant offset in γ is added to account for that. Finally, the right-most figure shows how the network EE trades off DE. Note that when transmitting in free space ($\alpha = 2$), the trade-off relation no longer holds.

1.2.5 Summary

In the previous four subsections, we have elaborated the definitions of the four fundamental trade-offs as well as their behavior predicted by the Shannon's capacity formula.

1.2 Insight from Shannon's capacity formula

Table 1.1. Simple calculations with the cell size shrinking ratio β. The constants k_1 and k_2 are $k_1 = A/\pi$, $k_2 = (2^{\eta_{SE}} - 1)/(\kappa_0 \eta_{SE})$, where A is the total area considered and η_{SE} is the target SE for cell-edge users. γ is the ratio of cost per base station against the initial value c_0.

Scenarios	Cell radius D	No. of cells N	Total cost C	Total E_b^t/N_0
Reference	d_0	$\dfrac{k_1}{d_0^2}$	$N \cdot c_0 = \dfrac{k_1 c_0}{d_0^2}$	$k_2 d_0^\alpha$
Shrinking ratio β	$d_0 \cdot \beta$	$\dfrac{k_1}{d_0^2} \cdot \dfrac{1}{\beta^2}$	$N \cdot \gamma c_0 = \dfrac{k_1 c_0}{d_0^2} \cdot \dfrac{\gamma}{\beta^2}$	$k_2 d_0^\alpha \cdot \beta^{\alpha-2}$

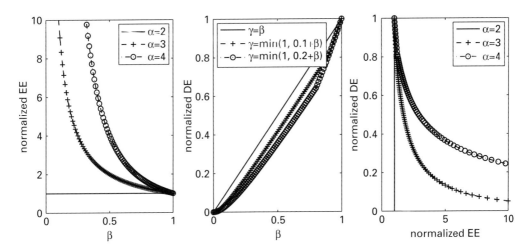

Figure 1.6 Illustration of the DE–EE trade-off derived from Shannon's formula. The left one shows the EE trend at different path-loss exponent α, while the middle one shows the DE behavior under different assumptions of γ, and the right one shows the trade-off relation between the two. The DE and EE values are normalized by that of the reference scenario. γ in the right-most figure is chosen as $\min\{1, 0.1 + \beta\}$.

In this subsection, we would like to summarize what we have learnt so far. Figure 1.7 gives the sketch of the trend curve for each of the trade-offs, which are monotonically decreasing as predicted by Shannon's formula.

Along each curve, we can identify the operation region with high power and that with low power, respectively, as shown by the shaded ellipses in the figures. The two large arrows in each sub-figure suggest two potential directions to improve EE or to reduce power. Having identified the operating regions with different power requirement, one possible direction for energy-oriented system optimization is to shift the operating point of the system along the trade-off curve, from the high-power region to the low-power region. This can, in general, be achieved by optimizing the key system parameters and adapting them to the dynamics of the system traffic requirement. For instance, when there is extra bandwidth available, aggregating them and then transmitting on the wider band also results in lower transmit power requirement (shifting the operating point along the curve from right to left as in Figure 1.7 (b)). Similarly, for a delay tolerant service,

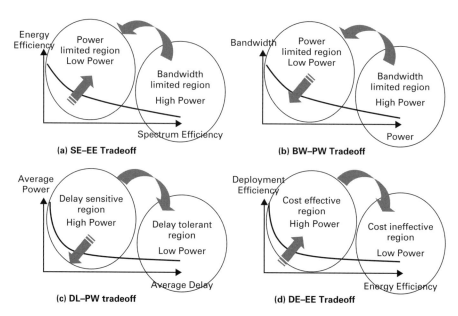

Figure 1.7 Summary of the sketch of the four fundamental trade-offs derived from Shannon's formula. Along the curves, high-power regions and low-power regions are identified and marked using shaded ellipses. The bold arrows suggest the potential direction of energy performance improvement.

lowering the modulation level to transmit more slowly can help to reduce the transmit power needed (shifting the operating point along the curve from left to right, as in Figure 1.7 (c), for PHY delay).

On the other hand, it is possible to improve the two ends of the trade-off simultaneously, i.e. pushing the trade-off curve outwards as for the SE–EE and DE–EE relations and pushing the curve inwards for the other two trade-offs. Multiple-input-multiple-output (MIMO) techniques, which can provide higher SE without increasing power for transmission, are potential candidates in this case. However, as we will see from the next section, it is not always good to have MIMO for energy-oriented design, because the power expenditure in other parts of the system (e.g. the electronic circuits) to support MIMO may degrade the gain obtained for the transmit power only.

1.3 Impact of practical constraints

In the results derived from Shannon's capacity formula, only the transmit power (radiated energy) is considered. In this case, we obtain a set of monotonically decreasing curves for each of the fundamental trade-offs. In fact, however, for any base station available today, the power radiated to the environment for signal transmission is only a portion of its total power consumption [13]. The ratio of the base station's total power consumption over its radiated power is called *base station efficiency*, which is far from the ideal value of 1. For

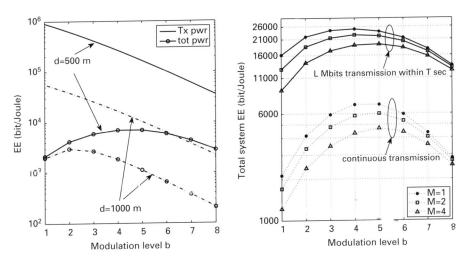

Figure 1.8 The SE–EE curves under practical constraints. The LHS figure show the EE at difference transmission distance with either transmit power or total power considered. The RHS figure compares the system EE ($d = 500$ m) with total power considered in both the full buffer case (continuous transmission) and the bursty traffic case (L bits before T sec.), given different numbers of beam-forming antennas M.

instance, if the base station efficiency is 10%, then to transmit 100 Watt for delivering information, about 900 Watt extra power is needed to keep the system working properly. As technologies evolve, the power efficiency has been improved greatly but is still far from the ideal (close to 1). Therefore, for an energy-efficient design of wireless networks, it is important to consider not only the radiated power, but also the overall system input power. Moreover, for the DE–EE trade-off from the whole network's aspect, it is also essential to have a correct model of the network deployment and operation cost.

In the following, we just give some examples of how the trade-off curves derived in Section 1.2 are impacted by the practical constraints elaborated above. We shall see that when the non-linear efficiency of the PA and the processing power and circuit power are considered, the trade-off relation usually deviates from the simple, monotonically decreasing curve.

As a comparison to Figure 1.3, we show in the LHS of Figure 1.8 the relation between the transmit EE and the modulation level of the uncoded M-QAM signals, which can be seen as the SE of the uncoded transmission. Similar issues were also investigated in [15, 16]. We further study the impact of multiple antennas and bursty transmission with sleep mode in the RHS of Figure 1.8. Transmit beamforming is assumed at the transmitter, which has been proved to reduce the transmit power by M times if the antenna array is equipped with M antennas [17]. However, when considering the total system power, we have the inverse observation, namely the beamforming gain may become negative and get worse with more antennas. This is because the transmit energy reduction becomes smaller than the circuit power increase brought by the increase of extra beamforming antennas. Also, it is easy to see that sleep mode helps to boost the energy efficiency of

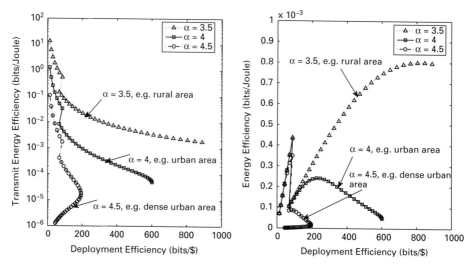

Figure 1.9 The DE–EE curves under practical concerns. The LHS figure considers only the transmit power of the network, while the RHS figure take the total input power into consideration. α is the path-loss exponent, whose value implies the service areas.

the system, since it directly reduces the circuit power and part of the processing powers in the system.

Another example is shown in Figure 1.9 for the DE–EE trade-off, based on the preliminary study in [18]. From the right-most plot, there might not always be a trade-off between DE and EE and the shape of a DE–EE curve depends on the specific deployment scenarios. For the suburb scenario, where the path-loss exponent is small (about 3.5), the network EE even increases with its DE. For the dense urban scenario, where the path-loss exponent is large (about 4.5), two different EE values may result in the same DE value, corresponding to very small and very large cell radii, respectively. The former is because of the huge increase in CapEx by increasing the number of sites; the latter is due to the sharply increased electricity bill in OpEx.

1.4 Latest research and future directions

1.4.1 SE–EE trade-off

The previous illustrations show that the SE–EE trade-off curves are highly relevant to the static circuit power consumptions. However, this is not the only factor that causes the different behavior of the SE–EE trade-off curves. Current literature also shows that the transmission technologies and the network architectures will affect the trade-off curve as well. In the current literature, the SE–EE trade-off relations has been extensively studied in the OFDMA and MIMO systems.

- *OFDMA systems*: The concept of EE in OFDM systems first appeared in [15] with the consideration of the circuit power consumption. In contrast to the traditional spectrally-efficient water-filling scheme that maximizes throughput under a fixed overall transmit power budget, the new scheme maximizes the overall EE by adjusting both the overall transmit power and its distribution among subcarriers. It is demonstrated that there is at least a 15% reduction in energy consumption when frequency selectivity is exploited. An energy-efficient design was also extended to general OFDMA networks [15]. For uplink transmission with flat fading channels, it was shown that using the optimal modulation, EE increases when the distance between the terminal and BS decreases. Moreover, by adapting both power and modulation order, the EE-oriented design always consumes less energy than the traditional fixed power schemes.

- *MIMO systems*: Although MIMO techniques have been proven to be effective in improving the network capacity of wireless systems, the power consumption also increases significantly. First of all, more circuit power is consumed due to the real implementation of the multiple transmit-receive radio processing chains. We can easily conclude that the EE of transmission with multiple antennas is lower than single antenna transmission if the incremental in the capacity cannot afford the additional expenditure of energy. Second, MIMO transmission triggers additional time/frequency resources for the signaling overhead. For example, in order to estimate the CSI and feed it back to the transmitter, pilots or training symbols need to be delivered before the data transmission. Since the number of channel coefficients increases with the product of the number of transmit antennas and that of receive antennas, a huge amount of signaling overhead is required for MIMO systems. As far as we are aware, the EE of MIMO systems is still unknown if all the signaling overhead is considered. Adaptively changing the number of active antennas at BS is proposed for 3GPP LTE to address the time-varying traffic issue in cellular networks. In [19], adaptive switching between MIMO with two transmit antennas and SIMO is addressed to save energy at mobile terminals.

In the existing SE–EE trade-off relationship investigation, most of the techniques focus on the uplink scenarios, where the terminal battery is the critical concern. However, there are few green technologies regarding the downlink transmission at the current stage. In addition, the impact of traffic types and statistics, the multi-user and multi-cell environment and other effects have not yet been taken into consideration in the green radio research. We have listed a few of the possible research opportunities, by no means complete, in the following part.

- *Role of traffic statistics*: The knowledge of traffic statistics is currently "blind" to the communication system designer since all the designs target the maximum throughput of the network. In the future wireless system, the dynamic adjustment of the system capability is a key feature which enables the exploitation of the time-varying nature of the traffic load variations. Hence, how to utilize the traffic statistics to improve the SE and EE trade-off relations is of great importance.
- *Energy-efficient schemes in multi-user and multi-cell scenarios*: In the multi-user and multi-cell environments, a single dimension of SE–EE trade-off relation is not sufficient to describe the complicated systems. The characteristics of the interference, the user selection policies, and the dynamics of active user distribution will have a joint

impact on the SE–EE trade-off relations. Thus, understanding the SE–EE trade-off in the multi-user and multi-cell environments is critical for the energy-efficient design of the practical system, and more research efforts will be allocated in this direction.

1.4.2 BW–PW trade-off

BW, as a sparse resource in the communication system, is fully utilized in the 2G wireless communication systems, such as GSM, where the fixed BW allocation scheme is adopted. With the recent advent of carrier aggregation technologies, UMTS, LTE and beyond, systems have come up with more flexibility to dynamically aggregate unutilized spectrums. In addition, spectrum re-farming,[1] cognitive [20] usage of white spectrum holes, provides enough freedom to dynamically adjust the BW according to the traffic requirement. Theoretical results [21] show that to obtain the optimal EE in the multi-user systems, BW and PW will be jointly optimized if the practical assumption is considered, which also triggers the practical resources allocation algorithm design in the licensed multi-band system and in unlicensed cognitive systems.

- *Multi-band systems*: The practical evaluation of the multi-band system contains many technical challenges, for example, the modeling of circuit power consumption with respect to BW and the joint BW–PW resources scheduling policies. In addition, energy-efficient solutions cannot be obtained in a straightforward way due to the non-convexity of the EE utility function. [22] gives a preliminary research result on this topic. It is shown that to achieve the same amount of data rate, the optimal transmit power should scale with the same direction of the optimal bandwidth and the required bandwidth should scale linearly with respect to the required data rate. Moreover, the optimal transmit power will be increased when the dynamic circuit power increases. Through a joint BW–PW allocation scheme, more than 30% improvement of the EE can be achieved when compared with the traditional throughput maximization algorithm.
- *Cognitive systems*: For the unlicensed spectrum, the resource allocation strategy is quite different from the licensed band case. This is because the secondary users have to sense the unlicensed band before they can use it and the sensing process usually consumes energy as well. Meanwhile, the variable bandwidth adjustment and modulation adaptation, the cognitive assignment strategy as well as the consideration of the channel occupancy probability, will also complicate the problem. [23] discusses ways to reduce the complexity of cognitive radio systems, particularly in the need for spectrum sensing by exploiting distributed artificial intelligence. By utilizing the available bandwidth and adjusting the modulation level correspondingly, the mean power efficiency can be improved by up to 80% compared with the fixed case.

The BW–PW trade-off is a crucial effect for designing a wireless system, since the bandwidth and the transmission power are the most important factors. With the evolution

---

[1] Spectrum re-farming is more like a government action to support more efficient use of wireless spectrum via reassigning 2G spectrum to 3G applications. For instance, it is now possible to deploy UMTS (3G system) on 900 MHz (2G spectrum).

of wireless technologies, such as software-defined radio and carrier aggregation, future communication systems are maturing to support the flexible use of BW. However, there are still many open issues that deserve further investigation.

One important issue is the development of novel network architecture. The deployment of advanced network architecture may have an influence on the BW–PW trade-off frontier. In particular, the deployment of cooperative and heterogeneous networks introduces additional infrastructure nodes into the network; consequently, the BW and PW planning will be different from the conventional network architectures. Hence, the BW–PW trade-off with advanced resource management algorithms under new network architectures deserves further research. In addition, with the combination of carrier aggregation and cognitive radio techniques, cross-layer approaches that jointly consider dynamic BW acquisition and BW–PW trade-off will certainly play important roles in future design.

1.4.3 DL–PW trade-off

From the pure information theory aspect, we already know from Shannon's capacity formula that prolonging the PHY layer transmission time for each bit (PHY delay) reduces the transmit power. Taking this one step further, since the wireless channel varies from time to time, if we can wait to transmit when the channel condition is good, the transmit energy can be reduced. This idea is similar to water-filling, which is also known as opportunistic transmission. However, these approaches are not aware of the backlogs in the MAC layer. As smart phones become popular, delay-sensitive applications such as VoIP and video streaming now account for a large portion of the mobile data volume. For delay-sensitive applications, it is important to take MAC delay into consideration, which is impacted by both the traffic arrivals and departures. In addition, when packet-based traffic flow is considered, the average delay per packet may replace the average delay per bit as the performance metric. Packet transmission scheduling that minimizes transmission energy subject to a deterministic deadline under deterministic packet arrivals was investigated in [24]. An online lazy scheduler was proposed and proved to achieve close-to-optimal MAC DL-transmit PW trade-off. Under the lazy scheduler, the system transmits packets for longer times when the backlog is low and for shorter times when the backlog is high, which outperforms the deterministic schedule applying fixed packet transmission time for all packets. In this work, however, the transmit power was assumed to be fixed, so that the dynamics of the wireless fading channel was not fully exploited.

Given the traffic dynamics and wireless channel uncertainties, the problem of characterizing the bound of MAC DL–PW trade-off is equivalent to minimizing the average energy consumption under all possible average delay constraints. Since the departure rates are closely related to the transmission schemes and the radio resources allocation strategies, solving the optimization problem implies finding the best power-rate allocation policy for the transmission for a given transmission scheme. Moreover, it is important to consider both the channel state information (CSI) and the queue state information (QSI) as system status in the optimization. In the literature, there are mainly three approaches [25].

- *Delay constraint to rate constraint conversion approach*, which converts the delay constraint into an average rate constraint using tail probability at a large delay regime (large derivation theory) and solves the optimization problem using information theoretical formulation based on the rate constraint [26]–[27]. While this approach allows a potentially simple solution, the control policy will be a function of CSI only and such control will be good only for a large delay regime.
- *Stochastic Lyapunov stability drift approach*, which utilizes the notion of Stochastic Lyapunov stability and establishes throughput optimal control policy (in a stability sense). The benchmark paper [28] depicted the behavior of the tail of the power-delay curve $P(D)$ as the buffer delay D over a point-to-point fading channel. In [29], the authors extended the bound in [28] to a multi-user context, demonstrating that any algorithm that yields average power within $O(1/V)$ of the minimum power required for network stability must also have an average queueing delay greater than or equal to $\Omega(\sqrt{V})$. Though analytical bounds can be characterized using this approach, the derived DL–PW bound under certain arrival models is tight only for heavy traffic loading and the derived policy is not optimal in general.
- *Markov decision process and stochastic learning approach*, which casts the average delay minimization problem into an infinite horizon average-cost Markov decision problem (MDP) [12]. This is a systematic approach to characterize the DL–PW bound. However, the solution to MDP has the well-known problem of the *curse of dimensionality*; brute-force value iteration and policy iteration will not lead to any practical solutions.

Delay is directly related to the user's experience of the service, so when trading off longer delay for lower power, the engineers need to be very careful. Though we have the theoretical results described above, it is still far from a design for practical systems. In order to fill the gap, the following topics are worth investigating.

- *Heterogeneous delay requirements*: In practice, there are most probably parallel service flows running in the system, each having a different delay requirement. These service queues may belong to one user or multiple users. For the downlink system, the departure process for each of the service queues may be coupled together through the common radio resource scheduler, which makes the characterization for DL–PW trade-off even more demanding for multi-user scenarios. Some results can be found in [30].
- *Low-complexity control policy*: In general, under dynamic traffic arrivals and fast channel variations, there is no closed-form expression available to show the direct relation between the average service delay and the system power consumption. Therefore, the investigation of simplified but approximate models is desirable to provide insights for practical system design for low-complexity control policy.

1.4.4 DE–EE trade-off

As the data traffic volume increases dramatically in the near future, if we keep the traditional macro-cell only deployment, it is highly probable that the network will not

1.4 Latest research and future directions

function properly at high load. Candidate solutions are: i) not increasing the number of base stations much but enhancing the capacity of each base station via a wider bandwidth or a larger number of antennas; ii) shrinking the cell size and thus deploying more base stations to cover the same area; iii) using multi-layer heterogeneous deployment and adding extra small cells where the capacity demand is high; iv) introducing collaboration between base stations to help each other boost capacity; v) brand-new network architectures. Which solution is best? This question is not easy to answer for two reasons. First, the answer depends on many factors, such as the type of areas, the network traffic load, the capability and cost of the small cells, the cost and efficiency of the collaboration, etc. Second, there is not yet a proper metric to evaluate all the key factors. Facing such a situation, the study of BS deployment problems in cellular networks has become a very active research area recently. Existing research is mainly exploring the following issues.

- *Network cost model*: The financial aspect of base station deployment and operation is closely related to the revenue of the operators. [31] provided a comprehensive view of the financial cost, including CapEx and OpEx, of heterogeneous deployments with micro/pico base stations as an extra layer on top of the traditional macro sites. It is shown that a complementary hot-spot layer of micro/pico cells on top of macro cells is the most cost-effective architecture for non-uniform spatial traffic. [32], on the other hand, analyzed the financial impact of home network deployment such as femto-cells.
- *Energy-efficient network deployment*: This study closely relates to the candidate solutions (ii) and (iii). For instance, it is shown in [33] that by shrinking the cell radius from 1,000 m to 250 m, the maximum EE of the HSDPA Network will be increased from 0.11 Mbits/Joule to 1.92 Mbits/Joule, respectively, corresponding to a gain of 17.5 times. Therefore, to minimize energy radiation, radio resource management engineers favor small cell-size deployment. [34]–[35] considered improving the network energy efficiency by deploying micro-cells on top of the macro-cell coverage. A varying number of micro base stations are assumed to be deployed in predetermined locations. Results show that power saving from the deployment of micro base stations strongly depends on the offset power consumption of both macro and micro base stations. Moreover, the benefit also depends heavily on the system load (users) and area traffic requirement: the higher the load or traffic demand, the more energy-saving gains we can get from deploying small cells. However, in this work the EE performance of stochastic deployment of pico cells was not investigated.
- *Energy-efficient base station collaboration*: The concept of collaboration between base stations may date back to the soft handover scheme in WCDMA systems, also known as macro diversity. In LTE/LTE+ systems, coordinated multi-point (CoMP) transmission is introduced to support more sophisticated cooperation. A general question is to what extent CoMP can actually improve energy efficiency, considering that the potential improvements due to interference cancellation and capacity boosting are compensated by a significantly increased complexity and extra pilots, backhauling, and processing cost. Some results are given in [36]–[37]. For instance, [36] shows that the bits-per-Joule performance of the system is only moderately affected by the use of CoMP

schemes, with potential gains of 10% and 20% for small and large site distances, respectively. In addition, increasing the cluster size of cooperation base stations to more than 3 brings marginal improvement.

Seen from the above works, the network DE and EE are usually treated separately. However, as both metrics are valued by mobile operators, characterizing the DE–EE trade-off relation is important to shed light on energy-efficient and economical network deployment. In the future, the following research aspects deserve further investigation.

- *Backhaul capacity constraint and power consumption*: As in much of the existing work, heterogeneous networks (HetNet) and cooperative network (CoopNet) are considered as promising network architectures in the future. Small cells with low equipment costs can be deployed with larger flexibility to boost the area throughput. Moreover, the closer distances between small cells in dense deployment favor network functions such as base station cooperation. However, the larger number of base stations makes the backhaul power consumption much more significant than that in traditional deployment. The complex cooperation between the small cells also adds a great burden to the backhaul transmission. From this aspect, investigations should be done to show the most economical and efficient way to use small cells, given the cost they add to the backhaul transmission and power consumption.
- *On-demand network management*: Today's network is designed for peak traffic load and the functions of coverage provision and capacity provision are closely coupled together. However, network traffic has a dynamic nature, which appears as spatial and temporal variations, as elaborated in [38]. One intuitive solution is to switch off some base stations when the capacity offered by them is not necessary, and quickly open more cells "on-demand." The efficiency of this scheme depends largely on the network architecture, which is efficient in nature for multi-tier heterogeneous networks. 3GPP is currently discussing the issue in [39], in which how to efficiently open the right cells is identified as the most difficult problem, which is still open for further study. [40] and [41] also suggest solutions for homogeneous deployment, where quick coverage compensation from neighboring cells is important.

1.5 Conclusion

In this chapter, a theoretical framework for green radio research based on four fundamental trade-offs has been proposed to address the future energy-consumption challenges for wireless systems. Although the basic trade-off relations for the point-to-point single cell scenario can be simply derived through the famous Shannon's formula, the practical system limitations in general break the monotonic curves as shown in Section 1.3. With the evolution of network architecture and transmission technologies, trade-off relations deserve further investigation in the sense that they will bring design insights for the improvement of the energy efficiency of future wireless systems. Moreover, how to use trade-off relations for system performance trade-off, and how to jointly design network

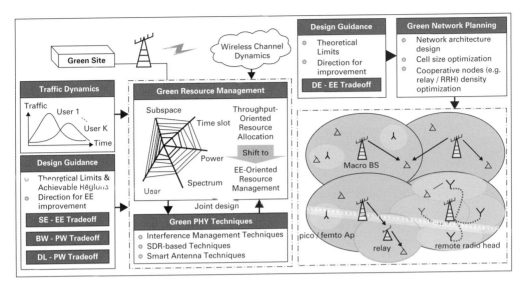

Figure 1.10 The application of the fundamental trade-offs in wireless design.

architecture based on trade-off relations, are expected to bring a green future for wireless networks. Finally, Figure 1.10 demonstrates a whole picture of how the proposed framework will impact the green design of future systems.

As the market develops, wireless networks will continue to expand in the future. Green evolution, as a result, will continue to be an urgent demand and inevitable trend for operators, equipment manufacturers, and other related industries. Progress in fundamental green research, as outlined in this chapter, will certainly help in making a green future.

References

[1] G. P. Fettweis and E. Zimmermann, "ICT energy consumption - trends and challenges," in *Proc. of 11th International Symposium on Wireless Personal Mulitimedia Communications*, Lapland, Finland, Sept. 2008.

[2] Cisco, "Cisco Visual Networking Index: Global Mobile Data Traffic Forecast Update, 2010˜C2015," in *White Paper*, Feb. 2011.

[3] www.vodafone.com/start/media_relations/news/group_press_releases/2007/01.html

[4] Huawei Technologies Co., Ltd., "Improving energy efficiency, lower CO_2 emission and TCO," in *Energy Efficiency Solution White Paper*. [Online] Available: www.huawei.com/en/static/hw-076768.pdf

[5] R. Irmer, "Evolution of LTE - operator requirements and some potential solutions," in *Proc. of 5th International FOKUS IMS Workshop*, Berlin, Germany, Nov. 2009.

[6] R. Tafazolli, "EARTH - energy aware radio and network technologies," in *Proc. of Next Generation Wireless Green Networks Workshop*, Paris, France, Nov. 2009.

[7] L. Correia, *et al.*, "Challenges and enabling technologies for energy aware mobile radio networks," *IEEE Communications Magazine*, vol. 48, no. 11, pp. 30–37, Nov. 2010.

[8] C. Han, *et al.*, "Green radio: radio techniques to enable energy-efficient wireless networks," *IEEE Communications Magazine*, vol. 48, no. 11, pp. 46–56, Nov. 2010.

[9] Y. Chen, *et al.*, "Fundamental trade-offs on green wireless communications," *IEEE Communications Magazine*, vol. 49, no. 6, Jun. 2011.

[10] C. E. Shannon, "A mathematical theory of communication," *Bell System Technical Journal*, vol. 27, pp. 379–423, 1948.

[11] L. Kleinrock, *Queueing Systems Volume I: Theory*. John Wiley & Sons, 1975.

[12] D. P. Bertsekas, *Dynamic Programming - Deterministic and Stochastic Models*. New Jersey, NJ, USA; Prentice Hall, 1987.

[13] O. Arnold, *et al.*, "Power-consumption modeling of different base station types in heterogeneous cellular networks," in *Proc. of Future Network and Mobile Summit*, 2010.

[14] S. Cui, A. J. Goldsmith, and A. Bahai, "Energy-constrained modulation optimization," *IEEE Trans. Wireless Commun.*, vol. 4, no. 5, pp. 2349–2360, Sept. 2005.

[15] G. Miao, *et al.*, "Cross-layer optimization for energy-efficient wireless communications: a survey," *Wiley Journal Wireless Communications and Mobile Computing*, vol. 9, pp. 529–542, Apr. 2009.

[16] G. Miao, *et al.*, "Interference-aware energy-efficient power optimization," in *Proc. of IEEE International Communications Conference (ICC)*, Dresden, Germany, Jun. 2009.

[17] Y. Chen, S. Zhang, and S. Xu, "Impact of non-ideal efficiency on bits per joule performance of base station transmissions," in *Proc. of IEEE Vechicular Technology Conference (VTC)*, May 2011.

[18] Y. Chen, S. Zhang, and S. Xu, "Characterizing energy efficiency and deployment efficiency relations for green architecture design," in *Proc. of IEEE International Communications Conference (ICC)*, Cape Town, South Africa, May 2010.

[19] H. Kim, *et al.*, "A cross-layer approach to energy efficiency for adaptive MIMO systems exploiting spare capacity," *IEEE Trans. Wireless Commun.*, vol. 8, no. 8, pp. 4264–4275, Aug. 2009.

[20] J. Mitola and G. Q. Maguire, "Cognitive radio: making software radios more personal," *IEEE Personal Commun. Mag.*, vol. 6, no. 4, pp. 13–18, Jun. 1999.

[21] S. Zhang, Y. Chen, and S. Xu, "Improving energy efficiency through bandwidth, power, and adaptive modulation," in *Proc. of IEEE Vechicular Technology Conference (VTC)*, Ottawa, Canada, Sept. 2010.

[22] S. Zhang, Y. Chen, and S. Xu, "Joint bandwidth-power allocation for energy efficient transmission in multi-user systems," in *Proc. of IEEE GlobeCom*, Nov. 2010.

[23] D. Grace, *et al.*, "Using cognitive radio to deliver green communications," in *Proc. of 4th International Conference on Cognitive Radio Oriented Wireless Networks and Communications (CROWNCOM)*, Jun. 2009.

[24] E. Uysal-Biyikoglu, B. Prabhakar, and A. E. Gamal, "Energy-efficient packet transmission over a wireless link," *IEEE/ACM Trans. Netw.*, vol. 10, no. 4, pp. 487–499, Aug. 2002.

[25] V. Lau, "Delay optimal cross-layer design for SDMA/OFDMA systems – stochastic decomposition and stochastic learning," *Invited Talk*, Stanford 2009. [Online]. Available: www.ece.ust.hk/~eeknlau/HKUST-Office-HomePage/Books_Papers_and_Patents_files/Delay-Optimal-Cross-Layer-Design-v1.0.pdf

[26] D. Wu and R. Negi, "Effective capacity: a wireless link model for support of quality of service," *IEEE Transactions on Wireless Communications*, vol. 2, no. 4, pp. 630–643, Jul. 2003.

[27] D. S. W. Hui, V. K. N. Lau, and H. L. Wong, "Cross-layer design for OFDMA wireless systems with heterogeneous delay requirements," *IEEE Transactions on Wireless Communications*, vol. 6, no. 8, pp. 2872–2880, Aug. 2007.

[28] R. A. Berry and R. Gallager, "Communication over fading channels with delay constraints," *IEEE Trans. Inf. Theory*, vol. 48, no. 5, pp. 1135–1148, May 2002.

[29] M. J. Neely, "Optimal energy and delay trade-offs for multi-user wireless downlinks," *IEEE Trans. Inf. Theory*, vol. 53, no. 9, pp. 3095–3113, Sept. 2007.

[30] V. Lau and Y. Chen, "Delay-optimal power and precoder adaptation for multi-stream mimo systems," *IEEE Trans. Wireless Commun.*, vol. 8, no. 6, pp. 3104–3111, Jun. 2009.

[31] K. Johansson, "Cost-effective deployment strategies for heterogeneous wireless networks," Ph.D. dissertation, KTH Information and Communication Technology, Stockholm, Sweden, Nov. 2007.

[32] H. Claussen, L. T. W. Ho, and L. G. Samuel, "Financial analysis of a pico-cellular home network deployment," in *Proc. of IEEE International Communications Conference (ICC)*, Glasgow, Scotland, Jun. 2007.

[33] B. Badic, *et al.*, "Energy efficiency radio access architectures for green radio: large versus small cell size deployment," in *Proc. of IEEE 70th Vehicular Technology Conference (VTC Fall)*, Anchorage, USA, Honolulu, USA, Dec. 2009.

[34] A. J. Febske, F. Richter, and G. P. Fettweis, "Energy efficiency improvements through micro sites in cellular mobile radio networks," in *Proc. of 2nd GreenCom Workshop, parallel with IEEE GLOBECOM*, Honolulu, USA, Dec. 2009.

[35] F. Richter, *et al.*, "Traffic demand and energy efficiency in heterogeneous cellular mobile radio networks," in *Proc. of IEEE VTC Spring*, Taipei, China, May 2010.

[36] A. Fehske, P. Marsch, and G. Fettweis, "Bit per joule efficiency of cooperating base stations in cellular networks," in *Proc. of IEEE Globecom, GreenComm Workshop*, Miami, Dec. 2010.

[37] M. A. Imran and R. Tafazolli, "Energy effiiency analysis of idealized cooperated multi-point communication system (CoMP)," in *Proc. of IEEE PIMRC Green Workshop*, Istanbul, Turkey, Sept. 2010.

[38] E. Oh, *et al.*, "Toward dynamic energy-efficient operation of cellular network infrastructure," *IEEE Communications Magazine*, vol. 49, no. 6, pp. 56–61, Jun. 2011.

[39] 3GPP draft TR 36.927, "Potential solutions for energy saving for E-UTRAN." [Online]. Available: www.3gpp.org/ftp/Specs/html-info/36927.htm

[40] Z. Niu, *et al.*, "Cell zooming for cost-efficient green cellular networks," *IEEE Communications Magazine*, vol. 48, no. 11, Jun. 2010.

[41] M. A. Marsan, *et al.*, "Optimal energy savings in cellular access networks," in *Proc. of IEEE Globecom, GreenCom Workshop*, Dec. 2009.

2 Algorithms for energy-harvesting wireless networks

Vinod Sharma, Utpal Mukherji, and Vinay Joseph

2.1 Introduction

Carbon emissions due to various man-made devices are one of the main reasons for global warming. Green communications is largely concerned with reducing the carbon emissions caused during communications. One of the main reasons for carbon emissions in communication is the emission in generating the electrical energy consumed in communication networks. Soon, the electrical energy consumed in information and communication technology (ICT) related activities will be about 14% of all the electrical energy consumed in the world [1]. A large part of this is consumed in wireless communication networks (where the base station consumes most of the energy). Thus, reducing the carbon emission due to the energy used in wireless communications will have a significant environmental impact. This can be done in two ways. One of course is that we should reduce the electrical energy consumed in communications. Recent work has shown that the energy consumed can easily be reduced to one-third [1, 2]. The other way is to generate the energy used in a way that reduces (or totally eliminates) carbon emission. If one generates electricity via coal or gas installations, typically 800–950 gram CO_2 (equivalent) is emitted per kwh. However, if one generates it via solar cells or wind turbines there is no emission at all [1, 2]. Therefore, there has been tremendous interest in using solar and/or wind energy at the base stations (BSs) of cellular systems [2]. This will not only reduce the carbon footprint due to wireless communications but also enable BSs to be installed at rural, remote, isolated places in the world (which currently use diesel gensets causing pollution and carbon emissions in addition to being expensive). Future growth of wireless communication will mostly be in such areas. BSs operated via solar/wind energy require extra initial investment but will have much lower operating costs, recovering the extra initial investments within a few years. Recently, solar cell operated cell phones have also become available.

One of the problems with generating electricity from solar and/or wind energy (as against the regular power supply or from a generator) is that the energy generated at any time is randomly changing with time [3, 4]. For example, a solar cell will not generate any electricity at night. Thus, the communication system will need to carefully use and conserve its energy such that it is able to satisfy its requirements and is not starved of energy at any time (called energy neutral operation in [4]). This is different from the usual scenario one comes across: either there is a regular power supply through a power point or a genset (at a BS) or there is a battery at a mobile phone. A battery has a fixed amount of energy (ignoring the recharging and leakage effects) and it goes dead once the energy is exhausted. A regular power supply can effectively provide any energy needed at any given time reliably. In the case of an energy-harvesting source, there is potentially an infinite amount of energy available over time, but at a given time only a random limited amount is available. A communication system with this setup has not been studied much except recently in the context of energy-harvesting sensor networks [1], [4]–[5]. The models studied in the context of sensor networks are applicable in the present scenario.

One other method recommended for reducing the energy consumed in a wireless system is that the BS and/or the cell phones are operated in energy-saving modes (sleep/OFF) when there is not enough traffic [5]. For example, these modes in a consumer premises equipment in a WiMAX system have been studied in [6]. Also, BSs with sleep mode are now commercially available.

In this chapter we study communication systems using energy-harvesting sources. We address the issue of energy management such that the requirements of the users are met while the system is not starved of energy at any time. We also develop algorithms that maximize system performance by exploiting the sleep mode judiciously. One of the features of our work is that using our approach, one can use the usual MAC and multi-hop techniques and not worry about the random unreliable supply of the energy harvester. Furthermore, our algorithms are optimal or close to optimal. Implementation of a green base station using solar energy and adaptively varying its power as a function of the energy in its battery is presented in [7].

Power control for energy-harvesting sensor nodes has been studied in [8] and [4]. The information capacity for a point-to-point Gaussian channel without leakage and processing energy is available in [9]. Energy-saving communications, modulation and MAC are studied in [10]–[11]. Routing protocols for such networks have been studied in [12]–[14]. Joint optimization of power and scheduling is considered in [15]. In [16], the data arrival rate and transmit power are jointly controlled for a single node to maximize the throughput subject to the stability of the data queue. Energy conservation in a wireline backbone internet has been recently studied in [17]–[18].

This chapter is organized as follows. In Section 2.2 we briefly explain the energy-harvesting technologies available today. In Section 2.3 we make a basic model of an energy-harvesting point-to-point channel. The energy harvested is stored in an energy queue (battery) and the data to be transmitted is stored in a data buffer. We first find the stability region of the data queue when the energy is spent only in transmission. Next, we consider the more realistic case when the energy is also spent in processing and other activities at the BS/nodes and there may be leakage in the battery storing the energy. Finally, we obtain the Shannon capacity of such a channel. These results are reported in [6], [19], [20], and [21]. Next in Section 2.4 we provide efficient algorithms for a multiple access channel (MAC) [22] when each user has an energy-harvesting source. This corresponds to the case in a cellular system when each cell phone is using a solar cell. Finally in Section 2.5 we consider a multi-hop wireless network with energy-harvesting nodes [23]. For point-to-point channel and multihop scenario we also study the system with optimal ON and OFF modes of the nodes. Section 2.6 concludes the chapter and discusses future directions.

2.2 Energy-harvesting technologies

Solar and wind power are the main renewable energy sources being considered for wireless communications. The advantages of solar/wind energy of course are that such energy is maintenance-free, pollution-free, requires no transport or storage of oil/diesel, and is immune to the cost escalation of fuel. In the following we briefly describe these two technologies in our context.

The power requirements of BSs have recently been cut down by powering down the BS outside of peak hours, shortening of cables to the power amplifier, overall better design of power amplifiers, and reducing the cooling costs, etc. Thus, now there are commercially available BSs that consume less than 200W. This makes solar or wind powered base stations a practical solution requiring fewer and smaller solar panels and wind turbines.

Solar power is generated from light via photo-voltaic cells. With modern technology, a 10 sq ft. solar panel can generate 100–200W at 12 volt. One can combine panels in parallel or series to increase the current and/or voltage. Today, Wi-Fi-mesh, WiMAX and cellular (3G) BSs are commercially available that are powered by solar energy. For example VNL has developed a small GSM base station which requires only 50–100W (as against the 3000W usually required) and hence can work with a single solar panel. It can support hundreds of users and is well suited to rural areas.

Three percent of all of the sun's energy to earth is converted to wind energy. The world's total current requirements will be met if 3% of that is converted to electrical energy. However, only 12–15% of the world's land mass has the potential for large wind power installations. But 60–65% of the land mass has potential for small-scale wind energy. For example 50W–5KW of energy can be generated at low wind speeds by a micro wind turbine. One can generate 10KW–50KW from wind speeds beyond 5 m/sec. Thus, there is a great potential to use micro wind turbines to power wireless BSs. Indeed, today BSs powered by wind turbines are already available.

A combination of solar and wind energy can also be used to build a more robust system (Figure 2.1). In countries like India, often there is sufficient sunshine (during the day), but in the rainy season sunlight may be insufficient due to cloud cover. Thus, wind energy can supplement solar energy on those days and of course at night. Storing the energy generated by wind and/or solar energy in a battery further mitigates the effects of random energy generation by these sources. Such green BSs are described in [24].

Efficient deployment of such a BS at a site requires an initial study of the site for solar and wind data as well as the expected traffic load. Furthermore, it is critical to use high efficiency rectifiers, backup batteries, inverters (to convert DC to AC), air conditioning unit, and an intelligent control unit, to judiciously mix energy from solar, wind and back-up batteries (Figure 2.1).

The energy spent in a BS has the following components:

1. Energy for baseband processing: modulation, coding and transmission interface processing.
2. Energy for transmission to the BSC/RNC.

2.2 Energy-harvesting technologies

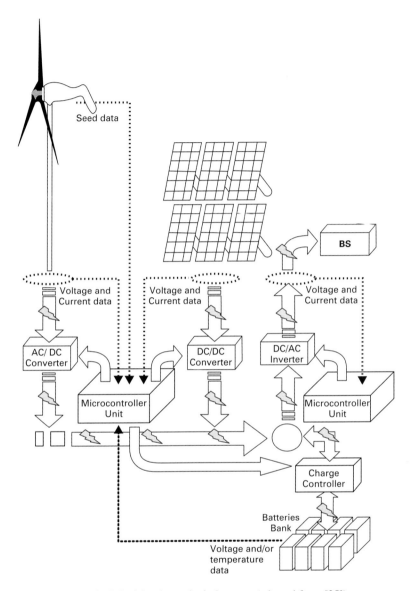

Figure 2.1 Block diagram of a BS with solar and wind energy (adapted from [25]).

3. Energy dissipated in cable feeders, AC/DC converters, etc.
4. Energy used in transmission of data and control signals.

We will combine the first three components into one (which will be assumed random but uncontrolled). The last, for transmission (which constitutes about 65% of the total energy spent in a BS), will be subject to energy management. This is justified via recent implementations of the green BS [7].

Our modeling and energy-management schemes in this chapter will illustrate that it is possible to design reliable systems which meet the QoS requirements of practical systems based on solar and wind energy.

2.3 Point-to-point channel

In this section we describe our basic model for a point-to-point channel with the transmitter powered by an energy harvester. The harvester could be using a solar, wind or any other energy source or a combination of them. This model can be relevant for downlink when a BS is powered by an energy harvester.

Section 2.3.1 provides the basic assumptions and notations used. We assume that all the energy is consumed in transmission only. Section 2.3.2 studies the stability of the data queue. Section 2.3.3 obtains delay optimal policies. Section 2.3.4 extends the model to also consider energy consumption in data processing and other activities at the transmitter and when the channel experiences fading. Section 2.3.5 compares the different energy-management algorithms obtained via simulations. Section 2.3.6 provides algorithms that improve the system performance by including the sleep option. Finally, Section 2.3.7 provides the information theoretic capacity of such a system.

2.3.1 Model and notation

We consider a transmitter (Figure 2.2) generating packets to be transmitted to a receiver node. The system is slotted. During slot k (defined as time interval $[k, k+1]$, i.e. a slot is a unit of time), X_k bits are generated. Although the transmitter may generate data as packets, we allow arbitrary fragmentation of packets during transmission. Thus, packet boundaries are not important and we consider bit strings (or just fluid). The bits X_k are eligible for transmission in the $(k+1)$th slot. The queue length (in bits) at time k is q_k. The transmitter is able to transmit $g(T_k)$ bits in slot k if it uses energy T_k. Initially, for simplicity, we assume that transmission consumes most of the energy at the transmitter and ignore other causes of energy consumption. This assumption is removed in Section 2.3.4. We denote by E_k the energy available in the node at time k. The transmitter is

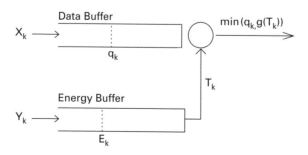

Figure 2.2 Model of an energy-harvesting point-to-point channel.

able to replenish energy by Y_k in slot k. The $\{Y_k\}$ sequence could possibly be generated by a solar cell, a wind turbine or a superposition of such energy-harvesting sources. We initially assume that $\{X_k\}$ and $\{Y_k\}$ are i.i.d. but will generalize this assumption later.

The processes $\{q_k\}$ and $\{E_k\}$ satisfy

$$q_{k+1} = (q_k - g(T_k))^+ + X_k, \qquad (2.1)$$

$$E_{k+1} = (E_k - T_k) + Y_k, \qquad (2.2)$$

where $T_k \leq E_k$. This assumes that the data buffer and the energy storage buffer are infinite.

The function g is assumed to be monotonically non-decreasing. An important such function is given by Shannon's capacity formula $g(T_k) = \frac{1}{2}log(1+\beta T_k)$ bits per channel use for Gaussian channels, where β is a constant such that βT_k is the SNR. This is a non-decreasing concave function. At low values of T_k, $g(T_k) \sim \beta_1 T_k$, i.e. g becomes a linear function. Thus in the following we limit our attention to linear and concave nondecreasing functions g. We also assume that $g(0) = 0$, which always holds in practice.

Many of our results (especially the stability results) will be valid when $\{X_k\}$ and $\{Y_k\}$ are (periodic) stationary, ergodic. These assumptions are general enough to cover most of the stochastic models developed for traffic (e.g. Markov modulated) and energy harvesting [1].

In Section 2.3.2 we study the stability of this queue and identify easily implementable energy-management policies that provide good performance.

2.3.2 Stability

We obtain a necessary condition for stability. Then we present a transmission policy that achieves the necessary condition, and show that it is also sufficient and is throughput optimal. The mean delay for this policy is not minimal. Thus, we obtain other policies that provide a lower mean delay. In the next section we consider delay optimal policies. These results are from [21] and [22].

Let us assume that we have obtained an (asymptotically) stationary and ergodic transmission policy $\{T_k\}$ which makes $\{q_k\}$ (asymptotically) stationary with the limiting distribution independent of q_0. Taking $\{T_k\}$ asymptotically stationary seems to be a natural requirement to obtain (asymptotic) stationarity of $\{q_k\}$. The following Lemma gives a necessary condition for the stability of such a policy. In the following $r.v.$ X will denote a generic $r.v.$ with the distribution of X_k. Similarly we denote the other $r.v.s$.

Lemma 1: Let g be concave nondecreasing and $\{(X_k, Y_k)\}$ be stationary, ergodic sequences. For $\{T_k\}$ to be an asymptotically stationary, ergodic energy-management policy that makes $\{q_k\}$ asymptotically stationary (jointly with $\{(T_k, X_k, Y_k)\}$) with a proper stationary distribution π, it is necessary that $E[X] < E_\pi[g(T)] \leq g(E[Y])$. ∎

Now, we present a policy that satisfies the above necessary condition. Let

$$T_k = min(E_k, E[Y] - \epsilon), \qquad (2.3)$$

where ϵ is a small positive constant with $E[X] < g(E[Y] - \epsilon)$. We show in Theorem 1 below that it is a throughput optimal policy. In other words, this policy keeps the queue stable if it is possible to do it by any (asymptotically) stationary, ergodic transmission policy.

Theorem 1: If $\{(X_k, Y_k)\}$ are stationary, ergodic, g is continuous, nondecreasing, concave, then if $E[X] < g(E[Y])$, (2.3) makes the queue stable (with $\epsilon > 0$ such that $E[X] < g(E[Y] - \epsilon)$), i.e. it has a unique, stationary, ergodic distribution and starting from any initial distribution, q_k converges in total variation to the stationary distribution. ∎

Thus, (2.3) is throughput optimal and we denote it by TO.

Let us consider a policy that does not store the harvested energy. Then, $T_k = Y_k$ for all k is a throughput optimal policy under this constraint. It provides stability of the queue if $E[X] < E[g(Y)]$. If g is linear then this coincides with the necessary condition in Lemma 1. If g is strictly concave then $E[g(Y)] < g(E[Y])$ unless $Y \equiv E[Y]$. Thus this policy provides a strictly smaller stability region. We will be forced to use this policy if there is no buffer to store the energy harvested. We will see in Section 2.3.5 that storing energy can also provide lower mean delays.

Although TO is a throughput optimal policy, it does not minimize the mean delay. The Greedy policy

$$T_k = min(E_k, f(q_k)), \qquad (2.4)$$

where $f = g^{-1}$, looks promising [22]. In Theorem 2, we will show that the stability condition for this policy is $E[X] < E[g(Y)]$ which is throughput optimal for linear g but strictly suboptimal for a strictly concave g. We will also show in Section 2.3.3 that when g is linear, equation (2.4) also minimizes the long-term mean delay.

In the next few results we assume that the energy buffer is finite, although large. For this case Lemma 1 and Theorem 1 also hold under the same assumptions.

Theorem 2: If the energy buffer is finite, i.e. $E_k \leq \bar{e} < \infty$ (but $\bar{e}$ is large enough) and $E[X] < E[g(Y)]$ then under the greedy policy (2.4), (q_k, E_k) has an ergodic set. ∎

The above result ensures that the Markov chain $\{(q_k, E_k)\}$ is ergodic and hence has a unique stationary distribution if $\{(q_k, E_k)\}$ is irreducible. A sufficient condition for this is $0 < P[X_k = 0] < 1$ and $0 < P[Y_k = 0] < 1$. In general, $\{(q_k, E_k)\}$ can have multiple ergodic sets. Then, depending on the initial state, $\{(q_k, E_k)\}$ converges to one of the ergodic sets and the limiting distribution depends on the initial conditions.

2.3.3 Delay optimal policies

In this section, we consider delay optimal policies. We choose T_k at time k as a function of q_k and E_k such that $E\left[\sum_{k=0}^{\infty} \alpha^k q_k\right]$ is minimized, where $0 < \alpha < 1$ is a suitable constant. This minimizing policy is called α-discount optimal. When $\alpha = 1$, we minimize $\lim_{n\to\infty} \sup \frac{1}{n} E\left[\sum_{k=0}^{n-1} q_k\right]$. This optimizing policy is called average cost optimal. By Little's law [27], an average cost optimal policy also minimizes mean delay. If for a given (q_k, e_k), the optimal policy T_k does not depend on the past values, and is time invariant, it is called a stationary Markov policy.

Theorem 3 [21]: Let $\{X_k\}, \{Y_k\}$ be i.i.d. with values in discrete sets or having probability densities. Also, let g be continuous and let the energy buffer be finite, i.e. $e_k \leq \bar{e} < \infty$. Then there exists an optimal α-discounted Markov stationary policy. If in addition, $E[X] < g(E[Y])$ and $E[X^2] < \infty$, then there exists an average cost optimal stationary Markov policy. The optimal cost v does not depend on the initial state. Also, then the optimal α-discount policies tend to an optimal average cost policy as $\alpha \to 1$. Furthermore, if $v_\alpha(q, e)$ is the optimal α-discount cost for the initial state (q, e) then $\lim_{\alpha \to 1}(1 - \alpha) \inf_{(q,e)} v_\alpha(q, e) = v$. ∎

In general one can compute a delay optimal policy numerically via Value Iteration or Policy Iteration. But this can be computationally intensive (especially for large data and energy buffer sizes).

In Section 2.3.2, we also provided a greedy policy (2.4) which is very intuitive, and is throughput optimal for linear g. However, for concave g (including the cost function $g(t) \frac{1}{2} log(1 + \beta t)$) it is *not* throughput optimal and provides low mean delays only for low load. However, we have the following.

Theorem 4 [21]: The Greedy policy (2.4) is α-discount optimal for $0 < \alpha < 1$ when $g(t) = \gamma t$ for some $\gamma > 0$. It is also average cost optimal. ∎

2.3.4 Generalizations

In this section, we make our model more realistic by relaxing several assumptions made in Section 2.3.1. In particular, we take into account energy inefficiency in storing energy in the energy buffer and energy leakage from the energy buffer. Next, we consider channel fading and the energy consumption in sensing and processing.

We assume that if energy Y_k is generated in slot k, then only energy $\beta_1 Y_k$ is stored in the buffer where $0 < \beta_1 < 1$ and that in every slot, energy β_2 leaks from the buffer, $0 < \beta_2 < \infty$. These seem to be realistic assumptions [4]. Then (2.2) becomes

$$E_{k+1} = ((E_k - T_k) - \beta_2)^+ + \beta_1 Y_k. \tag{2.5}$$

Now, Lemma 1 and Theorem 1 continue to hold with obvious changes and for energy neutral operation, our TO policy becomes $T_k = min(E_k, \beta_1 E[Y] - \beta_2 - \epsilon)$ for a small positive ϵ. A similar modification in (2.4) provides a greedy policy. Theorems 2, 3, and 4 continue to hold with these changes.

The modification (2.5) reduces the stability region to $E[X] < g(\beta_1 E[Y] - \beta_2)$. If the energy harvested Y_k in slot k is immediately used (instead of first being stored in the buffer) then (2.5) becomes $E_{k+1} = ((E_k + Y_k) - T_k - \beta_2)^+$ and the stability region increases to $E[X] < g(E[Y] - \beta_2)$.

We consider two further generalizations. First we extend the results to the case of fading channels and then to the case where the processing energy at the transmitter node is non-negligible with respect to the transmission energy. At a BS and/or a cell phone this will be true.

For fading channels, we assume flat fading during a slot. In slot k the channel gain is H_k. The sequence $\{H_k\}$ is assumed stationary, ergodic, and independent of $\{X_k\}$ and

$\{Y_k\}$. If T_k energy is spent in transmission in slot k, the $\{q_k\}$ process evolves as

$$q_{k+1} = (q_k - g(H_k T_k))^+ + X_k. \qquad (2.6)$$

If the channel state information (CSI) is not known to the transmitter, then T_k will depend only on (q_k, E_k). For the policy $T_k = min(E_k, E[Y] - \epsilon)$, the data queue is stable if $E[X] < E[g(H(E[Y] - \epsilon))]$. We will call this policy unfaded TO. If we use Greedy (2.4), then the data queue is stable if $E[X] < E[g(HY)]$.

If CSI H_k is available to both the transmitter and receiver then the following are the throughput optimal policies. If g is linear, then $g(x) = \alpha x$ for some $\alpha > 0$. Then, if $0 \leq H \leq \bar{h} < \infty$ and $P(H = \bar{h}) > 0$, the optimal policy is: $T(\bar{h}) = (E[Y] - \epsilon)/P(H = \bar{h})$ and $T(H) = 0$ otherwise. Thus if H can take an arbitrarily large value with positive probability, then $E[HT(H)] = \infty$ at the optimal solution.

If $g(t) = \frac{1}{2}log(1 + \beta t)$, then the water filling (WF) policy

$$T_k(H) = \left(\frac{1}{H_0} - \frac{1}{H}\right)^+ \qquad (2.7)$$

with the average power constraint $E[T_k] = E[Y] - \epsilon$, is throughput optimal because it maximizes $\frac{1}{2}E_H[log(1 + \beta HT(H))]$ with the given constraints. For a general concave g, we can obtain the optimal power control policy numerically with the average power constraint $E[T_k] = E[Y] - \epsilon$.

Until now we have assumed that all the energy that a node consumes is for transmission. However, the transmitter spends a significant amount of energy on various other activities. Now, we include this in our model. Later on we will also consider energy-saving modes that can improve the performance in the present setup.

We assume that Z_k is the energy consumed by the node for processing and other miscellaneous operations in slot k. Unlike T_k (which can vary according to q_k), $\{Z_k\}$ can be considered a stationary ergodic sequence. The rest of the system is as in Section 2.3.1.

Let c be the minimum positive constant such that $E[X] < g(c)$. Then if $c + E[Z] < E[Y] - \delta$ (where δ is a small positive constant), the system can be operated in energy neutral operation: If we take $T_k \equiv c$ (which can be done with high probability for all k large enough), the process $\{q_k\}$ will have a unique stationary, ergodic distribution and there will always be energy Z_k for sensing and processing for all k large enough. The result holds if $\{(X_k, Y_k, Z_k)\}$ is an ergodic stationary sequence. When the channel has fading, we need $E[X] < E[g(cH)]$ in the above paragraph.

2.3.5 Simulations

In this section, we compare the different policies we have studied via simulations. We provide results only for fading channels.

Figure 2.3 is for linear g and Figure 2.4 is for nonlinear g. The policies compared are unbuffered, Greedy (2.4), Unfaded TO and Fading, Greedy, Unfaded TO (2.4) and

2.3 Point-to-point channel

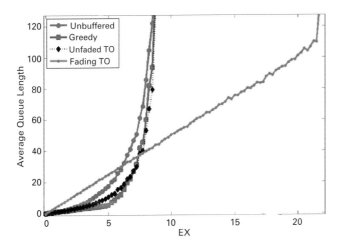

Figure 2.3 Comparison of policies with fading and linear g.

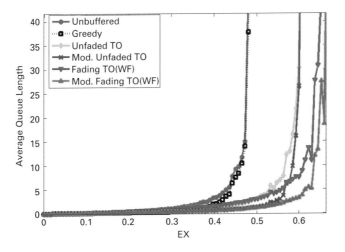

Figure 2.4 Comparison of policies with fading; $g(x) = \log(1+x)$.

Fading TO (WF) (2.7). In Figure 2.4, we have also plotted Modified Unfaded TO and Modified Fading TO (WF) which are improvements over the TO and WF policies. These are explained in [21].

In Figure 2.3, X and Y have hyperexponential distributions. The distribution of r.v. X is a mixture of 5 exponential distributions with means $E[X]/4.9, 2E[X]/4.9, 3E[X]/4.9, 6E[X]/4.9$ and $10E[X]/4.9$ and probabilities $0.1, 0.2, 0.2, 0.3,$ and 0.2, respectively. The distribution of Y is obtained in the same way.

$\{H_k\}$ is i.i.d. taking values $0.1, 0.5, 1.0,$ and 2.2 with probabilities $0.1, 0.3, 0.4,$ and 0.2, respectively. Now $E[Y] = 1$, $E[g(HY)] = 10$ and $E[g(HE[Y])] = 10$. We see that the stability region of fading TO is $E[X] < E[g(\bar{h}Y)]$ ($= 22.0$) while that of the other three algorithms is $E[X] < 10$. However, the mean queue length of fading TO is also larger from the beginning until almost 10. This is because in fading TO, we transmit only when $H = \bar{h} = 2.2$, which has a small probability ($= 0.2$) of occurrence. This example shows that for a linear g, CSI can significantly improve the stability region.

Figure 2.4 considers nonlinear $g(x) = log(1+x)$. We now consider a more realistic channel obtained by discretizing Rayleigh fading into ten values. X, Y are hyperexponential as in Figure 2.3; and $E[Y] = 1, E[g(HY)] = 0.51$ and $E[g(HE[Y])] = 0.64$. Now we see that the stability region of unbuffered and Greedy is the smallest, then of TO and MTO, while WF and MWF provide the largest region and are stable for $E[X] < 0.70$. MTO and MWF provide improvements in mean queue lengths over TO and WF.

In [21] we have applied these policies in a real-life setting, taking solar radiation from actual measurements and applied on a Mica2 sensor node. The g function for this sensor mote is almost linear.

2.3.6 Model with sleep option

In this section, we present our model for a single energy-harvesting node with two modes: Wake and Sleep. This option is increasingly becoming available at BSs and mobile phones. Sleep option is useful only when Z_k is non-negligible, which of course is always true in case of a BS or a mobile phone.

We will assume that the data buffer and energy buffer have infinite capacity. The sequence $\{(X_k, Y_k, Z_k)\}$ will be assumed to be (asymptotically) stationary, ergodic and sometimes even Markov. We can even take $Z_k = H(X_k)$, a deterministic function of X_k. If, in a slot, the energy $E_k < Z_k$, then, in that slot, the node cannot even do its basic operations of processing and hence will enter the sleep mode. Now, X_k is the number of bits that would be generated and Z_k is the amount of energy that would be used for processing if the node was awake in slot k. In sleep mode, the node shuts down all its operations except energy harvesting and thus consumes a negligible amount of energy (but does keep its data already stored in the buffer). The sleep mode can also be used at other times to optimize the overall performance.

We have shown in Section 2.3.4 that for energy neutral operation without the sleep option, it is necessary (and sufficient) that

$$E[X] < g\left((E[Y] - E[Z])^+\right).$$

If this condition is not satisfied, then one needs to introduce the sleep mode for energy neutral operation. We consider such a scenario in the following and develop optimal sleep-wake policies for a single node.

We define S_k to denote the mode of the node in slot k: $S_k = 0$ implies the node is awake and $S_k = 1$ implies it is sleeping.

The processes $\{q_k\}$ and $\{E_k\}$ satisfy

$$\begin{aligned} q_{k+1} &= \left(q_k - I_{\{S_k=0\}} g(T_k)\right)^+ + I_{\{S_k=0\}} X_k, \\ E_{k+1} &= \left((E_k - I_{\{S_k=0\}} T_k - I_{\{S_k=0\}} Z_k)\right) + Y_k, \end{aligned} \qquad (2.8)$$

where $T_k \leq E_k - Z_k$.

An optimal policy

We determine S_k and T_k as a function of q_k, E_k, X_{k-1}, Y_{k-1}, and Z_{k-1} such that the discounted cost

$$E\left[\sum_{k=0}^{\infty} \alpha^k \left(q_k + cI_{\{S_k=1\}}X_k\right)\right]$$

is minimized, where $0 < \alpha < 1$ and $0 < c < \infty$ are suitable constants. This cost takes into account two important performance measures: mean queue length and mean data loss rate (which happens when the node is sleeping). We also consider the average cost optimal policy minimizing

$$\lim_{n\to\infty} \sup \frac{1}{n} E\left[\sum_{k=0}^{n-1} \left(q_k + cI_{\{S_k=1\}}X_k\right)\right].$$

If, for a given $(q_k, e_k, x_{k-1}, y_{k-1}, z_{k-1})$, the optimal action S_k and T_k (when $S_k = 0$) is uniquely specified independent of k, the optimal policy is called a stationary policy.

For simplicity, in the following we consider these problems when $\{X_k\}$ and $\{Z_k\}$ are i.i.d. and $\{Y_k\}$ are modulated by a finite-state Markov chain.

Theorem 5: If g is continuous and the energy buffer is finite, i.e. $E_k \leq \bar{e} < \infty$, then there exist α-discount optimal and average cost optimal policies. Also, the optimal α-discount policy tends to the optimal average cost policy as $\alpha \nearrow 1$ and

$$\lim_{\alpha \to 1} (1-\alpha) \inf_{(q,e,y)} v_\alpha(q,e,y) = v,$$

where $v_\alpha(q, e, y)$ is the optimal α-discount cost for the initial state (q, e, y) and v is the optimal average cost that does not depend on the initial state. ∎

The optimal policy can be obtained via Value Iteration. However, it is computationally demanding. Next, we obtain two easily implementable suboptimal policies and compare their performance with that of the optimal policy.

Randomized sleep policy

In this policy, in every slot k with $E_k \geq Z_k$, the transmitter chooses to remain awake with probability $(1-p)$ independent of the other random variables, and in every slot that it remains awake it uses

$$T_k = \min\left\{\frac{E[Y]}{1-p} - E[Z] - \epsilon, E_k - Z_k\right\}, \quad (2.9)$$

where ϵ is a small positive constant. The p can be selected to have small average cost.

Energy threshold policy

In this policy, the sensor uses T_k given in (2.9) with $\epsilon = 0$ in every slot in which $S_k = 0$ where S_k is chosen as follows:

$$S_k = \begin{cases} 1, & \text{if } E_k \leq E_{T_1} \\ 0, & \text{if } E_k \geq E_{T_2}, \text{ and} \\ S_{k-1}, & \text{if } E_{T_1} < E_k < E_{T_2}, \end{cases}$$

where E_{T_1} and E_{T_2} are two energy levels such that $E_{T_1} \leq E_{T_2}$. We fix E_{T_1} and E_{T_2} so that the long-term fraction of sleep states is a given p. However, for a given p, the switching rate of this policy is considerably less than that of the randomized sleep policy for large enough $E_{T_2} - E_{T_1}$ (verified via simulations in [19]).

Simulation results

We compare average cost optimal, optimal randomized sleep and optimal energy threshold policies for $c = 5$ and 10 via simulations (obtained by optimizing p). We fix $E_{T_1} = 10$ units and $E_{T_2} = 60$ units. We take $\{X_k\}$ and $\{Y_k\}$ discrete valued i.i.d. random variables and $E[Y] = 2$ and $Z_k = 1$. From simulations, we found that average queue lengths and average throughputs, and hence optimal average costs, of optimal randomized sleep and optimal energy threshold policies are almost the same for different values of $E[X]$. Hence, in Figure 2.5 we compare only Optimal and Randomized Sleep Policies. We see that the average costs of the two policies are reasonably close to each other.

Throughput optimal policy

One disadvantage of the average cost optimal policy provided above is that it is computed numerically and we get minimal insight into the structure of the optimal policy. The suboptimal policies obtained, although useful, do pay a performance penalty. If throughput is the dominant criterion, we can, as in Section 2.3.4, obtain an explicit, easily computable policy which maximizes the throughput.

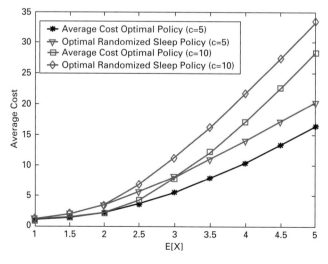

Figure 2.5 Comparison of average cost optimal policy and optimal randomized sleep (energy threshold) policy.

2.3 Point-to-point channel

As shown in Section 2.3.4, if

$$E[X] < g\left((E[Y] - E[Z])^+\right), \quad (2.10)$$

then the throughput optimal policy is: $S_k \equiv 0$ and $T_k = min\left\{(E_k - Z_k)^+, E[Y] - E[Z] - \epsilon\right\}$ where ϵ is an arbitrarily small positive constant. Now, we obtain the throughput optimal policy when (2.10) is not satisfied. From Lemma 1, we obtain that for $\{S_k, T_k\}$ to be an asymptotically stationary ergodic policy that is jointly stationary with $\{(q_k, E_k, X_k, Y_k, Z_k)\}$ and for which q_k has a proper distribution under stationarity, it is necessary that

$$E_\pi\left[X_k I_{\{S_k=0\}}\right] < E_\pi\left[g(T_k) I_{\{S_k=0\}}\right]$$
$$\leq (1-p) g\left(E_\pi[T_k | S_k = 0]\right), \quad (2.11)$$

where π is the stationary distribution and $p = P_\pi[S_k = 1]$, the probability of sleeping.

In [19], it is argued that the following is a good candidate for a throughput optimal policy:

In slot k, sleep with probability p^*, otherwise, sleep if $E_k < Z_k$. If awake, use

$$T_k = min\left\{E_k - Z_k, \frac{E_\pi[Y]}{1-p^*} - E_\pi[Z] - \epsilon\right\}$$

and admit X_k to the queue with probability $1 - \gamma$ and drop otherwise where p^* maximizes

$$(1-p) min\left\{E_\pi[X_k], g\left(\frac{E_\pi[Y]}{1-p} - E_\pi[Z_k] - \epsilon\right)\right\} \quad (2.12)$$

and γ is defined such that

$$(1-\gamma) E_\pi[X_k] = g\left(\frac{E_\pi[Y]}{1-p^*} - E_\pi[Z_k]\right) - \delta$$

for a positive δ. If δ is chosen to be small, we can have more throughput but we will also have larger queue lengths.

Energy threshold policy as a throughput optimal policy

The throughput optimal policy developed above corresponds to the randomized sleep policy developed above. It also has the drawback of high switching costs. Thus, it can be modified to obtain a policy corresponding to the Energy Threshold policy. Now the thresholds E_{T_1} and E_{T_2} should correspond to p^* obtained above. If $\{X_k\}$ and $\{Z_k\}$ are i.i.d., this will again provide a throughput optimal Policy (along with γ) but with much lower switching costs.

2.3.7 Fundamental limits of transmission

In this section we obtain the information theoretic capacity of a node with energy harvesting. We assume an additive white Gaussian channel (AWGN): if the node transmits

S_k, the receiver gets $S_k + W_k$ where W_k is a white Gaussian sequence with mean 0 and variance σ^2. Now, it is assumed that the data queue always has data to transmit and we obtain the maximum rate (capacity) at which the node can transmit reliably. All other assumptions and notation remain the same. The energy buffer has E_k energy at time k and hence in slot k the node can use only E_k for transmission as well as other activities. This makes this channel different from the usual AWGN channel with an average power constraint. The results reported in this section are from [20] and [21].

We provide the capacity of this system in the following scenario: (i) when the energy is consumed only in transmission; (ii) when non-negligible energy Z_k is consumed in processing and computations as well; (iii) Sleep option is available; and (iv) fading is experienced by the channel.

When energy is consumed only in transmission, we obtain the following.

Theorem 6: The capacity of the channel is $0.5 log(1 + E[Y]/\sigma^2)$. ∎

The above theorem says that there is no loss in capacity for the basic model due to the hard real-time energy constraint. The signaling scheme providing capacity is $S_k = Sgn(S'_K) min(\sqrt{E_k}, |S'_k|)$, where $\{S'_k\}$ is i.i.d. Gaussian with mean 0 and variance $E[Y]$, and $Sgn(x) = 1$ if $x \geq 0$ and $Sgn(x) = -1$ otherwise. S_k reflects the energy constraint at time k.

If Z_k energy is used for processing, then it is shown in [20] that a sleeping option can improve the capacity. Define

$$b(x) = \begin{cases} x^2 + \alpha, & \text{if } |x| > 0 \\ 0, & \text{if } |x| = 0. \end{cases}$$

The capacity of the system with the sleep option is given by

Theorem 7: For the system with processing energy,

$$C = \sup_{P_s : E[b(S)] \leq E[Y]} I(S, W+S) \qquad (2.13)$$

is the capacity of the system. ∎

The capacity-achieving signaling scheme for the system with the sleep option is a superposition of an ON-OFF code (OFF is the sleep condition with probability p) with an i.i.d. signal with density. Although we could not prove it, we have found the i.i.d. density is Gaussian with mean zero. The capacity (2.13) is plotted in Figure 2.6 for the optimal sleep probability along with other sleep probabilities p when $E[Z] = 0.5$ and $\sigma^2 = 1$. We observe that, when $E[Y] >> E[Z]$ there is no need to sleep. However, when $E[Y] < E[Z]$ then the capacity without sleep is zero. We can obtain a positive capacity with sleep. The figure also compares these results with those in [33].

In [20] achievable rates are provided when there is leakage in the energy storage and/or the efficiency of storing is less than 1 (e.g. if we try to store x energy we can only store $\beta x, \beta < 1$).

If the AWGN channel experiences flat fading H_k then the channel output is $H_k S_k + W_k$ where $\{H_k\}$ is an i.i.d. channel gain sequence independent of W_k. For this channel,

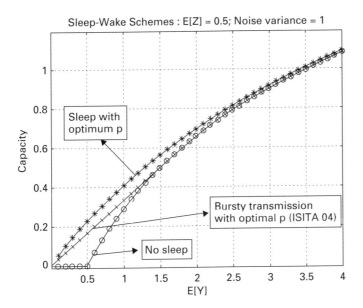

Figure 2.6 Comparison of sleep wake policies.

it is shown in [28] that when energy is consumed only in transmission, there are no energy-storage inefficiencies and perfect channel state information (CSI) is available at the transmitter and the receiver, the capacity is

$$\frac{1}{2} E_H \left[log \left(1 + \frac{HT^*(H)}{\sigma^2} \right) \right], \quad (2.14)$$

where $T^*(H)$ is the water-filling solution (2.7) subject to $E_H[T^*(H)] = E[Y]$. Again, as in Theorem 6 we see that there is no penalty to be paid for the hard energy constraint.

In [28] the effect of processing energy Z_k and/or energy storage inefficiencies on the capacity are also considered. It is further shown that having a battery to store energy increases the capacity of the system.

2.4 MAC policies

We consider the case where N nodes with data queues $Q_1, \ldots, Q_N$ are sharing a wireless channel. Each queue generates its traffic, stores in a queue, and transmits as in Section 2.3.1. Also, each node has its own energy-harvesting mechanism. The traffic generated at different queues and their energy mechanisms are assumed independent of each other. We do not consider the sleep option. The following results are from [22].

Let $\{X_k(i)\}$, $\{Y_k(i)\}$ and $\{Z_k(i)\}$ be the sequences corresponding to node i. For simplicity we will assume $\{X_k(i), k \geq 0\}$ and $\{Y_k(i), k \geq 0\}$ to be i.i.d. although these assumptions can be weakened as for a single queue. As mentioned at the end of Section 2.3.4, the energy consumption $\{Z_k(i)\}$ can be taken care of if we simply replace $E[Y(i)]$ by

$E[Y(i)] - E[Z(i)]$ in our algorithms. In the following we do that and write it as $E[Y(i)]$ only (and hence ignore $Z_k(i)$).

The N queues can share the channel in different ways. The stability region of $Q_1, Q_2, \ldots, Q_N$ and optimal (good) transmit policies depend upon the sharing mechanism used. We consider a few commonly used scenarios.

2.4.1 Orthogonal channels

The N nodes use TDMA/orthogonal CDMA/FDMA/OFDMA to transmit. Then the N queues become independent, decoupled queues and can be considered separately. Thus, the transmission policies developed in previous sections for a single queue can be used here. In the following we explain them in the context of TDMA.

If the queues have to use the channel in a TDMA fashion then the necessary conditions for stability of the N queues are: there exist $\alpha_1, \alpha_2, \ldots, \alpha_N, \alpha_i \geq 0$ and $\sum_{i=1}^{N} \alpha_i = 1$ such that

$$E[X(i)] < \alpha_i \, g_i \left(\frac{E[Y(i)]}{\alpha_i} \right), \quad i = 1, 2, \ldots, N, \tag{2.15}$$

where g_i is the energy to bit mapping for Q_i. A stable policy for each queue is, as in Section 2.3.2: Q_i is given α_i fraction of slots (on a long-term basis) and it uses energy $(E[Y(i)] - \epsilon)/\alpha_i$ whenever it transmits. For better delay performance, the slots allocated to different queues should be uniformly spaced.

It is possible that more than one set of $(\alpha_1, \ldots, \alpha_N)$ satisfies the stability condition (2.15). In that case one should select the values that minimize a cost function, (say) weighted sum of mean delays.

2.4.2 Opportunistic scheduling for fading channels: orthogonal channels

Now we discuss the MAC with fading. Let $\{H_k(i), k \geq 1\}$ be the channel gain process for Q_i. It is assumed stationary, ergodic, and independent of the fading process for $Q_j, j \neq i$. We first discuss opportunistic scheduling for the contention-free MAC. Later, we will study the CSMA-based algorithms.

If we assume that each of Q_i has infinite data backlog, then the policy that maximizes the sum of throughputs for $g(x) = log(1 + \beta x)$ and for symmetric statistics (i.e. each H_i has the same statistics and all $E[Y(i)]$ are the same) is to choose queue

$$i_k^* = arg\, max(H_k(i)) \tag{2.16}$$

in slot k and use $T_k(H)$ via the water-filling formula (2.7) with the average power constraint

$$E[T_k(H)] = N E[Y(i_k^*)] - \epsilon].$$

This is an extension of an algorithm in [29] for energy-harvesting nodes. A modification of this policy is available in [29] for the asymmetric case.

2.4 MAC policies

If g is linear, then for the symmetric case, a channel is selected only if it has the highest possible gain (for H bounded). If more than one channel is in the best state, select one of them with equal probability.

Although (2.16) maximizes throughput, it may be unfair to different queues and may not provide the QoS. Furthermore, an infinite backlog may not be a realistic assumption. Without this assumption, a throughput optimal policy (in the class of policies that use constant powers) is to choose queue

$$i_k^* = arg\,max\left(q_k(i)g_i\left(H_k(i)\left(\frac{E[Y(i)]-\epsilon}{\alpha(i)}\right)\right)\right) \quad (2.17)$$

and then use $T_k = (E[Y(i_k^*)] - \epsilon)/\alpha(i_k^*)$. Here, $\alpha(i_k^*)$ is the fraction of time that slots are assigned to i_k^*. However now we do not know $\alpha(i_k^*)$ but this may be estimated (see the end of this section). If $\alpha(i_k^*)$ is replaced with the true value, then the stability of the queues in the MAC follows from [30] if the fading states take values in a finite set and the system satisfies the following condition: Let there exist a function $f(r_k(1),\ldots,r_k(N))$ which picks one of the queues as a function of $(r_k(1),\ldots,r_k(N))$, where $r_k(i) = g_i\left(H_k(i)\frac{E[Y(i)]-\epsilon}{\alpha_i}\right)$, $\alpha(i) \triangleq E_\pi[1\{f(r_1,\ldots,r_N) = i\}]$, and $\pi(r_1,\ldots,r_N)$ is the stationary distribution of $(r_1,\ldots,r_N)$. Then if $E[X(i)] < \sum r_i 1\{f(r_1,\ldots,r_N) = i\}\pi(r_1,\ldots,r_N)$ for each i, the system is stable. This policy tries to satisfy the traffic requirements of different nodes but may not be delay optimal. But a Greedy policy

$$i_k^* = arg\,max\left(min\left(g_i\left(H_k(i)\left(\frac{E[Y(i)]-\epsilon}{\alpha(i)}\right)\right), q_k(i)\right)\right) \quad (2.18)$$

provides better mean delays. However, it is throughput optimal only for symmetric traffic statistics and when $E[Y(i)] = E[Y(j)]$, for all i, j. But it can be made throughput optimal (as (2.17)) while still retaining (partially) its mean delay performance as follows. Choose an appropriately large positive constant L. If none of $q_k(i)$ is greater than L, use (2.18); otherwise, on the set $\{i : q_k(i) > L\}$, use (2.17). We call this the *modified greedy policy*.

The policies (2.17) and (2.18) can be further improved if instead of using $T_k(i) = E[Y(i)] - \epsilon$, we use waterfilling in (2.17).

The policies (2.17), (2.18), and Modified Greedy provide good performance, require minimal information (only $E[Y(i)]$), are easy to implement, and have low computational requirements. In addition, they naturally adapt to changing traffic and channel conditions.

In (2.17)–(2.18) we need $\alpha(i)$ to obtain the energy T_k. But unlike for TDMA, $\alpha(i_k)$ is not available in these algorithms and depends on the algorithm used. Thus, we use a simple variant of the LMS (Least Mean Square) algorithm [31] to estimate $\alpha(i)$: Initially start with guess

$$\alpha_0(i) = \frac{1}{N}, \; i = 1,\ldots,N.$$

Run the algorithm for (say) L_1 number of slots. Each node i computes the fraction $\alpha'(i)$ of slots it gets and recomputes

$$\alpha_{n+1}(i) = \alpha_n(i) - \mu(\alpha_n(i) - \alpha'(i)), \quad (2.19)$$

where μ is a small positive constant.

2.4.3 Opportunistic scheduling for fading channels: CSMA

Since ZigBee and 802.11 use CSMA, we also discuss opportunistic scheduling for CSMA. As opposed to (2.17)–(2.18), this is a completely decentralized algorithm. The basic idea is ([32]) to make the back-off mechanism in a node a function of the channel state of that node.

Let f be a nonincreasing function with values in $[0, \tau_{max}]$, where τ_{max} is the maximum allowed back-off time in slots.

The Q_i uses the back-off time of

$$f\left(g_i\left(H_k(i)\frac{(E[Y(i)]-\epsilon)}{\alpha(i)}\right)\right). \qquad (2.20)$$

When a node gets the channel, it will transmit a complete packet and use energy per slot as

$$T_k(i_k^*) = \left(E[Y(i_k^*)]-\epsilon\right)/\alpha(i_k^*). \qquad (2.21)$$

Using the ideas in the last section, we can develop better algorithms than (2.20)–(2.21). Indeed, with (2.20), instead of (2.21), we can use waterfilling. We can also improve over (2.20) by using, for the back-off time of the ith node,

$$f\left(q_k(i)\ g_i\left(H_k(i)\frac{E[Y(i)]-\epsilon}{\alpha(i)}\right)\right), \qquad (2.22)$$

which takes care of the traffic requirements of different nodes.

We can also use the (modified) Greedy in (2.18). The $\alpha(i)$ in the above algorithms are computed via LMS in (2.19).

2.4.4 Simulations for MAC protocols

In this section we simulate the system under symmetric conditions, apply the different algorithms, and compare their performances. We use $g(x) = log(1+x)$. The $\{X_k(i), k \geq 1\}$ and $\{Y_k(i), k \geq 1\}$ are i.i.d. exponential. The LMS (2.19) was taken with $\mu = 0.01$ and the α_ks were updated after 30–50 slots.

For orthogonal channels, under symmetric conditions with 3 queues, average queue lengths are shown in Figure 2.7 for TO (2.17), Greedy (2.18), TDMA, Greedy with water-filling, and TDMA with water-filling policies. The $\{H_k(i), k \geq 1\}$ are i.i.d. exponential with mean 1. For symmetric conditions, Greedy is throughput optimal and hence Modified Greedy is not implemented. We see that TDMA becomes unstable much before

2.4 MAC policies

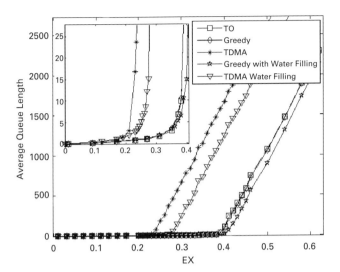

Figure 2.7 Orthogonal channels: symmetric, 3 queues.

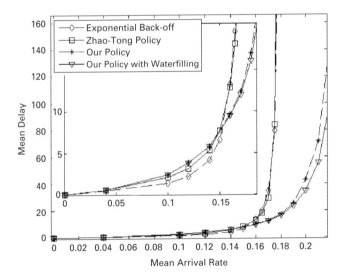

Figure 2.8 CSMA: mean delay, symmetric 10 queues.

the other policies, and that its average queue length is much worse even when it is stable. Greedy performs better than TO near the stability boundary, which is $E[X] = 0.39$. Water-filling improves the stability region of TDMA as well as Greedy.

For CSMA, Figure 2.8 shows mean delays under symmetric conditions with 10 queues and with normal exponential back-off, Zhao-Tong policy [32], our policy (2.22), and our policy with water-filling (with $f_{policy}(x) = \beta_{policy}/x$ and $Ef = 1.55$ at $EX = 0.17$; H assumes values 0.1, 0.5, 1.0, and 2.2 for time fractions 0.1, 0.3, 0.4, and 0.2). We simulated the 10 queues in continuous time. Also, $E[Y] = 1$ and the data packets of unit size arrive at each queue as Poisson streams. We see that opportunistic policies improve mean delays substantially.

2.5 Multi-hop networks

We consider a multi-hop wireless network with energy-harvesting nodes. The nodes generate packets to be transmitted to a fusion node. This can be a scenario for a sensor network [22] and can also be an uplink for a 4G cellular system (e.g. WiMAX, LTE) with relays. The following results are reported in [23]. A generalization of this setup with multiple source destination pairs with multi-casting and network coding is provided in [34].

The system is slotted, with each slot being of length T seconds. In any slot, a sensor node is in one of two modes: Wake or Sleep. In the sleep mode, a node can only generate energy via an energy-harvesting source. In the wake mode, it can process, receive and/or transmit data. We assume that the energy consumed by a sensor node in the sleep mode is negligible. In the wake mode, a node cannot transmit and receive in the same slot (i.e. it has a half duplex channel). For transmission, a node chooses its transmit power from a finite set of transmit powers (this holds in WiMAX and LTE).

For simplicity in notation, we assume that all the nodes choose their transmit powers from the same set consisting of K elements. We will denote the power consumed in the transmitter as e_t. The power consumption during reception is approximately constant and is denoted by e_r. Let e_s denote the power consumed by the node for sensing.

If sensor node n is awake in slot k (which denotes time interval $[(k-1)T, kT]$ seconds), $X_k(n)$ bits are generated in the slot. We assume that $\{X_k(n); k \geq 1\}$ is a stationary, ergodic random process. Then, node n when awake is generating data at a rate $d_n = E[X_k(n)]/T$. Further, in wake mode, we assume that a node consumes energy at a constant rate of e_w. This accounts for the energy consumed by the processor, the energy lost due to leakage, etc. Node n is able to replenish energy by $Y_k(n)$ in slot k. We assume that $\{Y_k(n); k \geq 1\}$ is a stationary, ergodic random process. Then energy is harvested at a rate $P_n^{av} = E[Y_k(n)]/T$ by node n. We assume that each node has infinite data and energy buffers.

We consider a network of N stationary (e.g. not mobile) energy-harvesting nodes with a fusion node. Let N denote the set of all nodes in the network. Due to the broadcast nature of wireless links, each node can potentially transmit to every other node, although the link quality will depend upon the distances between the nodes and possible shadowing and other physical obstacles in the way. Let $O(n)$ denote the set of outgoing links and $I(n)$ the set of incoming links of node n. There are $L = N^2$ links in the network (note that the fusion node doesn't have any outgoing links) and we denote the set of all links in the network by L. We assume that a node can communicate with at most one other node at any time.

We obtain jointly optimal power control, routing and scheduling policies for this network. We have seen above that for energy neutral operation in steady state, it is

2.5 Multi-hop networks

necessary and sufficient for the nth node to use average power $< P_n^{av}$. In the following, we will address the optimality problem for two scenarios: given the traffic requirements d_n of each sensor node n, obtain policies that satisfy these requirements. If it is not possible, then satisfy their requirements in a "fair" (to be explained below) way. Next, we exploit the energy-saving sleep-wake modes to obtain policies that can satisfy the traffic requirements better. We will find that the complexity of these problems is high and hence we have developed a computationally efficient suboptimal approach in [23] to obtain good solutions to this problem.

We define a mode of the network as a possible combination of active links along with their transmit powers, such that any node is communicating with only one other node over a half duplex channel. Thus, there can only be up to $N_A = \lfloor(N+1)/2\rfloor$ active links. Since each node chooses its transmit power from a finite set, the number of modes is finite. Let M denote the set of all modes. Corresponding to any transmission mode m, we have a link transmit power vector $\mathbf{P}^m = (P_1^m, \ldots, P_L^m)$ where the lth element, P_l^m, is the transmit power of link l in mode m.

Let the channel gain from the transmitting node of link k to the receiver node of link l be H_{kl}. This includes the effect of fading, shadowing, and attenuation due to distance. We assume slow fading, which is realistic for a network with stationary nodes. Hence, in the following, H_{kl} will be assumed constant for all $k, l \in L$. For any link transmit power vector $\mathbf{P}$, the SINR at the receiver of link l is

$$\gamma_l(\mathbf{P}) = \frac{h_{ll} P_l}{\sum_{k \neq l} h_{kl} P_k + \sigma_l^2},$$

where σ_l^2 is the noise power at the receiver of link l. The data rate, $R_l(\mathbf{P})$, over link l can be taken as a non-decreasing concave function of $\gamma_l(\mathbf{P})$. In this section, we use the Shannon formula

$$R_l(\mathbf{P}) = W \log_2(1 + \gamma_l(\mathbf{P})),$$

where W is the bandwidth of the wireless channel.

In the following, we consider scheduling of transmission modes for the network which satisfies the average power constraints, as well as the transmission requirements of different nodes, in a fair way.

2.5.1 Problem formulation

We consider each node as a source of a flow with the fusion node as the destination. Thus, we have N source–destination flows in the network. Denote the source of a flow f by $s(f)$ and the set of all flows by F. Also, let N_F denote the fusion node. Let r_f denote the rate at which the flow f is transmitted to the fusion node. To obtain the routing, we use a multi-commodity flow model [26] where each flow f corresponds to a commodity in the network. Let x_l^f denote the traffic assigned to link l by the routing scheme for flow f. The flow assignment given by the routing scheme must satisfy the flow conservation at each node n:

$$\sum_{l \in O(n)} x_l^f - \sum_{l \in I(n)} x_l^f = 0, \forall n \in N \setminus \{s(f)\}, \quad \forall f \in F,$$

$$\sum_{l \in O(s(f))} x_l^f = \sum_{l \in I(N_F)} x_l^f = r_f, \forall f \in F; \quad \sum_{l \in I(s(f))} x_l^f = \sum_{l \in O(N_F)} x_l^f = 0, \quad \forall f \in F.$$

These equations express the facts that if a node is not the fusion node or a source node for a flow, then the total input flow rate to it must equal the total output flow rate. If a node is a source node for a flow, then the difference between the output flow rate and the input flow rate equals the rate at which it generates the traffic corresponding to the flow.

The above equations can be compactly written as $\mathbf{A}\mathbf{x}^f = \mathbf{y}^f$, $\forall f \in F$ where A is the node-link incidence matrix, $\mathbf{x}^f$ is length L column vector with its lth element being x_l^f, and $\mathbf{y}^f$ is an $N+1$ column vector with $[\mathbf{y}^f]_{s(f)} = r_f$, $[\mathbf{y}^f]_{N_F} = -r_f$ and all other elements zero.

We initially consider a simplified model where the nodes are always awake. But, the network may not be able to carry all the data originating at each node. In such a case, we require a flow control mechanism at the nodes so that each node gets a *fair* share of the network resources. In order to ensure fairness, we maximize the minimum of the satisfied fraction of demand of the nodes in the network.

We have the following formulation for the optimization problem:

OPT-1: maximize λ subject to

$$\frac{r_f}{d_{s(f)}} \geq \lambda, \quad \forall f \in F; \; \mathbf{A}\mathbf{x}^f = \mathbf{y}^f, \quad \forall f \in F \qquad (2.23)$$

$$\sum_{f \in F} x_l^f \leq \sum_{m \in M} \alpha_m R_l(\mathbf{P}^m) - \delta, \quad \forall l \in L \qquad (2.24)$$

$$\sum_{m \in M} \alpha_m \left(\sum_{l \in O(n)} \left(P_l^m + e_t I(P_l^m > 0) \right) \right.$$

$$\left. + e_r \sum_{l \in I(n)} I(P_l^m > 0) \right) + e_s + e_w \leq P_n^{av} - \epsilon, \forall n \in N \qquad (2.25)$$

$$\sum_{m \in M} \alpha_m \leq 1, \qquad (2.26)$$

$$x_l^f \geq 0, \forall l \in L, \forall f \in F; r_f \geq 0, \forall f \in F; \alpha_m \geq 0, \forall m \in M,$$

where ϵ and δ are arbitrarily small positive constants and $I(\;)$ is the indicator function. This is a linear programming problem where we maximize the minimum, λ, of the satisfied fractions of demand $r_f/d_{s(f)}$ of the sensor nodes. The variable α_m is the fraction of time the network operates in mode m. The inequalities in (2.24) state that the traffic through any link should be less than its capacity. This is a necessary condition for stability of the data queue corresponding to link l. Under some weak conditions, it is also sufficient. The inequalities (2.25) state that the average power consumed for transmission, reception, and processing at each node n should be slightly less than P_n^{av}. This ensures energy neutral operation at each sensor node. We do not consider any power constraints on the fusion node whose main activity is reception (we assume that it has enough power to receive data in every slot).

By solving *OPT-1*, we obtain:

i) Optimal λ which we denote by λ^*. If $\lambda^* \geq 1$, then the network can carry all the traffic originating from the nodes to the fusion node. But, if $\lambda^* < 1$, there are nodes that are generating more data than what can be sent by the network to the fusion node under the condition of all nodes being awake all the time. These nodes will drop the excess data. In this case, using the energy-saving mode will be useful and this can possibly increase λ^*. This will be considered below.

ii) Optimal (mode scheduling) fractions of time $\{\alpha_m^*, m \in M\}$ that the network operates in the various valid modes.

iii) Optimal x_l^{*f} and the fair share of traffic r_f^* carried by the network for flow f.

At each node n, we have a queue corresponding to each outgoing link l and the fraction $x_l^{*f} / \sum_{i \in O(n)} x_i^{*f}$ of each flow f arriving at node n is buffered in this queue. Data in these queues can be served on a first-come first-served basis.

Now we formulate the optimization problem for the network with energy-harvesting nodes that have sleep and wake modes. Then *OPT-1* can be modified as

OPT-2: maximize λ subject to (2.23), (2.24), (2.26) and

$$\sum_{m \in M} \alpha_m \left[\mathbf{q}^m\right]_n + u_{sn} e_s + u_{wn} e_w \leq P_n^{av} - \epsilon, \quad \forall n \in N \qquad (2.27)$$

$$\frac{r_f}{d_{s(f)}} \leq u_{sn}, \ \forall n \in N; u_{sn} \leq u_{wn}, \quad \forall n \in N \qquad (2.28)$$

$$\sum_{m \in M} \alpha_m \left(\sum_{l \in O(n)} I(P_l^m > 0) + \sum_{l \in I(n)} I(P_l^m > 0) \right) \leq u_{wn}, \quad \forall n \in N \qquad (2.29)$$

$$u_{wn} \geq 0, \ \forall n \in N; u_{sn} \geq 0, \quad \forall n \in N. \qquad (2.30)$$

In the above formulation, we denote by u_{wn} and u_{sn}, the fraction of time that sensor node n is in wake mode and the fraction of time it is performing sensing, respectively. Thus, the node spends energy at the rate of e_w in u_{wn} fraction of time and at the rate of e_s in u_{sn} fraction of time. This is taken into account while writing inequalities (2.27). Since a sensor n is sensing and thus generating data only for u_{sn} fraction of time, $r_f/d_{s(f)}$ is upper bounded by u_{sn} and a sensor can perform sensing only when it is in the wake mode. Thus, we have the inequalities (2.28). Similarly, the fraction of time a node is transmitting or receiving is limited by the fraction of time it is in the wake mode, which is stated in inequalities (2.29).

Using the optimal solution, routing and scheduling with power control can be done as discussed in Section 2.5.1. The scheduling of the sleep slots can be done as follows.

Note that a node has to be awake when a mode is scheduled in which it has to transmit or receive. For the rest of the time, the sleep slots of the nodes can be scheduled in any manner, as long as the nodes are awake for the fraction u_{wn} and they sense for u_{sn}/u_{wn} fraction of time they are awake.

OPT-2 is also an LP which becomes computationally demanding as we increase N. Note that here we need to consider $L + 2N + 1$ modes to obtain the optimal solution (we need N additional modes for this problem because of additional constraints (2.29)).

2.5.2 Simulations

In this section, we will provide examples to gain some insights into the optimal solutions of the problem formulations discussed in the previous sections. These examples were considered in the context of sensor networks but indicate performance for other scenarios as well.

In the following, each node selects its transmit power from $\{0, 0.1, 1, 10\}$ mW. Channel gain over a link is taken as inversely proportional to the squared distance between its nodes. We take $e_t = 60$ mW, $e_r = 60$ mW, $e_s = 2$ mW, and $e_w = 20$ mW. The transmit powers, e_t and e_r correspond to some typical values in sensor node ADF7020 and e_s is a typical value seen in sensors like SHT1x/SHT7x. We take $\sigma_l^2 = 10^{-7}$ W and $W = 5$ MHz. These settings are applicable to all the examples in this section unless otherwise stated.

We first consider the solution of *OPT-2* for a simple network consisting of six nodes and a fusion node. Each node is generating data at the rate of 25 kbps and harvesting energy at the rate of 50 mW. In Figure 2.9, the net flows over each of the links in the network are shown. The net flow for a link l is obtained as $\sum_{f \in F} x_l^f$. In the optimal solution for this example, $\lambda = 0.88$, $u_{wn} = u_{sn} = 0.88 \ \forall n \in N$, fractions of time the nodes transmit and receive are [0.27, 0.27, 0.02, 0.40, 0.40, 0.02] and [0.17, 0.17, 0, 0.02, 0.02, 0],

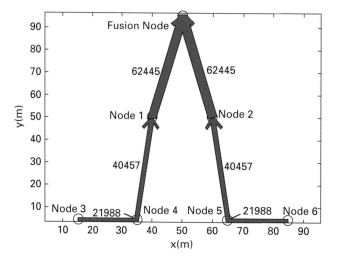

Figure 2.9 Optimal solution for $N = 6$.

2.5 Multi-hop networks

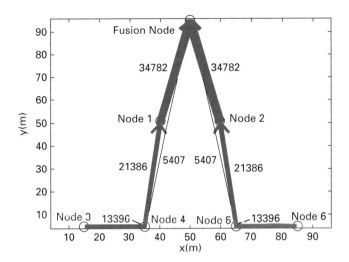

Figure 2.10 Optimal solution for $e_r = 300$ mW.

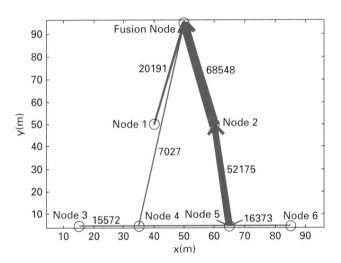

Figure 2.11 Optimal solution with Node 1 harvesting energy at a lower rate than others.

respectively. In Figures 2.9, 2.10, and 2.11, the thickness of a line representing a net flow is proportional to its rate (indicated against the arrow) and we have depicted only the significant net flows, ignoring the ones with smaller rates.

Figure 2.10 depicts the optimal net flows for the same network as Figure 2.9 using the same settings except for a higher power consumption during reception. We assume $e_r = 300$ mW. On increasing the power consumption required for reception, more energy is spent for the reception on routes with more hops. Thus, the increase in e_r results in an increase in the data flow over routes with fewer hops in the optimal solution. In the optimal solution, $\lambda = 0.54$, $u_{wn} = u_{sn} = 0.54$, $\forall n \in N$, fractions of time the nodes transmit and receive are [0.15, 0.15, 0.01, 0.44, 0.44, 0.01] and [0.09, 0.09, 0, 0.01, 0.01, 0] respectively. As expected, λ and the fractions of time nodes receive data have decreased. (Although e_r is usually not more than e_t, we took $e_r = 300$ mW to illustrate this point.) Further, we have also observed that the number of simultaneous transmissions in the significant modes increases with an increase in σ_l^2. For example, the fraction of time there are more than one simultaneous transmission is 0.67 when $\sigma_l^2 = 10^{-5}$ W, whereas it is only 0.37 when $\sigma_l^2 = 10^{-7}$ W.

Figure 2.11 depicts the optimal net flows for the same network as Figure 2.9 using the same settings except for a different energy-generation profile for node 1. It harvests energy at a rate of 22 mW. We can see that, in the optimal solution, other nodes avoid Node 1 for routing their data.

We also solved *OPT-1* for the settings discussed in Figures 2.9, 2.10, and 2.11 and in the respective optimal solutions, the values of λ were 0.82, 0.44, and 0.48. In the optimal solutions for *OPT-2*, the corresponding values of λ were 0.88, 0.54, and 0.65. Through these examples, we see the usefulness of having such sleep modes. The drastic reduction in performance in the *OPT-1* solution for the third case indicates that the sleep modes are more crucial for energy-deprived nodes than for the other nodes. *OPT-1* and *OPT-2*, even though they are linear programs, have a high computational complexity because the set M of all modes is very large. Thus, in [23] we have also developed efficient suboptimal algorithms that have much less complexity.

2.6 Conclusion

We have considered a wireless communication system powered by energy-harvesting sources. Initially, a point-to-point channel is considered. Throughput optimal and mean delay optimal energy-management policies are identified that can make the system work in energy neutral operation. It is found that having energy storage allows a larger stability region as well as lower mean delays. Also, often (when the energy to transmission rate map g is linear) a greedy policy studied here is throughput as well as mean delay optimal. This does not require energy-harvesting and traffic-generation statistics and is easy to implement.

The results have been extended to fading channels and when energy at the transmitter is also consumed in data processing. It is shown that if energy to transmission rate function g is linear, then knowing the channel state can significantly increase the stability region. We have also included the effects of leakage/wastage of energy when it is stored in the energy buffer and when it is extracted.

Next, we have considered the case where the transmitter also has a sleep mode to save energy. Optimal energy and sleep mode management policies have been obtained that minimize a weighted sum of mean queue length and mean data loss rate. Throughput optimal and easily computable efficient suboptimal policies are also provided. Fundamental communication limits of such systems when the channel is AWGN with fading have been also obtained. The effects of energy storage, processing energy, and leakage in the storage have also been studied.

Next, a MAC system with energy-harvesting nodes has been considered. Policies based on TDMA, opportunistic MAC with fading and CSMA have been provided and compared. Finally, we have considered joint power control, scheduling, and routing in multi-hop networks. Optimal and efficient suboptimal fair solutions have been provided.

In future, one may consider information theoretic fundamental limits for a MAC and multi-hop scenario. Also, it will be useful to obtain computationally efficient centralized and decentralized algorithms for a multi-hop scenario.

References

[1] W. Vereecken, *et al.*, "Overall ICT footprint and green communication technologies," in *Proc. of 4th Int. Symp. on Commun., Control and SP (ISCCP)*, 2010.

[2] A. Amanna, "Green Communication," *Annotated Literature Review and Research Vision*, Virginia Tech. [Online]. Available: http://filebox.vt.edu/users/aamanna/webpage/

[3] H. Aksoy, *et al.*, "Stochastic generation of hourly mean wind speed data," *Renewable Energy*, vol. 29, pp. 2111–2131, 2004.

[4] A. Kansal, *et al.*, "Power management in energy harvesting sensor networks," *ACM Trans. Embed. Comput. System*, vol. 6, 2007.

[5] E. Oh, *et al.*, "Toward dynamic energy efficient operation of cellular network infrastructure," *IEEE Commun. Magazine*, vol. 49, no. 6, pp. 56–61, 2011.

[6] K. D. Turck, *et al.*, "Performance of IEEE 802.16e sleep mode mechanism in the presence of bidirectional traffic," *1st Int. Workshop on Green Communications (GreenComm'09)*, 2009.

[7] D. Valerdi, *et al.*, "Intelligent energy managed service for green base stations," *IEEE GLOBECOM Workshop on Green Communications*, pp. 1453–1457, 2010.

[8] V. Raghunathan, S. Ganeriwal, and M. Srivastava, "Emerging techniques for long-lived wireless sensor networks," *IEEE Commun. Magazine*, vol. 44, no. 4, 2006.

[9] O. Ozel and S. Ulukus, "Information theoretic analysis of an energy harvesting communication system," *IEEE PIMRC*, 2010.

[10] S. Cui, A. Goldsmith, and A. Bahai, "Energy constrained modulation optimization," *IEEE Trans. Wireless Commun.*, vol. 4, pp. 2349–2363, 2005.

[11] P. Grover and A. Sahai, "Time division multiplexing for green communication," *IEEE Int. Symposium on Information Theory (ISIT)*, vol. 4, 2009.

[12] Z. A. Eu, H-P. Tan, and W. K. G. Seah, "Opportunistic routing in wireless sensor networks powered by ambient energy harvesting," *Computer Networks*, vol. 54, no. 17, 2010.

[13] E. Lattanzi, *et al.*, "Energetic sustainability of routing algorithms for energy harvesting wireless sensor networks," *Computer Commun.*, vol. 30, no. 14–15, pp. 2976–2986, 2007.

[14] L. Lin, B. Shroff, and R. Srikant, "Asymptotically optimal energy-aware routing for multihop wireless networks with renewable energy sources," *IEEE/ACM Trans. Networking.*, vol. 15, pp. 1021–1034, 2007.

[15] M. Gatzianas, L. Georgiadis, and L. Tassiulas, "Control of wireless networks with rechargeable batteries," *IEEE Trans. Wireless Commun.*, vol. 9, pp. 581–593, Feb. 2010.

[16] Z. Mao, C. Koksal, and N. Shroff, "Resource allocation in sensor networks with renewable energy," in *Proc. of 19th Int. Conf. on Computer Commun. and Networking (ICCCN)*, 2010.

[17] S. Singh and C. Yiu, "Putting the cart before the horse: merging traffic for energy conservation," *IEEE Commun. Magazine*, vol. 49, no. 6, pp. 78–82, 2011.

[18] C. Hu, *et al.*, "On the design of green reconfigurable router toward energy efficient internet," *IEEE Commun. Magazine*, vol. 49, no. 6, pp. 83–87, 2011.

[19] V. Joseph, V. Sharma, and U. Mukherji, "Optimal sleep-wake policies for an energy harvesting sensor node," in *Proc. of IEEE Int. Conf. on Commun. (ICC)*, 2009.

[20] R. Rajesh, V. Sharma, and P. Viswanath, "Information capacity of energy harvesting sensor nodes," in *Proc. of IEEE International Symposium on Information Theory (ISIT)*, 2011.

[21] V. Sharma, *et al.*, "Optimal energy management policies for energy harvesting sensor nodes," *IEEE Trans. Wireless Commun.*, vol. 9, pp. 1326–1336, 2010.

[22] V. Joseph, V. Sharma, and U.Mukherji, "Efficient energy management policies for networks with energy harvesting sensor nodes," in *Proc. of 46th Allerton Conf. on Commun., Control and Computing*, 2008.

[23] V. Joseph, V. Sharma, and U. Mukherji, "Joint power control, scheduling and routing for multihop energy harvesting sensor networks," in *Proc. of ACM MSWiM*, 2009.

[24] G. Schmitt, "The green base station," *4th International Conference on Telecommunication – Energy Special Conference (TELESCON)*, 2009.

[25] P. Ferreira, *et al.*, "Interfaces for renewable energy sources with electric power systems," in *Proc. of Environment 2010: Situation and Perspectives for the European Union, Porto, Portugal*, 2003.

[26] R. K. Ahuja, T. L. Magnanti, and J. B. Orlin, *Network Flows: Theory, Algorithms and Applications*. N.J., USA: Prentice Hall, 1993.

[27] S. Asmussen, *Applied Probability and Queues*. 2nd ed., N.Y.: Springer, 2003.

[28] R. Rajesh, V. Sharma, and P. Viswanath, "Capacity for fading gaussian channel with an energy harvesting sensor node," in *Proc. of IEEE Globecom*, 2011.

[29] R. Knopp and P. Humblet, "Information capacity and power control in single-cell multiuser communications," in *Proc. of IEEE Int. Conf. on Commun. (ICC)*, pp. 331–335, 1995.

[30] A. Eryilmaz, R. Srikant, and J.R. Perkins, "Stable scheduling policies for fading wireless channels," *IEEE/ACM Trans. on Networking.*, vol. 13, pp. 411–424, 2005.

[31] S. Haykin, *Adaptive Filtering Theory*. 3rd ed., N.J., USA: Printice Hall, 1996.

[32] Q. Zhao and L. Tong, "Opportunistic carrier sensing for energy efficient information retrieval in sensor networks," *EURASIP Journal on Wireless Commun. and Networking*, vol. 2, pp. 231–241, 2005.

[33] P. Y. Massad, M. Medard and L. Zheng, "Impact of processing energy on the capacity of wireless channels," *Proc. of ISITA*, Italy, 2004.

[34] V. Joseph, *et al.*, "Joint power control, scheduling and routing for multicast in multihop energy harvesting sensor networks," in *Proc. of ICUMI*, 2009.

3 PHY and MAC layer optimization for energy-harvesting wireless networks

Neelesh B. Mehta and Chandra R. Murthy

3.1 Introduction

In typical wireless communication systems, nodes are deployed with pre-charged batteries. The energy stored in the battery is used by the node for communication, sensing, and signal processing tasks. However, when the battery drains out, the node "dies" and is no longer available in the network. When a sufficient number of nodes die, the network itself becomes dysfunctional. Therefore, periodic maintenance and battery replacement are necessary to ensure that the network continues to operate. Such maintenance is often operationally challenging or even impossible in several network deployments. The alternative option of running power cables to the nodes is also often infeasible.

An emerging green alternative that circumvents this problem is the use of energy-harvesting (EH) functionality in the nodes [1]–[7]. An EH node harvests energy from the environment using solar, thermoelectric effects, vibration, and other phenomena. Unlike a conventional node, an EH node that drains out its battery can harvest energy later and, thus, again become "alive." Energy harvesting eliminates the need for periodic battery replacements and significantly decreases the network maintenance overhead. This also makes it an attractive alternative to wireless networks that are powered by external power cables. Consequently, EH networks are finding applications in monitoring systems in aerospace, automobile, and civil applications, environmental/habitat monitoring, intrusion detection, inventory management, etc.

Unlike a conventional battery-operated node, a potentially infinite amount of energy is available to an EH node, albeit over an infinite duration of time. Hence, the focus of the physical layer and multiple access (MAC) layer protocols shifts to judiciously utilizing the harvested energy and ensuring that the energy is available when required – to the extent possible. Reducing the energy consumption and improving the spectral efficiency now become secondary goals of the design of these protocols. This motivates a redesign of the physical and multiple access layers of the network. For example, increasing the transmit energy improves the reliability of transmissions by a node as it counters noise. However, it also drains the node's battery faster and lowers the odds that the node can transmit later. In a multi-node network, an additional multiple access problem that arises is the determination of which node(s) should transmit and when to transmit.

Several factors affect the performance of an EH wireless network. They are as follows:

- *Energy neutrality:* The energy neutrality constraint mandates that at any instant of time the total energy consumed by a node cannot exceed the total energy harvested by it. The energy neutrality constraint, which is nothing but the law of conservation of energy, fundamentally limits the end-to-end system performance of the network.
- *Energy-harvesting profile:* The energy profile models the energy harvested at a node as a function of time. It depends on the source from which energy is harvested. In solar and thermoelectric harvesting, the rate at which energy is harvested remains relatively constant over a long timescale (minutes). This can be exploited to predict the energy availability in a node in the near future. On the other hand, the harvested energy may also vary over a smaller timescale. Various analytical models have been proposed for the energy profile in the literature. For example, a leaky bucket model was proposed in [2]. In it, the exact rate at which energy is harvested is unknown, but lies within a known range. Interestingly, this model was motivated by internet traffic modeling. A simpler probabilistic Bernoulli injection model, in which a node harvests a fixed amount of energy with some probability in a slot or does not harvest at all, was considered in [8].
- *Storage buffer:* The availability of an energy buffer such as a battery or a super-capacitor helps alleviate the randomness in the energy-harvesting process. This is because the node can store excess harvested energy in the buffer and use it later. In practice, leakage currents in the energy buffer and inefficiencies in the storage process reduce the amount of energy that is used. Accounting for these inefficiencies leads to a more effective design of the physical and MAC layer protocols.
- *Channel and channel state information (CSI):* Since the nodes communicate over wireless channels, the path loss, shadowing, and fading encountered by the transmitted signals drive the overall performance. The transmitting node can adapt its transmission rate or even defer transmission depending on the current channel state. Also, if the node attempts transmissions over multiple channel realizations, it can exploit time diversity to transmit its packets at a lower power or more reliably.

In this chapter, we study the implications of energy harvesting on the design and optimization of the physical and MAC layers. We first consider the optimization of the physical layer of a single link scenario, in which a transmitter communicates with a receiver. Two different scenarios are investigated. In the first scenario (Section 3.2.1), the EH transmitter has no channel state information and uses retransmissions to increase the reliability of communication; the goal is to optimize the EH node's transmit power settings. In the second scenario (Section 3.2.2), the EH transmitter has channel state information and can adjust its transmit power as a function of the channel state. Further, the system design explicitly incorporates storage inefficiencies and the different timescales at which the energy harvesting and fading processes evolve. We then consider energy-efficient multi-hop communication using multiple EH cooperative relay nodes that transmit over a common channel (Section 3.3). The system brings out the inter-relationship between several aspects related to cross-layer system design for EH nodes, such as multiple access layer design, amplify-and-forward cooperative relaying to exploit spatial diversity to improve robustness to fading, and two-hop communication to improve energy efficiency. We conclude in Section 3.4.

3.2 Physical layer design

3.2.1 No CSI at transmitter and retransmissions

Consider an EH node that needs to transmit a measurement of δ data bits periodically once in every measurement of interval of duration T_m s to a destination node over a wireless fading channel [9]. The data bits are encapsulated in a packet of duration T_p sec. We hereafter use the term "slots" to refer to intervals of duration T_p, and discretize time in multiples of T_p. If the EH node does not receive an acknowledgment (ACK) from the receiver, then it retransmits the packet. The node retransmits the packet until it receives an ACK or it is time to transmit the next measurement. Once it receives an ACK, the node stops transmitting and just accumulates the energy harvested during the rest of the measurement interval. Let K denote the maximum number of transmissions in a measurement interval and let $T_m/T_p = K$.

No channel state information is assumed to be available at the transmitting EH node. Therefore, the node transmits its packets with a fixed power P, which is a static system parameter. The transmission model is illustrated in Figure 3.1. The energy required to transmit a packet equals PT_p. In the nth slot, the node transmits a packet if the available energy, B_n, in its battery exceeds PT_p. Otherwise, no packet transmission occurs.

A measurement *outage* occurs if the node fails to deliver the packet to the destination successfully within duration T_m. In a slot, the receiver may fail to decode for two reasons: (i) the EH transmitter does not have sufficient energy to transmit or (ii) the transmitted packet is corrupted by the channel and cannot be decoded by the receiver.

In order to gain analytical insights into the key trade-offs that occur in an EH link, we make the following simplifying assumptions. The packets are taken to be uncoded. We assume the bits to be on-off keying (OOK) encoded, as has been assumed in some sensor networks [10]. Each OOK symbol is of duration T_s. We focus on transmit energy-consumption since it is the dominant component when the inter-sensor distances are relatively large [11]. There are no inefficiencies in the storage of energy in the energy buffer.

We assume a Bernoulli energy-harvesting model in which an energy E_s is injected into the EH node at the beginning of every slot with probability ρ [8]. With probability

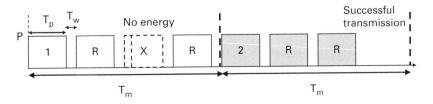

Figure 3.1 Periodic packet transmission model of the EH node.

$1-\rho$ no energy is injected into the node in the slot. A more refined model would allow for multiple levels of harvested energy and also model time correlations, if any. While the above model is simple, it captures the sporadic and random availability of energy at an EH node.

A block fading Rayleigh channel model is assumed. The channel gain remains the same during the entire measurement interval, T_m. Consequently, all retransmissions of a packet see the same channel.[1] Therefore, the instantaneous SNR, γ, at the receiver is an exponentially distributed random variable with mean $\left(\frac{d_0}{d}\right)^\kappa \frac{PT_s}{N_0}$, where d_0 is a reference distance, κ is the path loss exponent, and N_0 is the noise power. Let $P_e(\gamma)$ denote the probability that a transmitted packet cannot be decoded by the receiver. For an uncoded packet and OOK, it equals $1 - (1 - Q(\sqrt{\gamma}))^\delta$.

Energy-unconstrained regime

Consider first the scenario where the average energy harvested exceeds the average energy consumed. Then, after a sufficiently long time and with sufficient storage capacity, the probability that the node does not have energy to transmit will be zero. The node is said to be *energy unconstrained*. In such a case, the outage probability depends only on the transmit energy and not on E_s.

With infinite storage capacity, $B_{\max} = \infty$, the energy unconstrained regime can be characterized in the following simple manner.

Proposition 3.1. *An EH node is energy unconstrained when*

$$\rho \geq \frac{PT_p}{KE_s} \mathbf{E}\left[\frac{1-(P_e(\gamma))^K}{1-P_e(\gamma)}\right]. \tag{3.1}$$

Proof. In the energy-unconstrained regime, it is easy to see that exactly i transmissions occur in a measurement interval with SNR γ with probability $(1-P_e(\gamma))(P_e(\gamma))^{i-1}$, for $1 \leq i < K$. And, K transmissions occur with probability $(P_e(\gamma))^{K-1}$. Therefore, the average energy consumed in an interval equals $PT_p \mathbf{E}\left[(1-P_e(\gamma))\sum_{i=1}^{K-1} i(P_e(\gamma))^{i-1} + K(P_e(\gamma))^{K-1}\right]$, where the expectation is over the distribution of γ. Simplifying this expression and equating it with the average energy injected in a measurement interval, $K\rho E_s$, yields the desired expression. □

We now consider the regime in which the EH node is not energy unconstrained. Let

$$W = \frac{E_s}{PT_p}. \tag{3.2}$$

In general, W can be any positive real number and needs to be optimized by choosing the transmit power. The cases where $W = 1/L < 1$ and $W = L \geq 1$, where L is an integer, are analyzed separately below. For both these cases, the evolution of the battery energy at the beginning of a measurement interval constitutes a discrete time Markov chain (DTMC).

---

[1] The reader is referred to [9] for an analysis of the scenario in which different retransmissions see different channel fades.

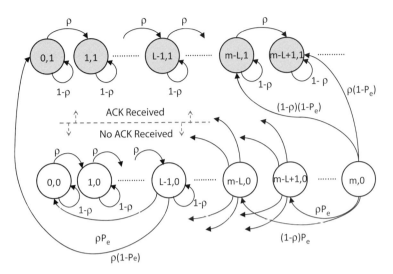

Figure 3.2 Discrete-time Markov evolution of battery energy (normalized with respect to E_s) and transmission states within a measurement interval when $\frac{E_s}{PT_p} = \frac{1}{L}$, $L \in \mathbb{N}$ (© [2009] IEEE).

Further, *within a measurement interval and given the SNR γ*, the battery energy coupled with a two-state variable that tracks whether an ACK has been received or not in the interval is another DTMC.

$E_s = PT_p/L, L \in \mathbb{N}$.
In this case, the energy stored in the battery of the EH node decreases by LE_s when it transmits. Recall that the channel state remains the same within a measurement interval and changes to an independent value thereafter. The DTMC for the evolution of the battery energy within a measurement interval is shown in Figure 3.2. The state (B_n, U_n) is defined by two variables: (i) the battery energy B_n (normalized with respect to E_s) and (ii) the feedback state, U_n, where $U_n = 0$ if no ACK has arrived in the measurement interval until slot n, and is 1 otherwise.

At the beginning of the mth measurement interval (i.e. time slot mK), the node is always in the state $(B_{mK}, 0)$ since an ACK is yet to arrive. The mth measurement interval consists of slots $mK, mK + 1, \ldots, m(K + 1) - 1$. Once an ACK arrives, the node transitions to states in which U_n is 1. Further, it no longer transmits in the remaining slots of the measurement interval. Irrespective of the feedback state at the end of the mth interval, the battery energy at the beginning of the $(m + 1)$th interval is the same as that at end of the mth interval, and the feedback state at the beginning of the $(m + 1)$th interval is always 0.

Outage occurs in the mth measurement interval if and only if the node is in feedback state 0 at the end of the interval. Since B_{mK} is independent of the channel state in $[mK, (m + 1)K]$, the outage probability, as a function of $K \geq 1$, is

$$P_{\text{out}}(K) = \sum_i \pi(i) \mathbf{E}[P_{\text{out}}(K|i, \gamma)], \tag{3.3}$$

where $\pi(i)$ is the stationary probability that the node has energy iE_s at the beginning of a measurement interval and $P_{\text{out}}(K|i, \gamma)$ is the outage probability conditioned on the SNR being γ and the battery state at the beginning of a measurement interval being iE_s. For a node that is not energy unconstrained, π can be shown to exist and is unique. From the Markov structure, $P_{\text{out}}(K|i, \gamma)$ obeys the following recursion relation:

$$P_{\text{out}}(K|i, \gamma) = \begin{cases} \sum_{r=0}^{K} \binom{K}{r} \rho^r (1-\rho)^{K-r} (P_e(\gamma))^{\left\lfloor \frac{i+r}{L} \right\rfloor}, & i \leq L-1 \\ \rho P_e(\gamma) P_{\text{out}}(K-1|i-L+1, \gamma) \\ + (1-\rho) P_e(\gamma) P_{\text{out}}(K-1|i-L, \gamma), & i \geq L. \end{cases} \quad (3.4)$$

This is because for $i \leq L-1$, r energy injections lead to $\left\lfloor \frac{i+r}{L} \right\rfloor$ packets being transmitted. And, $\binom{K}{r} \rho^r (1-\rho)^{K-r}$ is the probability of r energy injections in K slots. For $i \geq L$, a packet transmission will occur and will use up energy LE_s. Thereafter, the next state's energy is either $(m-L)$ or $(m-L+1)$ depending on whether an energy injection occurs or not in the same slot. We also have $P_{\text{out}}(0|i, \gamma) = 1$.

The stationary probability can be numerically computed as follows. Let the transition probability matrix of the intra-measurement interval Markov chain be denoted by $\mathbf{G}(\gamma)$. It contains $\Pr(B_{n+1}, U_{n+1}|B_n, U_n, \gamma)$ as its elements, which are specified in detail later.

1. Compute $\mathbf{G}^K(\gamma)$.
2. Compute the probability $\Pr\left(B_{(m+1)K} = jE_s|B_{mK} = iE_s, \gamma\right)$ that the EH node has energy jE_s at the end of the mth measurement interval given the SNR, γ, and $B_{mK} = iE_s$. It equals $\sum_{u=0}^{1} \Pr\left(B_{(m+1)K} = jE_s, U_{(m+1)K} = u|B_{mK} = iE_s, U_{mK} = 0, \gamma\right)$.
3. The stationary probabilities are obtained by solving the balance equations

$$\pi(j) = \sum_{i} \mathbf{E}\left[\Pr\left(B_{(m+1)K} = jE_s|B_{mK} = iE_s, \gamma\right)\right] \pi(i). \quad (3.5)$$

Transition probability matrix $\mathbf{G}(\gamma)$: The element of the state transition probability matrix $\mathbf{G}(\gamma)$ that corresponds to the transition from (i, r) to (j, s) equals

$$G_{ij}^{rs}(\gamma) = \Pr(B_{n+1} = jE_s, U_{n+1} = s|B_n = iE_s, U_n = r, \gamma),$$

where $i, j \in \{0, 1, \ldots, \infty\}$ and $r, s \in \{0, 1\}$. Notice that the transition probabilities depend on γ because the ACK probability does.

For $0 \leq i \leq L-2$ and $r = 0, 1$, or $i \geq L-1$ and $r = 1$:

$$G_{ij}^{rs}(\gamma) = \begin{cases} 1-\rho, & j=i, s=0 \\ \rho, & j=i+1, s=0 \\ 0, & \text{otherwise.} \end{cases} \quad (3.6)$$

3.2 Physical layer design

For $i = L - 1$ and $r = 0$, we have

$$G_{ij}^{rs}(\gamma) = \begin{cases} 1 - \rho, & j = L - 1, s = 0 \\ \rho(1 - P_e(\gamma)), & j = 0, s = 1 \\ \rho P_e(\gamma), & j = 0, s = 0 \\ 0, & \text{otherwise.} \end{cases} \quad (3.7)$$

For $i \geq L$ and $r = 0$, we have

$$G_{ij}^{rs}(\gamma) = \begin{cases} (1 - \rho)P_e(\gamma), & j = i - L, s = 0 \\ \rho P_e(\gamma), & j = i - L + 1, s = 0 \\ (1 - \rho)(1 - P_e(\gamma)), & j = i - L, s = 1 \\ \rho(1 - P_e(\gamma)), & j = i - L + 1, s = 1 \\ 0, & \text{otherwise.} \end{cases} \quad (3.8)$$

These transition probabilities follow by inspection and track the following two independent events: (i) whether energy was harvested in a slot, which happens with probability ρ, and (ii) whether the transmitted packet was decoded by the receiver, which happens with probability $1 - P_e(\gamma)$. These probabilities can be modified suitably for finite battery capacity.

$E_s = L P T_p, L \in \mathbb{N}$

Now the energy harvested in a slot is sufficient for L transmissions. The DTMC for the battery evolution within a measurement interval is shown in Figure 3.3.

As before, the state consists of the energy state B_n (normalized with respect to PT_p) and the feedback state U_n. The outage probability, as a function of $K \geq 1$, and conditioned

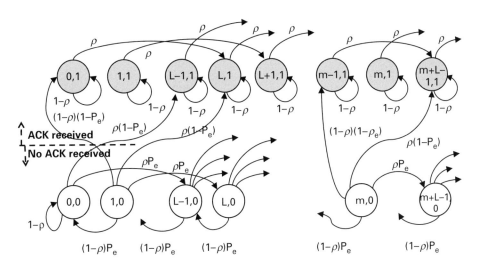

Figure 3.3 Discrete-time Markov evolution of battery energy (normalized with respect to PT_p) and transmission states within a measurement interval when $\frac{E_s}{PT_p} = L$ and $L \in \mathbb{N}$ (© [2009] IEEE).

on $B_{mK} = iE_s$ and γ, is given by the following recursive relation:

$$P_{\text{out}}(K|i,\gamma) = \begin{cases} \rho P_e(\gamma) P_{\text{out}}(K-1|L-1,\gamma) \\ \quad + (1-\rho) P_{\text{out}}(K-1|0,\gamma), & i = 0 \\ \rho P_e(\gamma) P_{\text{out}}(K-1|i+L-1,\gamma) \\ \quad + (1-\rho) P_e(\gamma) P_{\text{out}}(K-1|i-1,\gamma), & i > 0 \\ P_e^K(\gamma), & i \geq K. \end{cases} \qquad (3.9)$$

In addition, $P_{\text{out}}(K|i,\gamma) = 1$ for $K = 0$. The reasoning behind the above relations is similar to that used in the earlier case of $E_s = PT/L$.

Transition probability matrix $\mathbf{G}(\gamma)$: For $i \geq 1$ and $r = 0$, we have

$$G_{ij}^{rs}(\gamma) = \begin{cases} \rho P_e(\gamma), & j = i+L-1, s = 0 \\ \rho(1 - P_e(\gamma)), & j = i+L-1, s = 1 \\ (1-\rho) P_e(\gamma), & j = i-1, s = 0 \\ (1-\rho)(1 - P_e(\gamma)), & j = i-1, s = 1 \\ 0, & \text{otherwise.} \end{cases} \qquad (3.10)$$

For $i = 0$ and $r = 0$, we have

$$G_{ij}^{rs}(\gamma) = \begin{cases} 1-\rho, & j = 0, s = 0 \\ \rho P_e(\gamma), & j = L-1, s = 0 \\ \rho(1 - P_e(\gamma)), & j = L-1, s = 1 \\ 0, & \text{otherwise.} \end{cases} \qquad (3.11)$$

For $i \geq 0$ and $r = 1$, we have

$$G_{ij}^{rs}(\gamma) = \begin{cases} 1-\rho, & j = i, s = 1 \\ \rho, & j = i+L, s = 1 \\ 0, & \text{otherwise.} \end{cases} \qquad (3.12)$$

Numerical results and performance optimization

We now plot the analytical results and verify them with Monte Carlo simulations that use up to 10^7 samples. The parameters used are $T_m = 100$ msec, $\delta = 32$ bits, $T_s = 0.5$ msec, and $K = 4$. E_s is normalized, without loss of generality, by multiplying it with $\left(\frac{d_0}{d}\right)^K \frac{1}{N_0}$.[2] An additional 18 bits are included in a packet for a packet header, which is practically necessary for synchronization purposes [12]. While no information is conveyed by these bits, transmitting them does consume energy.

Figure 3.4 plots the outage probability as a function of ρ and W for $E_s = 35$ dB. The figure brings out several interesting facts. As ρ increases, P_{out} decreases since more energy is harvested on average. However, it undergoes a sharp transition to a fixed value when ρ lies in the energy-unconstrained regime. This is because P_{out} is now entirely determined by the packet decoding errors, which depend only on the transmit energy and not ρ.

---

[2] For example, $E_s = 35$ dB corresponds to an SNR of 18 dB when $P = 1.34$ mW, $T_s = 0.5$ msec, distance $d = 40 d_0$, a noise temperature of 300 K, and a bandwidth of 1 MHz.

3.2 Physical layer design

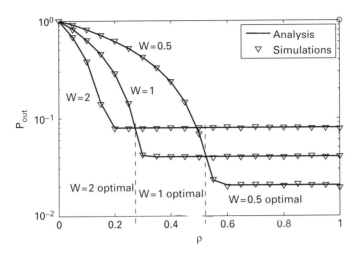

Figure 3.4 Outage probability vs. energy injection probability as a function of $W = \frac{E_s}{PT_p}$ for $E_s = 35$ dB and maximum battery capacity $50E_s$ (© [2009] IEEE).

Different values of W turn out to be optimal for different ρ. For example, $W = 2$ performs best for $0 \leq \rho \leq 0.275$. In other words, for lower injection probabilities, the best strategy is to set the transmit power such that multiple packet transmissions are possible. On the other hand, for $\rho \geq 0.525$, a lower value of W is better since communication reliability and not energy availability is the main obstacle. However, for W lower than $1/K$, the energy-unconstrained regime can never be reached. Hence, it will have a large outage probability for all values of ρ. Notice that the transmit power needs to be carefully optimized as a function of the energy profile and channel statistics.

3.2.2 Power management with channel state information

We now consider a different single link system where an EH sensor (EHS) is required to send data to a destination and has knowledge of the wireless channel to the destination. Such knowledge may be acquired, for example, using a known training signal from the receiver, when the channel is reciprocal. The EH node efficiently uses the harvested energy to maximize the average data rate by performing power control. Another aspect that will be explored is the difference in timescales between the channel-fading and energy-harvesting processes: the former typically changes significantly faster than the latter. For example, in a solar EHS, the harvested power stays roughly constant over, say, a half-hour period. On the other hand, the channel coherence time is typically of the order of a few tens or hundreds of milliseconds. Hence, it is reasonable to consider a two-stage power management algorithm:

1. In the first stage, we design the power control as a function of the channel gain to maximize the average data rate over a relatively long constant-power slot under a *constant available power* assumption.
2. We extend the results to an *outer stage* power management algorithm that works with the aim of maintaining energy neutrality of the node over a long time. It controls

the power used over the *constant available power* slots as a function of the harvested power. We obtain an optimal solution for the outer stage power allocation that maximizes the average data rate.

Together, these two stages provide a simple, yet powerful, framework for analyzing and optimizing various system parameters that determine the average data rate of EHS nodes. It will be seen that the design considerations and the corresponding solutions are fundamentally different from those obtained in Section 3.2.1.

Assumptions and notations

We make the following assumptions:

- Time is slotted, and at the start of every slot of duration T_s seconds the EHS node uses E_m J of energy to accurately measure the power gain/attenuation γ of the wireless channel to the destination. The slot duration T_s is greater than the coherence time of the channel, T_c.
- The wireless channel from the EHS node to the destination is assumed to be quasi-static and flat-fading, with the channel gain γ having a probability density function (PDF) $f_\gamma(x)$. When the channel undergoes Rayleigh fading, γ is an exponentially distributed random variable with PDF $f_\gamma(x) = \exp(-x)$, for $x \geq 0$.
- The data transmission duration is denoted T_τ, where $T_\tau \leq T_c$.
- The energy-harvesting process is also modeled as a quasi-static process that remains fixed for a *constant-power* (CP) slot of duration $T_p \gg T_c$. Let the PDF of the harvested power P_h be $f_p(P_h)$. We assume that the process is stationary and ergodic. Thus, our analysis applies to several energy-harvesting profiles considered in the literature, e.g. [2, 4, 8, 9, 13].
- In this chapter, we shall assume that the EHS only supports a constant bit rate (CBR) transmission. In order to connect the transmit power $P(\gamma)$ to the target data rate R_1, we use the well-known Shannon's capacity equation:

$$\log(1 + \gamma P(\gamma)) = R_1. \tag{3.13}$$

Practical codes can also be included in this framework by introducing a loss factor to model the gap between the Shannon capacity and the rate achieved by a given code [26]. Since the transmit power setting is based on an accurate measurement of the channel, individual packet acknowledgments are not required in this setup. The framework presented here can also be used with adaptive modulation and coding schemes (MCS). The interested reader is referred to [27] for details.
- If the EHS does not fully use the harvested energy in a given slot, it can save the excess energy in a battery and reuse it in a future slot. However, the battery is inefficient and only a fraction η of the energy input to it will be available for use later.

Note that, since the channel gain is modeled as Rayleigh distributed, the path-loss and shadowing are included in the signal powers themselves, i.e. they are the powers at the input to the receiver when the fading instantiation is unity. Also, E_m denotes the energy cost of measuring the channel reflected as received energy at the destination, relative to the noise power spectral density. Further, the noise power spectral density is normalized to 1. Next, we describe the equations that determine the average data rate of the EHS and its energy neutrality with the constant available power assumption. This will be extended to handle the variations in the harvested power in Section 3.2.2.

Inner stage: constant available power

As mentioned, the energy neutrality constraint mandates that the EHS should operate purely on the harvested energy without drawing on any external energy source. With a constant available power P_a at the node, we mathematically model energy neutrality by equating the *expected* energy deposited into the battery with the *expected* energy drawn from the battery:

$$\eta \int_0^\infty \left((P_a T_s - E_m) - (P(x) + P_f)T_\tau\right)^+ f_\gamma(x)\,dx$$
$$= \int_0^\infty \left((P(x) + P_f)T_\tau - (P_a T_s - E_m)\right)^+ f_\gamma(x)\,dx. \quad (3.14)$$

Here, $(x)^+$ equals x if $x \geq 0$ and is 0 otherwise, $P(\gamma)$ represents the transmit power of the EHS when the channel power gain is γ, and P_f represents the fixed power consumption of the transceiver circuitry. The *statistical energy neutrality (SEN)* constraint in (3.14) is similar to that in [2]. The term $\left((P_a T_s - E_m) - (P(x) + P_f)T_\tau\right)^+$ corresponds to energy being deposited in the battery, since it is greater than 0 in slots where, when the net available power, $P_a T_s - E_m$, exceeds the power consumed, $(P(x) + P_f)T_\tau$. Similarly, the term $\left((P(x) + P_f)T_\tau - (P_a T_s - E_m)\right)^+$ corresponds to slots where energy is drawn from the battery.

Note the η factor on the left-hand side of (3.14). It models the loss in energy when energy is stored in an inefficient battery and reused at a later time. The above equation is a simplification of the constraint that, at any point in time, the total energy deposited into the battery should equal or exceed the total energy drawn from the battery [1, 6, 28]. Nonetheless, (3.14) is a practically useful constraint when the maximum battery storage capacity is large compared to the variations in the battery level due to the imbalance between the energy consumed and energy deposited at any point in time. We shall assume unbounded battery capacity and data buffer.

Let R_a denote the average data rate achieved by the EHS. It is the data rate obtained after accounting for the slots when the EHS chooses not to transmit data. With CBR transmission, clearly, the optimum strategy to maximize the expected data rate over all channel realizations is one that always transmits data whenever the channel state is sufficiently good to guarantee reliable transmission at the chosen data rate. This is the truncated channel inversion (TCI) strategy [24]. In TCI, the transmission power is given by

$$P_t(\gamma) = \begin{cases} 0, & \text{if } \gamma < \gamma_m \\ \dfrac{P_0}{\gamma}, & \text{if } \gamma \geq \gamma_m, \end{cases} \quad (3.15)$$

where γ_m is the channel gain threshold for transmission and $P_0 \triangleq e^{R_1} - 1$ is the power required for successful transmission at rate R_1 when the channel gain is unity; this follows from (3.13).

Now, if the channel power gain in the current slot is γ, the energy E_c required to transmit at rate R_1 for T_τ seconds can be found from (3.13) as $E_c = P_0 T_\tau / \gamma$. For a Rayleigh fading channel, the energy neutrality condition in (3.14) is equivalent to

$$\rho_s T_s \left[\eta + (1-\eta)\left(e^{-\gamma_m} - e^{-\gamma_s}\right)\right] + P_0 T_\tau \left[(1-\eta)\text{Ei}(\gamma_s) - \text{Ei}(\gamma_m)\right] = 0, \quad (3.16)$$

where $\rho_s = (P_a T_s - E_m - P_f T_\tau)/T_s$, $\gamma_s = \frac{P_0 T_\tau}{\rho_s T_s}$, $\text{Ei}(x) \triangleq \int_x^\infty \frac{\exp(-t)}{t} dt$, $x > 0$, is the exponential integral function. When $\eta = 1$, (3.16) reduces to the simple form $\rho_s T_s = P_0 T_\tau \text{Ei}(\gamma_m)$.

Since the EHS achieves an instantaneous rate of R_1 whenever $\gamma > \gamma_m$ and is zero otherwise, the average data rate achievable over all the channel instantiations is

$$R_a = \frac{R_1 T_\tau}{T_s} e^{-\gamma_m}. \quad (3.17)$$

We can now use standard numerical techniques to determine the optimal values of the parameters such as T_τ, T_s, γ_m, etc., to maximize the average data rate R_a, which is given by (3.17), subject to the energy neutrality constraint (3.16). The interested reader is referred to [27, 29, 30] for more details and several specific case studies. One example design could choose $T_\tau = T_s = T_c$, and use (3.16) to set the value of γ_m given the P_a assigned by the outer stage. Note that R_a is implicitly related to P_a through the energy neutrality condition. It can be shown that the optimum R_a is a *concave* function of the available power P_a for most practically meaningful parameter settings. We will show this concavity and exploit it in the next section that deals with the design of the power allocation to the inner stage to maximize the long-term average data rate.

Outer stage design–variable harvested power

The goal in the outer stage is to find the optimum power allocation, P_a, for a given CP slot, *as a function of the harvested power,* P_h, to maximize the long-term average data rate, denoted R, while ensuring energy neutrality over the CP slots. It exploits the fact that there are a large number of T_s slots within a CP slot, by using R_a and P_a, the data rate and consumed power averaged over the channel distribution, as the inputs to the outer stage optimization.

The problem here is to choose the function $P_a(P_h)$, i.e. the mapping between the harvested power P_h and the allotted power P_a, for each CP slot, to maximize the average data rate, defined as

$$R = \int_0^\infty R_a(P_a) f_p(P_h) \, dP_h, \quad (3.18)$$

where R_a is determined from the previous subsection by maximizing the average data rate given by (3.17), subject to the energy neutrality constraint of (3.16). Since the inner

3.2 Physical layer design

stage uses TCI, for a given P_a, γ_m is determined by the energy neutrality condition in (3.16) when the other parameters are fixed. Hence, we can equivalently seek to find γ_m as a function of P_h. To clearly bring out the dependence of the TCI threshold γ_m on P_h, we shall henceforth write it as $\gamma_m(P_h)$. Thus, in a slot with harvested power P_h, the EHS transmits whenever $\gamma > \gamma_m(P_h)$.

Define $\gamma_T \triangleq P_0/(P_h - P_f)^+$. Two cases arise: $\gamma_m(P_h) < \gamma_T$ or $\gamma_m(P_h) \geq \gamma_T$. First, consider the case $\gamma_m(P_h) < \gamma_T$. Then, in channel slots when $\gamma < \gamma_m(P_h)$, the node does not transmit and deposits energy into the battery; when $\gamma_m(P_h) \leq \gamma < \gamma_T$, the node draws excess power from the battery; and when $\gamma \geq \gamma_T$, it deposits energy into the battery. Equating the expected power consumed by the node with P_a for a given CP slot results in

$$P_a = \int_{\gamma_T}^{\infty} \left[P_f + \frac{P_0}{x} \right] f_\gamma(x)\, dx + \int_{\gamma_m(P_h)}^{\gamma_T} \left(P_h + \frac{1}{\eta}\left(\frac{P_0}{x} + P_f - P_h \right) \right) f_\gamma(x)\, dx. \tag{3.19}$$

For the second case of $\gamma_m(P_h) \geq \gamma_T$, using similar arguments, we have,

$$P_a = \int_{\gamma_m(P_h)}^{\infty} \left[P_f + \frac{P_0}{x} \right] f_\gamma(x)\, dx. \tag{3.20}$$

Note that the above equations are in fact a restatement of the constraint given by (3.16). Also, in writing the above, we have incorporated E_m into the harvested power P_h by reducing it by E_m/T_s in every CP slot.

Now, for long-term energy neutrality, the expected harvested power should equal the expected consumed power. Hence, we can state the optimization problem as

$$\max_{P_a(P_h) \geq 0} R \triangleq \int_0^{\infty} R_a(P_a) f_p(P_h)\, dP_h, \tag{3.21}$$

subject to

$$E_p[P_h] = \int_0^{\infty} P_a(P_h) f_p(P_h)\, dP_h. \tag{3.22}$$

It can be shown that R_a is a concave function of P_a. Thus, the problem is to choose $\gamma_m(P_h)$ to maximize the average data rate, subject to the energy neutrality constraint (3.22). Introducing a Lagrange multiplier λ for the energy neutrality condition, we obtain the Lagrangian as

$$\int_0^{\infty} R_a f_p(P_h)\, dP_h + \lambda \left[E_p[P_h] - \int_0^{\infty} P_a(P_h) f_p(P_h)\, dP_h \right].$$

Setting the derivative of the Lagrangian with respect to P_a equal to zero, it can be shown that the optimum threshold $\gamma_m(P_h)$ satisfies

$$\gamma_m(P_h) = \frac{P_0}{\left[\frac{\eta}{\lambda} + (1-\eta) P_h - P_f \right]^+}, \tag{3.23}$$

with $\gamma_m(P_h) \geq \frac{P_0}{\left[\frac{1}{\lambda} - P_f\right]^+}$. This lower limit handles the case where $\gamma_m(P_h) \geq \gamma_T$. The Lagrange parameter λ is found numerically using (3.22).

Comments: The threshold rule (3.23) implies that $\gamma_m(P_h)$ is inversely proportional to P_h, which is intuitively satisfying. For the special case of $\eta = 0$ (no storage), the threshold becomes $\frac{P_0}{[P_h - P_f]^+}$, which ensures that the EHS never transmits at a power that exceeds the harvested power. At the other extreme of $\eta = 1$, we get $\gamma_m(P_h) = \frac{P_0}{\frac{1}{\lambda} - P_f}$, which is independent of P_h. This follows from the concavity of the rate function.

Thus far, we have designed the system given an on-rate, R_1. The choice of R_1 itself can be optimized to maximize the average data rate R. Recall that R_1 and the transmit power P_0 required for decodability are monotonically related via (3.13). On account of this relationship, optimizing R_1 to maximize R is equivalent to optimizing P_0 to maximize R.

3.2.3 Simulation results

We now present simulation results to illustrate the performance of the power-management solutions derived in this section. We assume a Rayleigh flat fading channel from the source to the destination. The noise variance at the destination is normalized to unity. Recall that the average data rate is the on-rate R_1 times the outage probability.

We first illustrate the performance of the inner stage design described in Section 3.2.2. Figure 3.5 illustrates the effect of the measurement energy, E_m, on the average data rate performance. The available power is taken to be unity, and fixed power losses in the transceiver circuitry are neglected. The average data rate for two values of R_1 are plotted, along with the rate achieved by using the best R_1 for each value of E_m, obtained via numerical optimization. The plot shows how the performance degrades as the percentage of the energy spent in measuring γ increases. We see that measurement of the CSI with

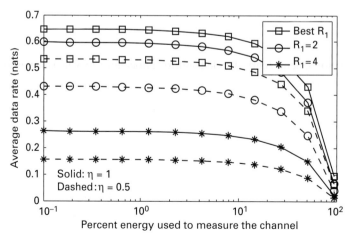

Figure 3.5 Average data rate as a function of the percentage energy used to measure the channel, i.e. $100 \times E_m/P_a T_s$.

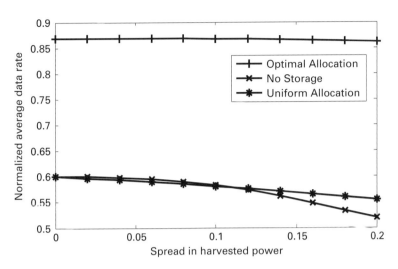

Figure 3.6 Average data rate (normalized with respect to R_1) for the uniform harvesting power distribution.

CBR transmission causes only a marginal loss in data rate, so long as the energy cost of measuring the channel is less than 15% of the total energy harvested. This is achievable for slow fading channels, which occur, for example, in fixed network deployments.

Next we consider the outer stage design of Section 3.2.2. We assume the following parameter values, which are typical in a low-power CC1000 radio running the IEEE 802.15.4 standard [14]. The transmit power, P_0, necessary to attain the fixed rate R_1 when the channel gain is unity, is assumed to be 50 mW. With a power amplifier efficiency of 20%, this corresponds to a radiated power of 13 dBm. The fixed power component P_f and the channel measurement cost E_m are neglected for simplicity.

The harvesting power distribution is taken to be uniform between $0.1 - a/2$ and $0.1 + a/2$. Figure 3.6 plots the average data rate (normalized with respect to R_1) achieved by the optimal allotment scheme derived in Section 3.2.2 as a function of the spread in harvested power, a. Also shown for comparison are: (i) an energy-efficient allocation (curve labeled no storage), where energy is used as and when it is available, and (ii) an equal allocation scheme (curve labeled uniform allocation), in which the power dissipated by the system is kept constant irrespective of the instantaneous harvested power in the slot. The optimal solution offers more than 40% improvement in the average data rate.

3.3 Cross-layer implications in a multi-node network

We now consider a network with multiple EH nodes. We study an interesting example from cooperative communications that involves interactions among multiple EH relays that transmit over a common multiple access channel [13]. The different wireless channels that different relays in the network observe allow the network to exploit an extra level of spatial diversity to improve its efficiency and reliability. By exploiting spatial

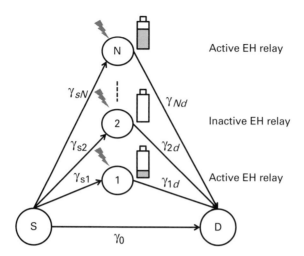

Figure 3.7 Voluntary energy-harvesting relays with battery storage capability that assist cooperative communication between a source and a destination.

diversity to combat fading and using multiple hops to combat path-loss, cooperative communications can significantly decrease the overall energy consumption. This makes it an important technology for green communication systems.

Consider a system with a source (S), a destination (D), and N energy-harvesting amplify-and-forward (AF) relays. The source and destination nodes are conventional nodes that want to communicate with each other. The channel gains from the source to relay i and from relay i to the destination are denoted by h_{si} and h_{id}, respectively. The various SR channels are assumed to be i.i.d. and so are the RD channels. The direct source-destination channel gain, which is independent of the SR and RD channels, is denoted by h_0. The reader is referred to [13] for an analysis of the general case in which the channel gains of the nodes are not statistically identical.

The cooperative data transmission occurs over two phases. In the first phase, S transmits with power P_s, which is received by the destination and the relays that have sufficient energy to forward the signal to the destination. A relay that has sufficient energy to transmit shall be said to be *active*. In the second phase, one of the active relays is selected. Selection avoids the practically challenging problem of ensuring symbol-level synchronization among multiple relays that are geographically separated. The selected relay then amplifies and forwards its received signal to the destination. At the end of the second time slot, the destination combines the signals it has received from the source and the selected relay in the two time slots. All channels are assumed to be frequency-flat, block-fading Rayleigh channels that remain constant over the two phases of cooperation.

Note that an EH relay may be active or inactive at different time instants because the energy stored in its battery depends on how often it has been selected and how much energy it has harvested in the past and when. The number of relays that are active at any time is a random variable. Since the relays are geographically separated from each other, they only have local channel knowledge and do not know a priori which relay is the best

3.3 Cross-layer implications in a multi-node network

one. Therefore, a multiple access selection algorithm becomes essential to select the best relay. Before we analyze and optimize the performance of the system, we comment on opportunistic multiple access selection algorithms, which are an integral component of the system.

3.3.1 Multiple access selection algorithms

Several different algorithms such as polling, splitting, and timer algorithms are available for implementing a selection. The simplest algorithm is polling, in which each node computes a real-valued metric as a function of its local channel knowledge. A larger metric implies that the node will be more useful as a relay. Each node sequentially transmits its metric to the source without any collisions. While polling is simple, it is not scalable as the time required by it to select the best relay increases linearly with the number of available relays.

- The splitting-based selection algorithm is a time-slotted multiple access contention algorithm in which each node autonomously decides whether or not to transmit in a certain time slot [15]–[16]. The algorithm works as follows. In each time slot, only those nodes whose metrics lie in between an upper and a lower threshold transmit. At the end of every slot, the coordinator, which can be the source or destination, feeds back a three-state outcome indicating idle (when no node transmitted), success (when exactly one node transmitted and was decoded successfully), or collision (when two or more nodes transmitted and none could be decoded). The nodes update their thresholds based on the feedback. The thresholds are set so that about one node, on average, transmits in every slot.

  The splitting algorithm is fast and scales well with the number of users. It can find the relay with the highest metric in just 2.467 slots, on average, even with a large number of relays. Contrast this with the polling algorithm, which would require N slots to poll N relays.

- The timer algorithm partially resembles the back-off mechanism used in the distributed coordination function (DCF) in wireless local area networks. Every node sets its timer as a function of its metric, and transmits a packet when its timer expires. However, unlike the random back-off mechanism of DCF, the metric-to-timer mapping is a deterministic monotone non-increasing function, which ensures that the node with the highest metric always transmits first [17]–[18].

  The timer algorithm is attractive because of its simplicity and its distributed nature. It requires no feedback during the selection process. However, the timer algorithm can occasionally fail to select the best node. This occurs, for example, when the timers of the node with the highest and second-highest metrics expire too close to one another. The optimal metric-to-timer mapping that maximizes the probability of selecting the node with the highest metric for a given duration of selection, and the optimal mapping that minimizes the average time required to select given a minimum probability of success requirement, are derived in [17].

3.3.2 Energy harvesting, storage, and usage model

The power harvested by a relay over time is assumed to be a stationary and ergodic process with mean P^{av} W. An EH relay stores its harvested energy in a lossless energy buffer. Among the the active relays, the relay that maximizes the end-to-end SNR of (3.25) is selected. It draws energy from its energy buffer to transmit. For simplicity, we assume that the energy and time overhead of selection is negligible. The interested reader is referred to [19] for an analysis of the system-level performance trade-offs associated with a selection mechanism.

Let α_i denote the signal amplification by a relay if it were to transmit. It is chosen to ensure that the transmit power of the relay (averaged over noise) is always P. Thus,

$$\alpha_i^2 = \frac{P}{P_s |h_{si}|^2 + N_0}, \quad (3.24)$$

where N_0 is the additive white Gaussian noise (AWGN) power. The SNR at the destination, γ_D, then equals [20]

$$\gamma_D = \gamma_0 + \frac{\gamma_{si} \gamma_{id}}{\gamma_{si} + \gamma_{id} + 1} \approx \gamma_0 + \frac{\gamma_{si} \gamma_{id}}{\gamma_{si} + \gamma_{id}}, \quad (3.25)$$

where $\gamma_0 = |h_0|^2 \frac{P_s T_s}{N_0}$, $\gamma_{si} = |h_{si}|^2 \frac{P_s T_s}{N_0}$, and $\gamma_{id} = |h_{id}|^2 \frac{P T_s}{N_0}$. Let $\mathbf{E}[\gamma_0] = \bar{\gamma}_0$, $\mathbf{E}[\gamma_{si}] = \bar{\gamma}_s$, and $\mathbf{E}[\gamma_{id}] = \bar{\gamma}_d$.

3.3.3 Energy neutrality implication

Let $\xi \geq 0$ denote the steady state probability that a relay is active. We define the following two important terms that depend on ξ:

- *Energy-constrained relay:* A relay is said to be energy constrained when $\xi < 1$.
- *Energy-unconstrained relay:* A relay is energy unconstrained when $\xi = 1$.

An energy-unconstrained relay is, thus, always available for relaying. Intuitively, this should occur when the rate at which a relay harvests energy exceeds the average rate at which it consumes energy.

3.3.4 Performance analysis

The important result below shows that whether a relay is energy constrained or not depends on the other EH relays in the system. Let

$$\rho \triangleq \frac{2 P^{\text{av}}}{P}. \quad (3.26)$$

Proposition 3.2. *When $\rho < \frac{1}{N}$, the relays are energy constrained and the probability that a relay is active, ξ, is*

$$\xi = 1 - (1 - N\rho)^{\frac{1}{N}}. \quad (3.27)$$

When $\rho \geq \frac{1}{N}$, all the relays are energy unconstrained ($\xi = 1$).

Proof. First consider the case where the relays are energy constrained, i.e. a relay consumes all the energy it harvests. Let Pr(Relay i sel.) denote the probability that Relay i is selected (sel.). Therefore, it consumes energy at an average rate of $\frac{1}{2}$Pr(Relay i sel.)P. The factor of $\frac{1}{2}$ arises because the relay transmits in one of the two phases of the cooperation protocol and both phases are of equal duration. From the energy neutrality constraint, this must equal the rate at which the node harvests energy, P^{av}. Equating the two, we obtain

$$\text{Pr(Relay } i \text{ sel.)} = \frac{2P^{\text{av}}}{P} = \rho. \qquad (3.28)$$

Using the law of total probability, the probability that a relay is selected can be written as

$$\text{Pr(Relay } i \text{ sel.)} = \sum_{r=0}^{N-1} \text{Pr(Relay } i \text{ sel.|Relay } i \text{ active, } r \text{ other active relays)}$$
$$\times \text{Pr}(R_i \text{ is active, } r \text{ other active relays).} \qquad (3.29)$$

By symmetry, Pr(Relay i sel.|Relay i active, r other active relays) $= \frac{1}{r+1}$. To evaluate Pr(R_i is active, r other active relays), we make the *decoupling approximation* that the event that a relay node is active is independent of whether other relay nodes are active or not. Hence,

$$\text{Pr(Relay } i \text{ active, } r \text{ other active relays)} \approx \xi \binom{N-1}{r} \xi^r (1-\xi)^{N-1-r}. \qquad (3.30)$$

Substituting (3.29) and (3.30) in (3.28), we get

$$\rho = \xi \sum_{k=0}^{N-1} \frac{1}{r+1} \binom{N-1}{r} \xi^r (1-\xi)^{N-r-1} = \frac{1-(1-\xi)^N}{N}.$$

Rearranging the terms yields (3.27). The derivation above also shows that $\rho N < 1$ when $\xi < 1$. When $\xi = 1$, the system harvests more energy than it can use. Therefore, $\rho N \geq 1$. □

We now derive an expression for the fading-averaged end-to-end symbol error rate (SER) for MPSK. For this, we first determine SER$_k$, which is the SER given that k out of the N relays are active. Let

$$\Lambda_i = \frac{\gamma_{si}\gamma_{id}}{\gamma_{si}+\gamma_{id}}. \qquad (3.31)$$

Let $f_{\Lambda_i}(.)$ and $F_{\Lambda_i}(.)$ denote the PDF and cumulative distribution function (CDF), respectively, of Λ_i for an arbitrary relay i. They are given by [21]

$$F_{\Lambda_i}(x) = 1 - \frac{2x}{\sqrt{\nu}} K_1\left(\frac{2x}{\sqrt{\nu}}\right) \exp\left(-\frac{\mu}{\nu} x\right), \quad x > 0, \tag{3.32}$$

$$f_{\Lambda_i}(x) = \left(\frac{4x}{\nu} K_0\left(\frac{2x}{\sqrt{\nu}}\right) + \frac{2x\mu}{\nu\sqrt{\nu}} K_1\left(\frac{2x}{\sqrt{\nu}}\right)\right) \exp\left(-\frac{\mu}{\nu} x\right), \quad x > 0, \tag{3.33}$$

where $\nu = \bar{\gamma}_s \bar{\gamma}_d$, $\mu = \bar{\gamma}_s + \bar{\gamma}_d$, and $K_l(\cdot)$ is the modified Bessel function of the second kind of order l [22].

We treat the cases of $1 \leq k \leq N$ and $k = 0$ separately below.

Proposition 3.3. *For $1 \leq k \leq N$, SER_k can be written as*

$$SER_k = k \int_0^\infty \psi(x) f_{\Lambda_i}(x) F_{\Lambda_i}^{k-1}(x) \, dx, \tag{3.34}$$

where $\psi(x) = \frac{1}{\pi} \int_0^{\frac{M-1}{M}\pi} \exp\left(-\frac{x}{\beta \sin^2(\phi)}\right) \left(1 + \frac{\bar{\gamma}_0}{\beta \sin^2(\phi)}\right)^{-1} d\phi.$

Proof. From (3.25), $\gamma_D = \gamma_0 + \Lambda_{[1]}$, where $\Lambda_{[1]} \triangleq \max_i \Lambda_i$. Recall from (3.31) that $\Lambda_i = \frac{\gamma_{si}\gamma_{id}}{\gamma_{si}+\gamma_{id}}$. From [23], SER_k for MPSK takes the form

$$SER_k = \frac{1}{\pi} \int_0^{\frac{M-1}{M}\pi} M_{\gamma_D}\left(-\frac{\sin^2(\pi/M)}{\sin^2(\phi)}\right) d\phi, \tag{3.35}$$

where $M_{\gamma_D}(s)$ is the moment-generating function (MGF) of γ_D. Since the SD link is independent of the SR and RD links, we have $M_{\gamma_D}(s) = M_{\gamma_0}(s) M_{\Lambda_{[1]}}(s)$. Since γ_0 is exponentially distributed with mean $\bar{\gamma}_0$, $M_{\gamma_0}(s) = \frac{1}{1-\bar{\gamma}_0 s}$. Furthermore, from elementary order statistics, the PDF of $\Lambda_{[1]}$ is given by $f_{\Lambda_{[1]}} = k f_{\Lambda_i}(x) F_{\Lambda_i}^{k-1}(x)$. Therefore, $M_{\Lambda_{[1]}}(s) = \int_0^\infty k f_{\Lambda_i}(x) F_{\Lambda_i}(x)^{k-1} e^{sx} dx$ and

$$SER_k = \frac{1}{\pi} \int_0^{\frac{M-1}{M}\pi} \int_0^\infty \frac{1}{1+\bar{\gamma}_0 \frac{\sin^2(\pi/M)}{\sin^2(\phi)}} k f_{\Lambda_i}(x) F_{\Lambda_i}^{k-1}(x) \exp\left(-x \frac{\sin^2(\pi/M)}{\sin^2(\phi)}\right) dx \, d\phi. \tag{3.36}$$

Equivalently, $SER_k = \int_0^\infty \psi(x) k f_{\Lambda_i}(x) F_{\Lambda_i}^{k-1}(x) dx$, $\psi(x)$ with as defined before. □

3.3 Cross-layer implications in a multi-node network

The interested reader is referred to [13] for closed-form expressions for $\psi(x)$. Using Gauss–Laguerre quadrature [22], (3.34) can be further simplified to

$$\text{SER}_k \approx k \sum_{n=1}^{M} w_n \left(\frac{4a_n v}{\mu^2} K_0 \left(\frac{2a_n \sqrt{v}}{\mu} \right) + \frac{2a_n \sqrt{v}}{\mu} K_1 \left(\frac{2a_n \sqrt{v}}{\mu} \right) \right)$$

$$\times \psi \left(\frac{v}{\mu} a_n \right) \left(1 - \frac{2a_n \sqrt{v}}{\mu} K_1 \left(\frac{2a_n \sqrt{v}}{\mu} \right) \exp(-a_n) \right)^{k-1}, \quad (3.37)$$

where a_n and w_n, for $1 \leq n \leq M$, are the M Gauss–Laguerre abscissa and weights, respectively. The accuracy of the approximation improves as M increases.

Proposition 3.4. *When $k = 0$ relays are active,*

$$\text{SER}_0 = \frac{1}{2} - \frac{1}{2\sqrt{1 + \csc^2\left(\frac{\pi}{M}\right)/\bar{\gamma}_0}} - \frac{\arctan\left(\sqrt{\csc^2\left(\frac{\pi}{M}\right) - 1}/\sqrt{1 + \frac{\csc^2\left(\frac{\pi}{M}\right)}{\bar{\gamma}_0}}\right)}{\pi \sqrt{1 + \frac{\csc^2\left(\frac{\pi}{M}\right)}{\bar{\gamma}_0}}}$$

$$+ \frac{\arctan\left(\sqrt{\csc^2\left(\frac{\pi}{M}\right) - 1}\right)}{\pi}. \quad (3.38)$$

Proof. Now, only the direct SD Rayleigh fading link matters. Therefore, $\text{SER}_0 = \frac{1}{\pi} \int_0^{\frac{M-1}{M}\pi} \left(1 + \bar{\gamma}_0 \frac{\sin^2(\pi/M)}{\sin^2(\phi)} \right)^{-1} d\phi$. Making the variable substitution $x = \csc^2(\phi)$ and simplifying yields (3.38). $\square$

From Propositions 3.2, 3.3, and 3.4, the final SER expression for a system with N EH relays easily follows, as given below.

Result 3.1.

$$\text{SER} \approx (1-\xi)^N \text{SER}_0 + N\xi \sum_{n=1}^{M} w_n \left(\frac{4a_n v}{\mu^2} K_0 \left(\frac{2a_n \sqrt{v}}{\mu} \right) + \frac{2a_n \sqrt{v}}{\mu} K_1 \left(\frac{2a_n \sqrt{v}}{\mu} \right) \right)$$

$$\times \psi \left(\frac{v}{\mu} a_n \right) \left(1 - \frac{2a_n \xi \sqrt{v}}{\mu} K_1 \left(\frac{2a_n \sqrt{v}}{\mu} \right) \exp(-a_n) \right)^{N-1}, \quad (3.39)$$

where SER_0 is given by (3.38).

3.3.5 Numerical results

We illustrate the results for $\bar{\gamma}_{si} = P_s T_s$, $\bar{\gamma}_{id} = PT_s$, and $\bar{\gamma}_0 = \frac{P_s T_s}{4}$, where all relays transmit with the same power P. The noise power is normalized to $N_0 = 1$.

Figure 3.8 plots the SER as a function of the energy spent per symbol by a relay, PT_s, for four different values of ρ. When $\rho < \frac{1}{N} = 0.25$, the relays are energy constrained.

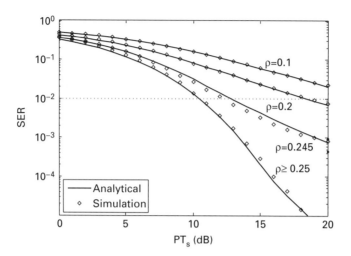

Figure 3.8 SER as a function of the transmit energy per symbol (PT_s) for different values of ρ for $P_s = P$, QPSK, and $N = 4$ relays (© [2010] IEEE).

Notice that the three SER curves for $\rho < 0.25$ become parallel to each other for large PT_s, which implies that they have the same diversity order. When $\rho \geq 0.25$, all the relays are energy unconstrained. This makes the diversity order increase to $N + 1 = 5$, and is reflected in the steeper slope for the corresponding curve.

Figure 3.9 plots the SER as a function of PT_s when $2P^{av}T_s = 6$ dB and 12 dB. When $2P^{av}T_s$ increases from 6 dB to 12 dB, the optimum value of PT_s increases from 12 dB to 18 dB. The curves can be understood as follows. For $PT_s < 2P^{av}T_s/N$, all the relays are energy unconstrained, which makes the two curves coincide for small PT_s. As PT_s increases, the SER decreases as expected. However, once PT_s exceeds $2NP^{av}T_s$, all the relays become energy constrained. Therefore, the SER becomes sensitive to $2P^{av}T_s$. It increases as PT_s increases because fewer relays are active. For large PT_s, since the relays are inactive for most of the time, the SER is determined primarily by the SD channel and is dependent only on $P_s T_s$.

3.4 Conclusion

The incorporation of energy-harvesting technology in wireless networks is an attractive and green solution that tackles the problem of limited network lifetime. However, the rules that govern the operation of energy-harvesting wireless networks compared with conventional nodes that are deployed with pre-charged batteries or nodes powered by the mains are different. The fundamental rule that governs the operation of an energy-harvesting network is the energy neutrality constraint. While energy conservation and spectral-efficiency maximization continue to remain desirable design objectives, they become secondary goals. The primary design focus changes to judiciously using all the harvested energy and ensuring that energy is available for consumption when required.

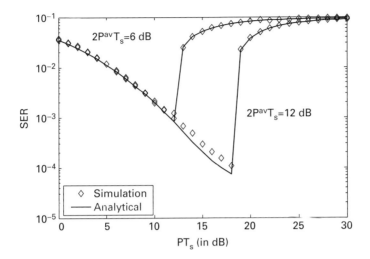

Figure 3.9 Optimizing the relay transmit energy per symbol (PT_s) on SER as a function of the average energy harvested by a relay, when $P_s T_s = 15$ dB, QPSK, and $N = 4$ relays (© [2010] IEEE).

Being too conservative in consuming the harvested energy simply wastes the harvested energy, while being too aggressive leads to the node unnecessarily being starved of the energy it requires at later time instants. Thus, the physical and multiple access layers of an EH network need to be redesigned.

The several design examples that have been considered in this chapter brought out the important factors that ought to be considered in the design, such as energy-harvesting profile, availability or unavailability of channel state information, and energy storage capability. We have observed that the system performance is quite sensitive to the transmit power settings. Altogether, green energy-harvesting wireless networks are a promising and emerging area of research in wireless communications, and pose several novel and interesting challenges for researchers and practitioners alike.

References

[1] V. Sharma, *et al.*, "Optimal energy management policies for energy harvesting sensor nodes," *IEEE Trans. Wireless Commun.*, vol. 9, pp. 1326–1336, Apr. 2008.

[2] A. Kansal, *et al.*, "Power management in energy harvesting sensor networks," *ACM Trans. Embedded Comput. Syst.*, vol. 7, pp. 1–38, Sept. 2007.

[3] P. S. Khairnar and N. B. Mehta, "Power and discrete rate adaptation for energy harvesting wireless nodes," in *Proc. of IEEE ICC*, Jun. 2011.

[4] D. Niyato, E. Hossain, and A. Fallahi, "Sleep and wakeup strategies in solar-powered wireless sensor/mesh networks: performance analysis and optimization," *IEEE Trans. Mobile Comput.*, vol. 6, pp. 221–236, Feb. 2007.

[5] A. Seyedi and B. Sikdar, "Energy efficient transmission strategies for body sensor networks with energy harvesting," *IEEE Trans. on Commun.*, vol. 58, pp. 2116–2126, Jul. 2010.

[6] O. Ozel, et al., "Transmission with energy harvesting nodes in fading wireless channels: optimal policies," *IEEE J. Sel. Areas Commun.*, vol. 29, pp. 1732–1743, Sept. 2011.

[7] M. Gatzianas, L. Georgiadis, and L. Tassiulas "Control of wireless networks with rechargeable batteries," *IEEE Trans. on Wireless Commun.*, vol. 9, pp. 581–593, Feb. 2010.

[8] J. Lei, R. Yates, and L. Greenstein, "A generic model for optimizing single-hop transmission policy of replenishable sensors," *IEEE Trans. on Wireless Commun.*, vol. 8, pp. 547–551, Feb. 2009.

[9] B. Medepally, N. B. Mehta, and C. R. Murthy, "Implications of energy profile and storage on energy harvesting sensor link performance," in *Proc. of IEEE Globecom*, Nov. 2009.

[10] J. Ammer and J. Rabaey, "Low power synchronization for wireless sensor network modems," in *Proc. of IEEE WCNC*, pp. 670–675, Mar. 2005.

[11] S. Cui and A. J. Goldsmith, "Cross-layer design in energy-constrained networks using cooperative MIMO techniques," *EURASIP Signal Process. J., Special Issue on Advances in Sig. Proc.-based Cross-layer Designs*, vol. 86, pp. 1804–1814, Aug. 2006.

[12] J. A. Paradiso and M. Feldmeier, "A compact, wireless, self powered pushbutton controller," in *Proc. of Int. Conf. Ubiquitous Comput.*, pp. 299–304, 2001.

[13] B. Medepally and N. B. Mehta, "Voluntary energy harvesting relays and selection in cooperative wireless networks," *IEEE Trans. on Wireless Commun.*, vol. 9, pp. 3543–3553, Nov. 2010.

[14] *IEEE standard 802, part 15.4: wireless medium access control (MAC) and physical layer (PHY) specifications for low rate wireless personal area networks (WPANs)*, 2003.

[15] V. Shah, N. B. Mehta, and R. Yim, "Splitting algorithms for fast relay selection: generalizations, analysis, and a unified view," *IEEE Trans. Wireless Commun.*, vol. 9, pp. 1525–1535, Apr. 2010.

[16] X. Qin and R. Berry, "Opportunistic splitting algorithms for wireless networks," in *Proc. INFOCOM*, pp. 1662–1672, Mar. 2004.

[17] V. Shah, N. B. Mehta, and R. Yim, "Optimal timer-based selection schemes," *IEEE Trans. Commun.*, vol. 58, pp. 1814–1823, Jun. 2010.

[18] A. Bletsas, et al., "A simple cooperative diversity method based on network path selection," *IEEE J. on Sel. Areas Commun.*, vol. 24, pp. 659–672, Mar. 2006.

[19] V. Shah, N. B. Mehta, and R. Yim, "The relay selection and transmission trade-off in cooperative communication systems," *IEEE Trans. Wireless Commun.*, vol. 9, pp. 2505–2515, Aug. 2010.

[20] A. Ribeiro, X. Cai, and G. B. Giannakis, "Symbol error probabilities for general cooperative links," *IEEE Trans. Wireless Commun.*, vol. 4, pp. 1264–1273, May 2005.

[21] P. A. Anghel and M. Kaveh, "Exact symbol error probability of a cooperative network in a Rayleigh-fading environment," *IEEE Trans. on Wireless Commun.*, vol. 3, pp. 1416–1421, Sept. 2004.

[22] M. Abramowitz and I. Stegun, *Handbook of Mathematical Functions with Formulas, Graphs, and Mathematical Tables*. Dover, 9th ed., 1972.

[23] M. K. Simon and D. Divsalar, "Some new twists to problems involving the Gaussian probability integral," *IEEE Trans. on Commun.*, vol. 46, pp. 200–210, Feb. 1998.

[24] A. J. Goldsmith and P. P. Varaiya, "Capacity of fading channels with channel side information," *IEEE Trans. Inf. Theory*, vol. 43, no. 6, pp. 1986–1992, Nov. 1997.

[25] S. Cui, A. J. Goldsmith, and A. Bahai, "Energy-constrained modulation optimization," *IEEE Trans. on Wireless Commun.*, vol. 4, no. 5, pp. 2349–2360, Sept. 2005.

[26] T. Starr, J. M. Cioffi, and P. J. Silverman, *Understanding Digital Subscriber Line Technology*. 1st ed., Prentice-Hall, 1999.

[27] C. Murthy, "Power management and data rate maximization in wireless energy harvesting sensors," *Intl. J. Wireless Inf. Netw.*, vol. 16, no. 3, pp. 102–117, Jul. 2009.

[28] C. K. Ho and R. Zhang, "Optimal energy allocation for wireless communications powered by energy harvesters," in *Proc. of IEEE ISIT*, Austin, TX, USA, Jun. 2010.

[29] V. Shenoy and C. Murthy, "Throughput maximization of delay-constrained traffic in wireless energy harvesting sensors," in *Proc. of IEEE ICC*, Cape Town, South Africa, May 2010.

[30] S. Reddy and C. Murthy, "Profile-based load scheduling in wireless energy harvesting sensors for data rate maximization," in *Proc. of IEEE ICC*, Cape Town, South Africa, May 2010.

4 Mechanical relaying techniques in cellular wireless networks

Panayiotis Kolios, Vasilis Friderikos, and Katerina Papadaki

4.1 Introduction

The tendency to devise more sophisticated network-management solutions is expected to sharpen in the near future due to the need to reduce energy-consumption levels of large-scale deployed cellular networks. With the adoption of smart phones that enable full access to different internet applications, mobile users will continue to require more for less. In light of these emerging trends in the use of mobile phones, network operators need to increase network capacity to fulfil the demand, while at the same time keeping capital and operational expenditure low to maintain a competitive edge. Traditional approaches to provide energy efficiency focus on the different elements of the network such as reducing the energy consumption at the component (silicon) level. But it is becoming increasingly apparent that energy consumption should be considered as an architectural element in the design of the network. This more holistic (architectural) view calls for a system-level approach on issues related to energy consumption for the sustainable proliferation of emerging and future wireless networks.

In this chapter, we detail a relaying technique that can be used in general heterogenous wireless access network scenarios whereby mobile relay nodes are allowed to store information while on the move before they engage in transmission with another node or the base station (BS) at a later time instance. This is in fact the fundamental principle of communication in delay-tolerant networks (DTNs), where the silent assumption is that there is no end-to-end path between the source and destination nodes, either due to the sparsity of the communicating nodes or other communication constraints imposed by the specific nature of the network. However, contrary to the DTN setup, infrastructure nodes in cellular networks provide an almost 100% coverage and thus are able to maintain full connectivity across large geographical areas. Therefore, while there is ubiquitous connectivity in cellular networks, message communication can be *deliberately* delayed for a future time instance with favorable networking conditions. Communicating only at the best locations within the cell and always under the message delivery deadlines imposed by the initiated service, significant reductions in communication energy-consumption can be achieved. Mechanical relaying (MR) aims to achieve exactly that, by enabling mobile nodes to operate under a store-carry and forward paradigm within cellular networks.

The rest of the chapter details the operation of mechanical relaying and is organized as follows. In Section 4.2, a background literature review in the scope research area is presented. The case of mechanical relaying is detailed in Section 4.3, which includes a detailed investigation of the applicability and the benefits of the different varieties of the proposed scheme. Section 4.4 quantifies the potential energy-efficiency gains that can be achieved via MR through a set of real-world measurements. Section 4.5 details the standardization efforts that are currently taking place and that are relevant to the proposed MR concept. Finally, the chapter concludes with Section 4.6 where future avenues of research are also discussed.

4.2 Background

4.2.1 DTN architecture

The area of delay-tolerant networking has its origins in the initial efforts to provide novel networking techniques that are able to sustain the significant packet delays that are observed in space communications. Based on this prerequisite of providing communication over such challenged environments, a set of architectural designs have been proposed in [1] under the umbrella of the IETF Inter-Planetary Internet Research Group. This effort led to a formal DTN network architecture [2]–[3] that can be used for a variety of different network scenarios where the underlying network topology is not connected and arbitrary long periods of link disconnection may occur between communicating entities. In essence, the fundamental assumptions of internet protocols, such as the existence of end-to-end paths and packet delays of the order of a few hundred milliseconds, are relaxed in a DTN network. The above are not requirements in DTNs and nodes are allowed to store an information message or its segment for an arbitrary amount of time before relaying it to another node.

Examples of networks where frequent topology partition can take place and hence DTN mechanisms can be used are outer-space networks, wireless ad hoc networks, mobile wireless sensor networks (such as environmental sensing and animal tracking) and vehicular networks. The key characteristic of all DTN protocols is that they employ a store-carry and forward (SCF) message propagation mechanism. In DTNs, segments of the information message (or sometimes the complete information message) are relayed between hosts and the hosts are allowed to store the segments before forwarding them in future time intervals under some predefined rules (or policies). Therefore, by exploring opportunistic contacts, DTNs can deliver information messages between end-to-end communicating nodes, even though the delay can – in the general case – be arbitrarily high.

Delay-tolerant networking introduces a new layer in the protocol stack, called the "bundle layer," which is situated above the transport layer in the TCP/IP protocol architecture. Hence, a DTN is a message-oriented architecture (i.e. information is exchanged in the form of bundles) and can also be considered as an overlay architecture that sits on top of the traditional transport layer. The definition of the bundle protocol in terms of the service description for the exchange of messages (bundles) in DTNs together with all other supported functionalities is detailed in [4].

Table 4.1. Routing schemes in delay-tolerant networks

Routing	Single copy?	Energy aware?	Characteristics
Data mules [5]	N	N	Opportunistic non-optimized routing
Message ferrying [6]	N	N	Assumes controlled mobility
Epidemic [7]	N	N	Flooding of messages, replication upon contact
Spray and wait [8]	N	N	Source "sprays" up to L copies of the message
SMART [9]	N	N	Fixed number of copies, utilize encountering history
PROPHET [10]	N	N	Prediction-based routing using encountering history
Deterministic [11]	Y	N	Adapted Dijkstra's SP algorithm, minimize the end-to-end delay

4.2.2 Routing in DTNs

A large number of routing protocols have recently been proposed for DTNs (please refer to Table 4.1 for a list of major schemes in the literature). The characteristic and common assumption of these routing schemes is that the network topology is a disconnected one. More importantly, for all those cases depicted in the table, there is no direct communication between the source and destination nodes, as is the case explored in this work. This issue is explained more clearly in Figure 4.1 where the possible forwarding paths of a classic DTN topology and the proposed MR networking paradigm are illustrated.

Routing schemes for DTNs are mainly based on a (controlled) probabilistic replication of information messages with the main objective to reduce the overall communication overhead in the network while at the same time aiming to achieve a good performance in terms of end-to-end message delivery delay. Since more than one copy of the information message is disseminated in the network, it is expected that the energy consumption for these schemes will inevitably be high. The authors in [11] proposed a set of routing schemes that involve single copy routing in DTNs. However, energy efficiency is not the primary research objective in that work either. Instead (and in most cases within the DTN literature), the challenge is to cope with the dynamic underlying networking environment.

Further, due to the very specific characteristics of networks that DTNs can be used for, it comes as no surprise that the applicability of DTN services has been very limited to date. To this end, the work in [12] addresses this issue and provides use case scenarios where DTNs could go mainstream to either extend the reach of services beyond the connected world or providing a social message dissemination mechanism in isolated/remote areas. This is in contrast to the objectives of this work, where the perceived user experience within the cell is not to be deteriorated in any way and system performance should be maintained and even enhanced. It is important to note at this point that via the proposed MR extension to cellular networks, a user can still access the network instantaneously and use its full potentials. However, for those cases where message delivery delays can

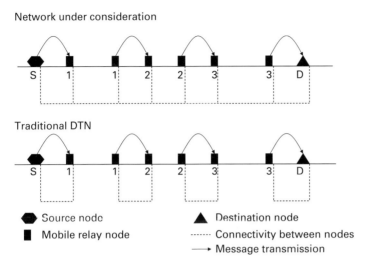

Figure 4.1 Comparison between traditional delay-tolerant networks and store-carry and forward schemes considered in this work. As can be seen in the figure, in DTN there is only connectivity between nodes during the message transmission times whereas in the scenario in this work, which resembles a cellular network, the source and destination are within communication range but SCF relaying is enforced in order to reduce energy consumption.

be tolerated, communication is deferred at a later time with better networking conditions. As shown in the sequel, there are a great number of services in the "connected" world that can in practice tolerate significant delays at no expense to the perceived user experience. By capitalizing on the delay tolerance of such services, the store-carry and forwarding paradigm can be realized within the cell to enhance the performance of current mobile networks.

4.3 Mechanical relaying

In the ongoing evolution of cellular networks, aggregate transmission rates on both the downlink and uplink have been the driving source of innovation to increase subscriber growth and support multimedia services to the users. Moreover, due to the fixed operating energy costs of both the infrastructure nodes and the mobile terminals, higher-capacity mobile networks have been shown to provide increased energy-efficiency gains as communicating nodes can enter sleep modes faster and for longer periods of time to conserve energy. Due to the best-effort nature of IP networks, continuous flow of data traffic is rarely achieved and thus idle periods can be observed between message/packet transmissions. As such, switching between transmit and sleep modes during data transmissions means that energy can be conserved. Standardization bodies have already provided enhancements to current protocol implementations that take advantage of this feature of full IP-based networks for discontinuous transmission and reception (DTX and DRX) of messages during active sessions [13]. However, while faster transmission links might improve the overall energy-efficiency gains over short-range communications, it would require unreasonably high power-consumption to provide high data rates over long distances.

For this reason, to support higher data rates, pico- and femto-cell deployments have attracted considerable attention from the research community as the means to provide high-speed, energy-efficient communication over short ranges to both indoor and outdoor users. Such shrinking-cell-size solutions however are mainly applicable to densely populated areas where the infrastructure is in place to support the cheap installation and maintenance of such topologies. As an indication, it is estimated that approximately 20% of all BS installations are placed in urban areas, whereas the rest of those units are deployed in suburban and rural areas where shrinking-cell-size solutions are simply not economical. Mechanical relaying is purposefully envisioned to achieve the desired short-range transmissions in cellular networks by capitalizing on the actual mobility of nodes and the elasticity of internet traffic services. It is interesting to note here that although the initial designs of cellular networks presumed the existence of only voice-centric applications with real-time delivery requirements, current data traffic mix studies indicate a significant amount of delay-tolerant traffic, coming mainly from internet applications such as email and P2P. The fact is that with the gradual merge of the internet and mobile domains, voice communication has become one of the many alternative services offered on the go. Internet services experience a broad range of message delivery delay requirements allowing the possibility of taking that feature into account for network optimization. MR benefits from this key observation to trade-off message delivery delay for increased energy-efficiency gains by allowing mobile terminals to postpone message transmissions for future time instances with favorable channel gains.

An illustrative example of the proposed MR scheme is shown in Figure 4.2. In addition to the traditional direct link transmission (as shown by link 1 in Figure 4.2) employed by current cellular deployments and the basic multi-hop scheme already proposed in the literature (link 2 in Figure 4.2) [14], mechanical relaying can be realized within the cell (link 3 in Figure 4.2). A mobile user (assumed here to be a user terminal within a vehicle) acting as a source node can simply postpone engaging in transmission until it is closer to the BS of the serving cell, thus reducing the transmission distance either for the uplink or downlink of information. The same mobile node however can act as a relay for other source nodes within the cell. In this case a static node (which can be either a user terminal or the BS in the uplink or downlink, respectively) can postpone the transmission to a mobile relay node while the relay moves closer to its vicinity. The relay node in turn can store and carry the received data while in transit before forwarding it to another relay or the destination at a later time instance. An example of such a scenario is illustrated via the path of links $3a \mapsto 3b \mapsto 3c$ in Figure 4.2 for the uplink of information messages. As shown in the figure, via MR localized transmissions can be achieved throughout the cell coverage area, minimizing in that respect the required communication energy-consumption.

It is important to note here that the timescale of operation of MR is in the order of a few seconds to a few tens of seconds, which is the time required for the displacement of nodes under realistic travelling speeds. Such mobile scenarios include urban commute

4.3 Mechanical relaying

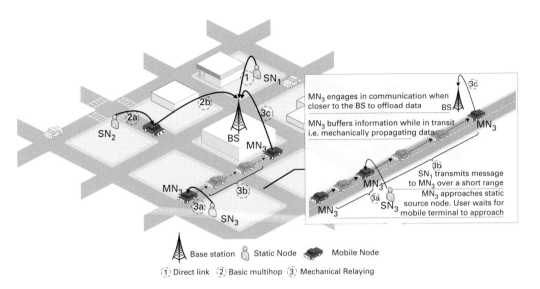

Figure 4.2 Mechanical relaying enables mobile relay nodes to postpone the transmission of information messages while in transit and engage in communication at a later time when the mobile node moves to a better forwarding location within the cell. The illustrative example here includes the traditional direct link communication, the basic multi-hop paradigm and the proposed MR scheme.

as illustrated in Figure 4.2, highway vehicular traffic and high-speed train scenarios, all of which make up a substantial source of generated data traffic. Further, as discussed in the sequel, there is a considerable amount of data traffic that can tolerate delays in the order of few seconds to tens of seconds at no cost to the perceivable user experience. As such, MR provides the opportunity to achieve substantial energy-efficiency gains over true mobile scenarios and complement other green radio techniques in reducing the total energy-consumption of cellular networks.

4.3.1 Mobile internet traffic mix

Traditionally, network operators have focused on the various mobile voice and messaging services as the primary offerings of their service portfolio. To increase their competitive edge, attract an even greater number of subscribers and increase the actual usage per customer, mobile operators have been increasingly willing to open up their networks to third-party services that have already been flourishing over the internet. Evidently, email clients now come pre-installed as a basic phone feature not only on high-end smartphones but also on standard low-end handsets. In addition, news readers and social networking applications are consistently appearing as big contributors to the total aggregated data traffic consumption. Moreover, on-demand video is expected to dominate the total number of information bits shifted over the air in the near future as more content becomes available for mobile platforms and as more terminals become capable of supporting video streams. Web browsing, file downloads (either through ftp

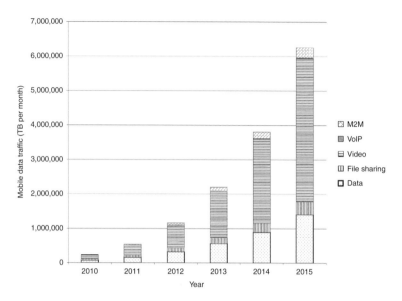

Figure 4.3 Mobile internet traffic growth.

or peer-to-peer channels), and voice-over IP (VoIP) traffic complete the league table of the most popular internet traffic services over mobile networks.

Figure 4.3 depicts the projections in data traffic growth as estimated by Cisco's visual networking index [15]. Similar trends are also reported by other independent studies such as [16]–[18]. In the latter report, US mobile network operator AT&T already notes an 8000% increase in data traffic over the four-year period since 2007. It is interesting to note here that such a demand growth has been a result of the spectacularly fast pace of adoption and usage of mobile internet services. While real-time voice services retain their popularity it is estimated that VoIP services will account for less than 1% of the total aggregated data traffic by 2015, as shown in Figure 4.3.

An important observation to be made here, based on the above discussions, is that unlike interactive voice applications, internet-type services (including email, news feeds, social networking, etc.) require a very broad range of delivery constraints. For example the quality-of-service requirements (QoS) including bandwidth, latency, and packet error rate of different data traffic classes are in fact substantially different. Real-time gaming, one such possible service in the network, has significantly tighter QoS constraints as opposed to those imposed by an email service. As such, in the current LTE standard 9 QoS class identifiers have already been defined [19]. In addition, the quality-of-experience (QoE) of the different service categories is another important differentiator that has mostly remained unexplored. Real-time gaming would require an immediate start-up time and minimum response delay from the system for smooth user interaction. However, it would not impact the perceivable user experience if an email client on a mobile terminal is synchronized instantaneously when an automated update request is made to the server or if some time has passed before this update request is serviced. Once again, it should be emphasized that a user is not restricted from having an immediate access to the network and the services provided. On the contrary, a user can have access and interact with any service at any time. Acknowledging however that numerous data services are in fact elastic to message delivery delays, provides the potential to realize novel optimization strategies that promise considerable network performance improvements. Below, the most prominent application categories are identified and use cases are provided.

Background operations

With the introduction of the "app" world, mobile phones now offer a personalized, always connected on-line experience. Rarely are application programs a one-off download experience. Email clients, for example, provide notifications of newly received emails while operating at the background. News readers and weather forecasts apps periodically check for new entries from the respective online repositories to maintain the latest content available at the terminal. Social networking applications interrogate the respective online sites to inform of status updates, profile changes, and comment notifications. Even the actual apps check and inform for newer versions of the running software to ensure that the best user experience is maintained. Notably, all such updates are conducted wirelessly. Over the air (OTA) firmware updates of the running operating system (OS) are also becoming a common practice due to the ease of implementation and economical appeal. OTA updates are expected to increase in frequency to be able to control and manage the increasing complexity and security threads of mobile handsets.

It is important to note that all such update operations are indeed delay insensitive. Any of those update operations could be postponed for short time periods without any impact on the perceived user experience. Allowing for this time window in data delivery could in fact conserve the residual terminal's battery lifetime, reduce the required transmit power at the infrastructure side, and achieve the desired green operation of cellular systems.

File sharing

File sharing has always been one of the most popular operations over the fixed internet (accounting for almost 20% of the total traffic in the network [17]) and continues to be a favorable service over mobile networks as well. As indicated in Figure 4.3, file sharing alone currently accounts for 15% of the total aggregated traffic. File sharing, that is downloading a file via an ftp server, collecting files via a peer-to-peer network, or uploading documents, pictures and video clips online, is inherently of a best-effort nature. As such, download/upload requests can handle message delays without significant impact on the user experience. It could be the case, for example, that a movie file is downloaded on the terminal to watch at a later time or mobile users may wish to upload a newly captured picture from their device to their social network profile. Postponing the file delivery of such operations for short periods of time would not severely impact the perceived user experience, especially if significant savings in energy-consumption could be achieved for the battery-operated terminal.

Video-on-demand

Due to the demanding data requirements of video playback, it is estimated that more than 60% of the mobile data traffic in the near future will be associated with video content delivery. Even though video streaming sets stringent QoS requirements on the network, it operates in a fashion that could in fact be considered very elastic. When viewing on-demand video, the media server sends consecutive fragments of the video to be played back at the terminal. As long as the terminal's buffer contains a file of duration longer than the playback time, a smooth viewing experience is guaranteed. The key observation to be made here is that message delivery can be postponed for the time difference between the buffer duration and the playback time while smooth playback is maintained. In this way for example, a media server can send enough data to ensure uninterrupted playback at the terminal for some time and then defer the transmission of the rest of the file until the mobile terminal moves to a location within the cell with favorable channel gains. In doing so, mechanical forwarding strategies can in fact be realized for energy-efficient video content delivery in cellular networks.

4.3.2 Mechanical relaying strategies

Mechanical relaying enables mobile nodes to postpone the transmission (or reception) of information messages while on the move and engage in communication at a later time when found at better forwarding (or receiving) locations. This action allows for a great flexibility in generating decision policies to maximize the energy-efficiency gains while maintaining and improving the system performance both in terms of the radio resource efficiency and the system utilization. The benefit of transmitting at the best locations within the cell is that not only is significantly less power required for successful communication but further the radio resources are used more efficiently (i.e. higher-order modulation and coding schemes can be employed over the allocated radio slots). On top of that, reducing the transmit power over an active link will have as a side effect, a reduction of the co-channel inter-cell interference imposed to neighboring cells, which can further enhance the signal-to-interference-plus-noise ratio (SINR) at the receivers across the whole system. In the sequel, the envisioned implementations of MR are described for both single and multi-cell scenarios.

Both single-hop and multi-hop solutions can be realized within the cell as shown in Figures 4.4(a) and 4.4(b), respectively.

Mechanical relaying within a single cell
Consider the simple scenario illustrated in Figure 4.4(a). In this case, a mobile node with knowledge of its short-term mobility pattern and the delay tolerance of the initiated service can simply postpone the transmission of information messages until it moves closer to the BS, thus reducing the physical communication distance. A similar strategy can also hold for the downlink scenario, whereby a BS with knowledge of the approximate location and direction of travel of an active user can delay the transmission of data until the mobile node moves within its vicinity. Transmitting over the best channel locations reduces the required transit power for successful communication since the main RF power expenditure is to combat the path-loss effects.

In the case where mobile nodes are allowed to act as relays for other nodes within the cell (as shown in Figure 4.4(b)) a great number of innovative forwarding strategies can be realized. A static node for instance could forward its data to the nearest mobile node, which in turn can carry the messages while in transit before forwarding it to another

4.3 Mechanical relaying

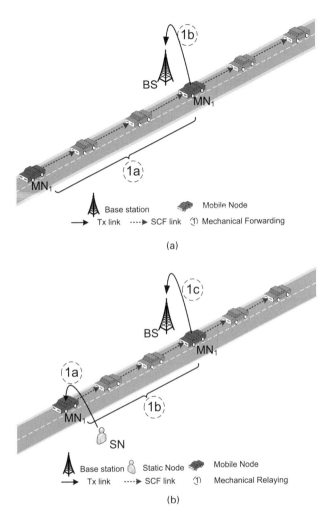

Figure 4.4 Delaying message transmissions allows for the physical propagation of information messages towards the destination.

node or the destination. Clearly in this case, the source and destination nodes could equally be a static user or the BS in either the uplink and downlink case of transmission. Alternatively, a mobile user moving away from the BS could forward its data traffic to a relay node travelling in the opposite direction, and towards the BS, to assist in achieving an energy-efficient communication strategy. The important point to make here is that these delay-tolerant relaying opportunities postulate that localized transmissions can be achieved throughout the cell.

Mechanical relaying in multi-cell topologies
Under the current networking paradigm in cellular networks, an active user is restricted to communicate only with the serving BS (i.e. the BS whose service area the node is within).

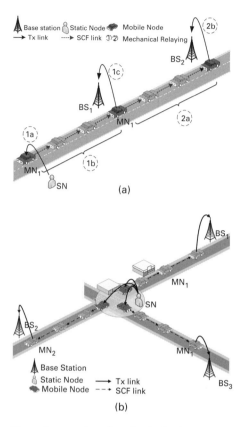

Figure 4.5 Example scenarios of mechanical relaying strategies in multi-cell topologies.

Enabling mobile nodes to store and carry data while in transit could potentially relax this fundamental assumption in cellular networks. As shown in Figure 4.5(a), a mobile node can defer the transmission of information and physically propagate data towards a target cell before engaging in transmission (link 2 in Figure 4.5(a)). Such strategies could effectively achieve a load-balancing effect across a number of cells to improve system utilization. For example, the delay-tolerant traffic of mobile nodes can be moved from hotspot areas to nearby cells that have significantly lower utilization. This allows the scarce available capacity in the hotspot areas to be used for delay-critical traffic such as VoIP. Alternatively, via MR, data traffic can be redirected towards target locations (as shown in Figure 4.5(b)) to allow under-utilized cells to suspend their operation and conserve energy. Cell switch-off mechanisms are considered a promising approach to save energy at the infrastructure side. Shifting load away from under-utilized cells could trigger such mechanisms into action much sooner, in effect maximizing the energy gains achieved. Further, MR strategies could be employed to provide connectivity in the dead-spot areas where the serving BSs have been switched off.

As illustrated in Figure 4.5(b), a static node with knowledge of the load conditions at the serving BS could decide (or even be encouraged by the serving BS) to redirect its traffic to one of the neighboring cells via MR. An important point to note here is that every terminal within the network has direct communication to an infrastructure node and thus sending or receiving the information message upon its delivery deadline, can be guaranteed. Therefore, even if a terminal diverges from its estimated path, frequent re-assessments of the forwarding opportunities will ensure that the best decision policies are followed. In effect, unlike all the challenges faced by highly dynamic infrastructure-less networks (including mobile ad hoc networks, DTNs, and vehicular networks to name just a few) the presence of BS entities ensures the timely, secured delivery of information messages from and to mobile nodes, while via MR the benefits of the actual mobility of nodes come into play.

4.4 Real-world measurements

As discussed in the previous section, allowing mobile nodes to postpone engaging in transmission for future time instances that experience favorable communication opportunities can result in significant performance gains. From its inception, mechanical relaying has been conceived as the means to combat the detrimental effects of the wireless channel. Due to the nonlinear relationship of the transmit power to the communication distance, allowing mobile nodes to physically propagate information towards the destination promises considerable energy-efficiency gains by combating the path-loss component in the wireless communication link. Further, due to the build-out in urban, suburban, and even rural areas, shadowing can cause severe weakening of the communicating signal levels. Employing mechanical relaying mechanisms to exit such shadowing regions could considerably improve the communication conditions.

In this section, the potential improvements that can be achieved in real-world cellular deployments are demonstrated. To that end, a series of radio measurements were conducted in the area near King's College, London and the findings are presented and discussed below. The Nokia Energy Profiler[1] application has been used for recording and logging of the terminal and signal characteristics, including the following parameters:

- 3G signal strength: captures transmit and received power levels in dBm.
- Data rate: captures downlink and uplink data rate speeds.
- Terminal power-consumption and current level from the battery, CPU activity, and RAM usage.
- IEEE 802.11 signal strength: a similar functionality is provided as for the the case of 3G signals.

In this chapter, the results related to the received signal strength and terminal transmit power levels are presented. The area under investigation is illustrated in Figure 4.6. The area layout as shown in the figure was extracted from Google maps and the location information of the BS was retrieved from Sitefinder[2] (an Ofcom service for base station

---

[1] www.forum.nokia.com/Library/Tools_and_downloads/Other/Nokia_Energy_Profiler
[2] Accurate cellular network deployments and BSs positions can be found at http://www.sitefinder.ofcom.org.uk/.

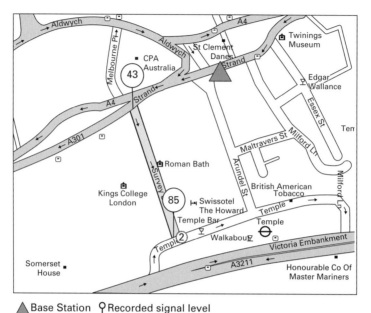

▲ Base Station ♀ Recorded signal level

Figure 4.6 The area near King's College, London is used as a toy example for the potential channel gain improvements that can be achieved via the proposed message postponement algorithms. The significant drop in the received power level moving away from the BS and towards shadowing regions within the cell are illustrated along the marked roadway.

audits in the UK). The figure also indicates the recorded received signal strength values in negative dBm units.

As shown in the figure, there is significant variation in the received signal strength values as the terminal moves from location 1 (as shown in Figure 4.6) to location 2; with the values dropping abruptly from -43 dBm to -85 dBm. Further, it is important to note that such a drop occurs over very short time intervals; in the order of few seconds, i.e. the time needed to move from location 1 to 2. Postponing the transmission of elastic data services to those locations within the cell with favorable channel gains and while on the move, proves to be an effective and efficient method to achieve the required energy-efficiency gains.

Figure 4.7(a) further illustrates how the drop in received power levels is experienced across the marked roadway in Figure 4.6 for a number of independent trials. It is evident that over very short time intervals, the channel quality changes significantly. Such location-specific signal attenuations persist in value across different times of the day, as illustrated in Figure 4.7(a). Clearly, this is a fundamental property of physical layer propagation constraints and is the basis of practically tested cell-planning procedures. This indication however suggests that simple mechanical forwarding strategies (i.e. single-hop message postponement schemes) can in fact be implemented today at no additional cost. By simply knowing the location and direction of motion, together with information on the expected channel attenuations along its path, a mobile node can target

4.4 Real-world measurements

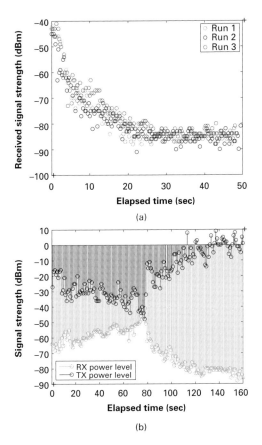

Figure 4.7 The received power levels moving from location 1 to 2 along the marked roadway of the test area are illustrated in (a). Further, the transmit power values in comparison to the received power levels when crossing the BS are illustrated in (b).

the best forwarding locations to engage in transmission by enabling simple distributed algorithms to conserve energy consumption. On top of that, mechanical relaying decision policies can in turn be constructed when such information is disseminated across the interested mobile terminals. Later on in this chapter we make explicit reference to current standardization efforts that are taking place to help build such networking schemes.

Finally, Figure 4.7(b) illustrates the received and transmit signal levels recorded on the device when crossing the BS entity. As expected, the transmit power levels follow the received signal strength quality to achieve an acceptable communication performance. The swing in transmit level from moving towards and away from the serving BS as shown in the figure, changes from approximately -50 dBm near the cell center to almost 10 dBm at the worst transmit locations. This indication strongly suggests that many-fold improvements in energy-consumption gains can be realized via MR strategies in cellular networks.

4.5 Related standardization efforts

The increasing complexity associated with efficiently operating and maintaining cellular systems mandates for the automated, self-optimizing, and self-organizing operation of common routine tasks. It is becoming increasingly apparent that to do so and achieve a high level of reliability requires additional intelligence on the underlying networking conditions.

One such key knowledge component that has been meticulously studied in cellular networks is location information of the mobile nodes. Set aside the anonymity and security issues that need to be resolved, location information provides the possibility of improving the operation of several mechanisms across the whole communication protocol stack. While it is beyond the scope of this book chapter to list all such work that benefits from user location information, a few indicative examples are presented. The work in [20], for instance, detailed the benefits of providing location-oriented *sensor hints* to adaptation algorithms implemented at different layers of the communications suite, in order to improve their performance. The work in [21] further studied how the QoS of video content delivery methods can be improved when in possession of location-specific bandwidth maps in cellular networks. The authors showed that by gaining prior knowledge of the available bandwidth along the travelling path of mobile nodes, the utilization of the downlink channel can be maximized. In [22] the concept of *coverage maps* was discussed, whereby a mobile node with knowledge of its location can predict the existence or not of different radio access technologies and assess the possibility of turning on a network interface to gracefully switch to a more energy-efficient alternative wireless access network. Similarly, the work in [23] and [24] investigated how the handover performance (and further the call blocking and dropping probabilities) can be improved with knowledge of user mobility trajectories. Moreover, the work in [25] illustrated the importance of user location information on improving the performance of different multi-hop routing schemes. The authors built a location information collection and dissemination mechanism by exploiting the ubiquitous connectivity provided by the cellular infrastructure. It was further shown that implementing such a mechanism consumes a minimum of system resources. At the application layer, location-based services are expected to significantly enhance the user's interaction with the device and the surrounding environment. In both [26] and [27], numerous such services were detailed, which vary from restaurant bookings to shopping assistance and tourist guidance. Emergency response services, such as the eCall service within the European Union [28] are expected to become mandatory in the near future and require onboard terminals to be capable of obtaining position information in assisting the rescue mission.

In light of all these (and associated) studies that take advantage of user location information, standardization procedures have been defined by 3GPP to efficiently acquire such location information from the possible alternative position estimation methods [29]. On top of that, mobile network operators have seen the opportunity to add further intelligence to the cellular systems by standardizing the procedures by which in addition to location information, mobile terminals scan the different frequencies of the wireless channel, record their channel states at a given location, log such data, and report it to the infrastructure for further processing. With such knowledge of the underlying networking conditions at hand, the time-consuming and costly procedure of driving around to identify coverage dead-spots is eliminated. The logging and reporting procedures for such terminal operations were detailed in [30]. Given this information, implementing MR strategies in cellular networks becomes significantly simpler. The only remaining requirement will then be to identify the conditions under which an initiated service request can possibly tolerate any delay and if yes, the duration of the delay that can be tolerated.

Some initial work towards that goal has recently emerged from Apple and its proposed HTTP rate-adaptive video content delivery method [31]. The protocol, proposed as an IETF standard, unlike other video content delivery methods keeps track of the buffered video on the terminal and postpones the request of new content until the buffer falls below a suggested level. This is done to save on data usage and system bandwidth in case a user decides to stop the video playback. With current video delivery methods (such as standard progressive download), the complete video file is delivered to the user as soon as possible, resulting in a significant waste of resources if the user decides not to view the content.

Further, the research work in [32] illustrated the energy-efficiency gains that can be achieved by serving delay-tolerant traffic requests (including email and news feeds) as a batch. They exploit the fact that every time the network is accessed, the high-power transmit state is maintained at the terminal even when the last message has been received. Reducing the times the network is accessed minimizes this inactivity duration and thus results in energy savings. However, in the latter work the authors do not consider how to quantify the delay tolerance of each service request and simply assume the user's input. It is envisioned that dipper integration between application programming interfaces and a middleware platform that can i) identify the origins of a user request, and ii) decide on the possible message delivery delay tolerance, may be the way forward. However, other approaches are not to be excluded. Nevertheless, as indicated from the above discussion a good starting point would be to identify the origin of the data request. For example, an automated update request triggered at the terminal while the screen is off can be used as an indication that the user is not interacting with the device and hence it could be possible to postpone the update request. However, if an update request was made explicitly by the user within the application, such a request would be serviced instantaneously.

4.6 Conclusion

In this chapter, architectural aspects, applicability issues, and a set of different schemes for providing store-carry and forward relaying (mechanical relaying) in cellular networks have been detailed and discussed. It has been argued that significant energy savings can be acquired by deliberately postponing the transmission of various delay-insensitive internet applications while the terminals move to better forwarding locations within the cell.

A number of future research threads and avenues for consideration can be envisioned. As indicated in the previous section, identifying and establishing the rules and conditions for delaying data transmissions is detrimental to the potential energy-efficiency gains that can be achieved and the perceived quality of experience. Both the application and networking layers should work synergetically to achieve the maximum performance gains. Further, message forwarding decision policies can be made at the BS in a centralized manner, in a distributed way among the participating nodes, or in a hybrid approach. Detailed investigation of the advantages and drawbacks of each solution at the time of implementation will reveal the best alternative.

One issue that can be further optimized in mechanical relaying is the size of the message to be transmitted in a store-carry and forward manner. The bundle protocol allows for fragmentation of the bundle to take place in order to allow message forwarding for small contact times between nodes. This feature can be used to derive optimal bundle sizes for mechanical relaying within the cell from an energy-consumption perspective.

Acknowledgment

The work reported in this chapter has formed part of the Green Radio Core 5 Research Programme of the Virtual Centre of Excellence in Mobile & Personal Communications, Mobile VCE, www.mobilevce.com. This research has been funded by EPSRC and by the Industrial Companies who are Members of Mobile VCE.

References

[1] V. Cerf, *et al.*, "Interplanetary Internet (IPN): architectural definition," InterPlaNetary Internet Research Group, memo IPNRG architecture, 2001.
[2] V. Cerf, *et al.*, "Delay-tolerant networking architecture," IETF RFC 4838, Apr. 2007.
[3] K. Fall and S. Farrell, "DTN: an architectural retrospective," *IEEE Journal on Selected Areas in Communications*, vol. 26, no. 5, pp. 828–836, June 2008.
[4] K. Scott and S. Burleigh, "Bundle protocol specification," IETF RFC 5050, Nov. 2007.
[5] R. C. Shah, *et al.*, "Data mules: modelling a three-tier architecture for sparse sensor networks," in *Proc. of IEEE International Workshop on Sensor, Networks and Applications*, pp. 30–41, May 2003.
[6] W. Zhao, M. Ammar, and E. Zegura, "A message ferrying approach for data delivery in sparse mobile ad hoc networks," in *Proc. of ACM International Symposium on Mobile Ad Hoc Networking and Computing*, pp. 187–198, May 2004.
[7] W. Mitchener and A. Vadhat, "Epidemic routing for partially connected ad hoc networks," Duke University, Technical Report Computer Science-2000-06, 2000.
[8] T. Spyropoulos, K. Psounis, and C. S. Raghavendra, "Spray and wait: an efficient routing scheme for intermittently connected mobile networks," in *Proc. of ACM SIGCOMM Workshop on Delay-Tolerant Networking*, pp. 252–259, Aug. 2005.
[9] L. Tang, *et al.*, "SMART: a selective controlled-flooding routing for delay tolerant networks," in *Proc. of IEEE Conference on Broadband Communications, Networks and Systems*, pp. 356–365, Sept. 2007.

[10] A. Lindgren, A. Doria, and O. Schelen, "Probabilistic routing in intermittently connected networks," in *Proc. of ACM SIGMOBILE Mobile Computing Communications Review*, vol. 7, no. 3, pp. 19–20, Jul. 2003.

[11] S. Jain, K. Fall, and R. Patra, "Routing in a delay tolerant network," in *Proc. of ACM SIGCOMM*, pp. 145–158, Aug. 2004.

[12] A. Lindgren and P. Hui, "The quest for a killer app for opportunistic and delay tolerant networks," in *Proc. of ACM Workshop on Challenged Networks*, Sept. 2009.

[13] "DRX mechanism for power saving in LTE," *IEEE Communications Magazine*, vol. 47, no. 6, pp. 48–55, Jun. 2009.

[14] R. Pabst *et al.*, "Relay-based deployment concepts for wireless and mobile broadband radio," *IEEE Communications Magazine*, vol. 42, no. 9, pp. 80–89, Sept. 2004.

[15] Cisco, "Visual networking index: global mobile data traffic forecast update, 2010-2015," White Paper, Feb. 2011.

[16] J. Seymour, "The path to 4G: LTE and LTE-Advanced," 4G World, Alcatel Lucent, Oct. 2010.

[17] Sandvine, "Global Internet phenomena report: spring 2011," Sandvine, Revision: 2011-05-12, May 2011.

[18] AT&T Inc., "Acquisition of T-Mobile USA, Inc. by AT&T Inc., Description of Transaction, Public Interest Showing and Related Demonstrations," Filed with the Federal Communications Commission, Apr. 2011.

[19] S. Sesia, I. Toufik, and M. Baker, *LTE: The UMTS Long Term Evolution From Theory to Practice*. 1st ed., John Wiley & Sons Ltd, pp. 33, 2009.

[20] L. Ravindranath, *et al.*, "Improving wireless network performance using sensor hints," NSDI, Mar. 2011.

[21] J. Yao, S. Kanhere, and M. Hassan, "Improving QoS in high-speed mobility using bandwidth maps," *IEEE Transaction of Mobile Computing*, to appear.

[22] K. Holger, "An overview of energy-efficiency techniques for mobile communication systems," Report of AG Mobikom WG7, Sept. 2003.

[23] S. Glisic and B. Lorenzo, *Advanced Wireless Networks: Cognitive, Cooperative & Opportunistic 4G Technology*. 2nd ed., Wiley Press, pp. 65–69, June 2009.

[24] T. Jansen *et al.*, "Handover parameter optimization in LTE self-organizing networks," in *Proc. of IEEE Vehicular Technology Conference*, Sept. 2009.

[25] I. Lequerica, P. M. Ruiz, and V. Cabrera, "Improvement of vehicular communications by using 3G capabilities to disseminate control information," *IEEE Network*, vol. 24, no. 1, pp. 32–38, Jan. 2010.

[26] S. Wang, J. Min, and B. Yi, "Location based services for mobiles: technologies and standards," in *Proc. of IEEE International Conference on Communications*, Tutorial, May 2008.

[27] Google, *The mobile movement: understanding smartphone users*, Google/IPSOS OTX MediaCT, Apr. 2011, [Online]. Available: www.google.com/think/insights.

[28] M. Werner *et al.*, "Cellular in-band modem solution for eCall emergency data transmission," in *Proc. of IEEE Vehicular Technology Conference*, Apr. 2009.

[29] ETSI 3rd Generation Partnership Project, "LTE; Evolved Universal Terrestrial Radio Access Network (E-UTRAN); stage 2 functional specification of user equipment positioning in E-UTRAN," ETSI Technical Specification 136 305 V9.0.0 Release 9, Oct. 2009.

[30] ETSI 3rd Generation Partnership Project, "Technical specification group radio access network study on minimization of drive-tests in next generation networks," ETSI TR 36.805 v9.0.0, Dec. 2009.

[31] R. Pantos and W. May, *HTTP live streaming*, IETF, Internet Draft draft-pantos-http-live-streaming-05, [Online]. Available: http://tools.ietf.org/html/draft-pantos-http-live-streaming-05.

[32] N. Balasubramanian, A. Balasubramanian, and A. Venkataramani, "Energy consumption in mobile phones: a measurement study and implications for network applications," *CM SIGCOMM conference on Internet measurement*, Nov. 2009.

Part II

Physical communications techniques for green radio networks

5 Green modulation and coding schemes in energy-constrained wireless networks

Jamshid Abouei, Konstantinos N. Plataniotis, and Subbarayan Pasupathy

5.1 Introduction

The past decade has witnessed many significant advances in the physical layer of wireless communication systems in both theory and implementation. Traditionally, the design of existing cellular networks has focused on increasing the spectral efficiency, throughput, and transmission reliability, while minimizing the latency. The recent research focus has also included studying the *energy efficiency* in next generation wireless networks; associated with this shift is a new point of view that wireless communications are becoming ubiquitous and that the energy consumption of embedded devices is gradually increasing. Of interest is the next generation of mobile technologies, where the energy resources are scarce and have to be conserved, in particular when the replenishment of the energy resource is not easy. On the other hand, in indoor or short-range communications such as pico-cellular networks and femtocells, or in dense wireless networks when a large number of mobile nodes is deployed over a region, the circuit energy-consumption is comparable to or even dominates the transmission energy due to the short distance between nodes. Thus, minimizing the total energy-consumption in both circuits and signal transmission compared to the current level should be considered primarily as an important requirement in the different layer design of future wireless networks. These requirements and realizations have led to a push towards *green wireless communications* and have created inter-disciplinary research challenges in hardware and protocols in different layers of the wireless network.

Towards green communication radios, the standardization processes for future wireless systems should target power control on circuit components as well as power-management algorithms for mobile nodes with sleep-mode processes, where the nodes only transmit a finite number of packets in a duty-cycling fashion. Central to this study is to find energy-efficient modulation and coding schemes in the physical layer of an energy-constrained wireless system. But, there is one major question: what types of modulation/coding techniques look promising? Generally, energy-efficient modulation/coding schemes should be simple enough to be implemented by state-of-the-art low-power technologies, but still robust enough to provide the desired service. Furthermore, since some wireless devices frequently switch from sleep mode to the active mode, modulation and coding circuits should have fast start-up times along with the capability of transmitting packets during a pre-assigned time slot before new sensed packets arrive. In addition, a wireless network needs a powerful channel coding scheme (when the distance between nodes exceeds a certain threshold level) to protect transmitted data against the unpredictable and harsh nature of channels. We refer to these low-complexity and low-energy-consumption approaches in an energy-constrained wireless network providing proper link reliability as *green modulation/coding* (GMC) schemes.

In the past decade, intense research has focused on adaptive modulation and coding algorithms in traditional wireless networks to provide the desired quality-of-service (QoS). For energy-constrained networks, however, adaptive approaches impose some additional system complexity and processing delay due to the multi-level modulation/coding formats plus the channel state information fed back from the receiver to the corresponding transmitter. In addition, existing protocols use various methods to overcome the packet loss and decoding error concerns in wireless networks. The automatic repeat request (ARQ), for instance, requires many retransmissions in the case of poor channel conditions, resulting in a high latency in the network. Furthermore, most of the pioneering work on energy-efficient modulation/codings has ignored the effect of the bandwidth and transmission time duration; in particular when some nodes use duty-cycling transmission processes. More recently, the attention of researchers has been directed toward deploying rateless codes (e.g. Luby transform (LT) code [1]) in wireless networks due to their significant advantages in erasure channels. However, investigating the energy efficiency of rateless codes in energy-constrained wireless systems with low-energy modulations over realistic fading channel models has received little attention.

The main goal of this chapter is to develop some theoretical concepts of energy efficiency along with the system view emphasis on the green modulation/coding. We demonstrate how the concepts are applied in actual energy-constrained wireless networks. This chapter consists of two parts:

- The first part of this chapter deals with a comprehensive analysis (supported by numerical results) of the energy efficiency of popular modulation designs considering the effect of the channel bandwidth, the *active mode duration*, and according to the realistic parameters in IEEE standards. The main focus is to find the distance-based green modulation in a wireless network when nodes perform a sleep-mode operation. For this purpose, we describe the system model according to a flexible duty-cycling process used in practical proactive devices. We start the analysis based on a Rayleigh flat-fading channel with path loss. Then, we evaluate numerically the energy efficiency of sinusoidal carrier-based modulations operating over the more general Rician model which includes a strong direct line-of-sight (LOS) path.
- In the second part, we address the analysis of the energy efficiency of LT codes with the green modulation introduced in the first part. We investigate how channel coding affects the circuit and RF signal energy-consumptions in the system. We study the effects of the LT code rate and the corresponding coding gain on the energy efficiency of the network. This study uses the classical BCH and convolutional codes (as reference codes), used in IEEE standards, for comparative evaluation. Numerical results, supported by some experimental setup on the computation energy, show that the optimized LT coded frequency-shift keying (FSK) scheme is the most energy-efficient scheme in low-power wireless networks with mobile environments. Also,

some design guidelines for using LT codes in practical wireless system with mobile applications are presented.

5.2 System model and assumptions

This chapter considers a dense wireless system consisting of a couple of pairs of nodes, where the term pair is used to describe a transmitter and its corresponding receiver. It is assumed that each transmitter node sends an equal amount of data during a pre-assigned time slot to a designated receiver node. The transmitter and receiver nodes synchronize with one another and operate in a real-time process known as *duty-cycling*, as depicted in Figure 5.1.

During the *active mode* period T_{ac}, the analog signal is first digitized by an analog-to-digital converter (ADC), and an N-bit message sequence $M_N \triangleq (m_1, m_2, \ldots, m_N)$ is generated, where N is assumed to be fixed, and $m_i \in \{0, 1\}, i = 1, 2, \ldots, N$. The bit stream is then sent to the channel encoder. The encoding process begins by dividing the message M_N into blocks of equal length denoted by $B_j \triangleq (m_{(j-1)k+1}, \ldots, m_{jk}), j = 1, \ldots, N/k$, where k is the length of any particular B_j, and N is assumed to be divisible by k. Each block B_j is encoded by a pre-determined channel coding scheme to generate a coded bit stream $C_j \triangleq (a_{(j-1)n+1}, \ldots, a_{jn}), j = 1, \ldots, N/k$, with block length n, where n is either a fixed value (e.g. for block and convolutional codes) or a random variable (e.g. for rateless codes). The coded stream is then modulated by a pre-determined modulation scheme and transmitted to the receiver. Finally, the transmitter returns to the sleep mode, and all the circuits are powered off for the sleep-mode duration T_{sl} for energy saving. We denote by T_{tr} the transient mode duration, consisting of the switching time from sleep mode to active mode (i.e. $T_{sl \rightarrow ac}$) plus the switching time from active mode to sleep mode (i.e. $T_{ac \rightarrow sl}$), where $T_{ac \rightarrow sl}$ is short enough to be negligible. Under the above considerations, each pair of nodes process one entire N-bit message M_N during a fixed period $T_N \triangleq T_{tr} + T_{ac} + T_{sl}$, where $T_{tr} \approx T_{sl \rightarrow ac}$. Note that T_{ac} is an influential factor in choosing the energy-efficient modulation, since it directly affects the total energy consumption as we will show later.

5.2.1 Performance metric

In a typical dense wireless network, the distance between nodes is normally short. Thus, the total circuit power-consumption, defined by $P_c \triangleq P_{ct} + P_{cr}$, is comparable to the

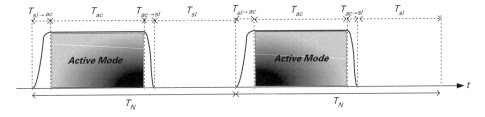

Figure 5.1 A duty-cycling process in a proactive wireless system.

RF transmit power-consumption denoted by P_t, where P_{ct} and P_{cr} represent the circuit power-consumptions for the transmitter and the receiver nodes, respectively. Taking these into account, the total energy consumption during the active-mode period, denoted by E_{ac}, is given by

$$E_{ac} = (P_c + P_t)T_{ac}, \qquad (5.1)$$

where T_{ac} is a function of N and the channel bandwidth as we will show in Section 5.3. Note that the power consumption during the sleep-mode duration T_{sl} is much smaller than the power consumption in the active mode (due to the low sleep-mode leakage current). The *energy efficiency*, referred to as the performance metric of the proposed system, can be measured by the total energy consumption in each period T_N corresponding to N-bit message M_N as follows:

$$E_N = (P_c + P_t)T_{ac} + P_{tr}T_{tr}, \qquad (5.2)$$

where P_{tr} is the circuit power-consumption during the transient-mode period. We use (5.2) to investigate and compare the energy efficiency of various uncoded modulation schemes and coded green modulation for various channel coding schemes in the subsequent sections.

5.2.2 Channel model

The choice of low transmission power in dense wireless networks results in several consequences to the channel model. It is a well-known fact that a low transmission power implies a small range. For short-range transmission scenarios, the root mean square (rms) delay spread is in the range of nanoseconds [2]. Thus, it is reasonable to expect a flat-fading channel model for a dense wireless network. In addition, many transmission environments include significant obstacles such as walls, doors, and furniture which lead to reduced LOS components. This behavior suggests a Rayleigh fading channel model for our system. Under the above considerations, the channel model between the transmitter and its corresponding receiver is assumed to be Rayleigh flat-fading with path loss. For this model, we assume that the channel is constant during the transmission of a codeword, but may vary from one codeword to another. We denote the fading channel coefficient corresponding to symbol i as h_i, where the amplitude $|h_i|$ is Rayleigh distributed with probability density function (pdf) $f_{|h_i|}(r) = (2r/\Omega)e^{-r^2/\Omega}$, $r \geq 0$, where $\Omega \triangleq \mathbb{E}[|h_i|^2]$ [3].

To model the path loss of a link where the transmitter and the receiver are separated by distance d, we denote by P_t and P_r the transmitted and the received signal powers, respectively. In this case, the gain factor L_d for a ηth-power path-loss channel is expressed as

$$L_d \triangleq \frac{P_t}{P_r} = M_l d^\eta L_1, \quad \text{with } \eta_{min} \leq \eta \leq \eta_{max}, \qquad (5.3)$$

where M_l is the gain margin which accounts for the effects of hardware process variations and background noise, and $L_1 \triangleq \frac{(4\pi)^2}{G_t G_r \lambda^2}$ is the gain factor at $d = 1$ meter which is specified by the transmitter and the receiver antenna gains G_t and G_r, and wavelength λ [4], and η is the path-loss exponent.

As a result, when both fading and path loss are considered, the instantaneous channel coefficient becomes $G_i \triangleq h_i/\sqrt{L_d}$. Denoting $x_i(t)$ as the RF transmitted signal with energy E_t, the received signal at the receiver node is given by $y_i(t) = G_i x_i(t) + n_i(t)$, where $n_i(t)$ is additive white Gaussian noise (AWGN) at the receiver with two-sided power spectral density given by $N_0/2$. Under the above considerations, the instantaneous signal-to-noise ratio (SNR) corresponding to symbol i can be computed as $\gamma_i = |G_i|^2 E_t/N_0$. Under the assumption of a Rayleigh fading channel model, γ_i is chi-square distributed with two degrees of freedom and with probability density function (pdf) $f_\gamma(\gamma_i) = \frac{1}{\bar{\gamma}} e^{-\frac{\gamma_i}{\bar{\gamma}}}$, where $\bar{\gamma} \triangleq \mathbb{E}[|G_i|^2] E_t/N_0 = \Omega E_t/(L_d N_0)$ denotes the average received SNR.

5.3 Energy consumption of uncoded scheme

One challenge faced by energy-constrained wireless networks is to design a small-sized and low-power transceiver which operates efficiently over the assigned frequency band. With this observation in mind, finding the energy-efficient modulation with low-complexity implementation is a crucial task in the design of small-sized wireless devices. In this section, we analyze the energy and bandwidth efficiency of three popular sinusoidal carrier-based modulations, namely MFSK, M-ary quadrature amplitude modulation (MQAM), and offset quadrature phase-shift keying (OQPSK), over a Rayleigh flat-fading channel with path loss. MFSK is used in many low-complexity and energy-constrained wireless systems and some IEEE standards (e.g. [5]), whereas MQAM is used in modem and digital video applications. Also, OQPSK is used in the IEEE 802.15.4 standard [6], which is the industry standard for wireless sensor networks (WSNs). In the sequel and for simplicity of the notation, we use the superscripts "FS," "QA," and "OQ" for MFSK, MQAM, and OQPSK, respectively.

5.3.1 M-ary FSK

An M-ary FSK modulator with $M = 2^b$ orthogonal carriers benefits from using the direct digital modulation (DDM) approach, meaning that it does not need the mixer and the digital-to-analog converter (DAC). This property gives MFSK a faster start-up time than the other modulation schemes. Let us denote $\Delta f = 1/(\zeta T_s^{FS})$ as the minimum carrier separation with the symbol duration T_s^{FS}, where $\zeta = 2$ for coherent and $\zeta = 1$ for non-coherent FSK [7, p. 114]. In this case, the channel bandwidth is obtained as $B \approx M \times \Delta f$, where B is assumed to be fixed for all sinusoidal carrier-based modulations. Denoting B_{eff}^{FS} as the *bandwidth efficiency* of MFSK defined as the ratio of data rate $R^{FS} = b/T_s^{FS}$ (b/s) to the channel bandwidth, we have

$$B_{eff}^{FS} \triangleq \frac{R^{FS}}{B} = \frac{\zeta \log_2 M}{M}, \text{ b/s/Hz.} \qquad (5.4)$$

It can be seen that using a small constellation size M avoids losing more bandwidth efficiency in MFSK. To address the effect of increasing M on the energy efficiency, we first derive the relationship between M and the active-mode duration T_{ac}^{FS}. Since, we have b bits during each symbol period T_s^{FS}, we can write

$$T_{ac}^{FS} = \frac{N}{b} T_s^{FS} = \frac{MN}{\zeta B \log_2 M}. \tag{5.5}$$

Recalling that B and N are fixed, an increase in M results in an increase in T_{ac}^{FS}. However, the maximum value of T_{ac}^{FS} is bounded by $T_N - T_{tr}^{FS}$ as illustrated in Figure 5.1. Thus, $M_{max} \triangleq 2^{b_{max}}$ in MFSK is calculated by the following non-linear equation:

$$\frac{M_{max}}{\log_2 M_{max}} = \frac{\zeta B}{N}(T_N - T_{tr}^{FS}). \tag{5.6}$$

At the receiver side, the received MFSK signal can be detected coherently to provide an optimum performance. However, the MFSK-coherent detection requires the receiver to obtain a precise frequency and carrier phase reference for each of the transmitted orthogonal carriers. For large M, this would increase the complexity of the detector, which makes a coherent MFSK receiver very difficult to implement. Thus, most practical MFSK receivers use non-coherent (NC) detectors. To analyze the energy efficiency of NC-MFSK, we first derive E_t^{FS}, the transmit energy per symbol, in terms of a given average symbol error rate (SER) denoted by P_s. It is shown in [8, Lemma 2] that the average SER of NC-MFSK is upper bounded by

$$P_s = 1 - \left(1 - \frac{1}{2 + \bar{\gamma}^{FS}}\right)^{M-1}, \tag{5.7}$$

where $\bar{\gamma}^{FS} = \Omega E_t^{FS}/(L_d N_0)$. Thus, the transmit energy-consumption per each symbol is obtained from the above P_s as

$$E_t^{FS} \triangleq P_t^{FS} T_s^{FS} = \left[\left(1 - (1 - P_s)^{\frac{1}{M-1}}\right)^{-1} - 2\right] \frac{L_d N_0}{\Omega}. \tag{5.8}$$

As a result, the output energy-consumption of transmitting N-bit during T_{ac}^{FS} of an uncoded MFSK is computed from (5.5) as follows:

$$P_t^{FS} T_{ac}^{FS} = \frac{T_{ac}^{FS}}{T_s^{FS}} E_t^{FS} = \left[\left(1 - (1 - P_s)^{\frac{1}{M-1}}\right)^{-1} - 2\right] \frac{L_d N_0}{\Omega} \frac{N}{\log_2 M}. \tag{5.9}$$

On the other hand, the total circuit energy-consumption of the transmitter/receiver devices during T_{ac}^{FS} is obtained from $(P_{ct}^{FS} + P_{cr}^{FS})T_{ac}^{FS}$. For the transmitter node with the MFSK modulator, we denote the power consumption of frequency synthesizer, filters and power amplifier as P_{Sy}^{FS}, P_{Filt}^{FS} and P_{Amp}^{FS}, respectively. In this case,

$$P_{ct}^{FS} = P_{Sy}^{FS} + P_{Filt}^{FS} + P_{Amp}^{FS}. \tag{5.10}$$

5.3 Energy consumption of uncoded scheme

It is shown that the relationship between P_{Amp}^{FS} and the transmission power of an MFSK signal is $P_{Amp}^{FS} = \alpha^{FS} P_t^{FS}$, where α^{FS} is determined based on the type of the power amplifier. For instance for a class B power amplifier, $\alpha^{FS} = 0.33$ [4]. For the circuit power-consumption of the receiver side, we use the fact that the optimum NC-MFSK demodulator consists of a bank of M matched filters, each followed by an envelope detector [9]. In addition, we assume that the receiver node uses a low-noise amplifier (LNA), which is generally placed at the front-end of an RF receiver circuit, an intermediate-frequency amplifier (IFA), and an ADC, regardless of the type of deployed modulation. Thus, denoting P_{LNA}^{FS}, P_{Filr}^{FS}, P_{ED}^{FS}, P_{IFA}^{FS}, and P_{ADC}^{FS} as the power consumption of LNA, filters, envelope detector, IF amplifier, and ADC, respectively, the circuit power-consumption of an NC-MFSK receiver is obtained as

$$P_{cr}^{FS} = P_{LNA}^{FS} + M \times (P_{Filr}^{FS} + P_{ED}^{FS}) + P_{IFA}^{FS} + P_{ADC}^{FS}. \quad (5.11)$$

Moreover, it is shown that the power consumption during the transient-mode period T_{tr}^{FS} is governed by the frequency synthesizer in both transmitter/receiver nodes [4]. Thus, the energy consumption during T_{tr}^{FS} is obtained as $P_{tr}^{FS} T_{tr}^{FS} = 2 P_{Sy}^{FS} T_{tr}^{FS}$ [2]. Substituting (5.5) and (5.9) in (5.2), the total energy consumption of an uncoded NC-MFSK scheme for transmitting N-bit information in each period T_N for a given P_s is obtained as

$$E_N^{FS} = (1 + \alpha^{FS}) \left(\left[1 - (1 - P_s)^{\frac{1}{M-1}} \right]^{-1} - 2 \right) \frac{L_d N_0}{\Omega} \frac{N}{\log_2 M}$$
$$+ (P_c^{FS} - P_{Amp}^{FS}) \frac{MN}{B \log_2 M} + 2 P_{Sy}^{FS} T_{tr}^{FS}, \quad (5.12)$$

with the fact that $L_d = M_l d^\eta L_1$. Thus, the optimization goal is to determine the optimum constellation size M, such that the objective function E_N^{FS} can be minimized, i.e.

$$\hat{M} = \arg \min_M E_N^{FS}$$

$$\text{subject to} \begin{cases} 2 \leq M \leq M_{max} \\ \eta_{min} \leq \eta \leq \eta_{max} \\ d > 0, \end{cases} \quad (5.13)$$

where M_{max} is derived from (5.6). To solve this optimization problem, we prove that (5.12) is a monotonically increasing function of M for every value of d and η. It is seen that the second term in (5.12) is a monotonically increasing function of M. Also, from the first term in (5.12), we have

$$\left(\left[1 - (1 - P_s)^{\frac{1}{M-1}} \right]^{-1} - 2 \right) \frac{1}{\log_2 M} = \left(\left[1 - e^{\frac{1}{M-1} \ln(1-P_s)} \right]^{-1} - 2 \right) \frac{1}{\log_2 M}$$
$$\stackrel{(a)}{\approx} \left(\left[1 - e^{-\frac{P_s}{M-1}} \right]^{-1} - 2 \right) \frac{1}{\log_2 M}$$
$$\stackrel{(b)}{\approx} \left(\frac{M-1}{P_s} - 2 \right) \frac{1}{\log_2 M}, \quad (5.14)$$

where (a) comes from the approximation $\ln(1-z) \approx -z$, $|z| \ll 1$, and the fact that P_s scales as $o(1)$. Also, (b) follows from the approximation

$$e^{-z} = \sum_{n=0}^{\infty} (-1)^n \frac{z^n}{n!} \approx 1 - z, \quad |z| \ll 1. \tag{5.15}$$

It is concluded from (5.14) that the first term in (5.12) is also a monotonically increasing function of M. As a result, the minimum total energy-consumption E_N^{FS} is achieved at $\hat{M} = 2$ for all values of d and η.

5.3.2 M-ary QAM

For M-ary QAM with the square constellation, each $b = \log_2 M$ bits of the message are mapped to the symbol S_i, $i = 0, 1, \ldots, M-1$, with the symbol duration T_s^{QA}. Assuming that the raised-cosine filter is used for pulse shaping, the channel bandwidth of MQAM is given by $B \approx 1/(2T_s^{QA})$. Thus, using the data rate $R^{QA} = b/T_s^{QA}$, the bandwidth efficiency of MQAM is obtained as $B_{eff}^{QA} \triangleq R^{QA}/B = 2\log_2 M$, which is a logarithmically increasing function of M. To address the impact of M on the energy efficiency, we derive the active-mode duration T_{ac}^{QA} in terms of M as follows:

$$T_{ac}^{QA} = \frac{N}{b} T_s^{QA} = \frac{N}{2B \log_2 M}. \tag{5.16}$$

It is seen that an increase in M results in a decrease in T_{ac}^{QA}. Also compared to (5.5) for the NC-MFSK, it is concluded that $T_{ac}^{QA} = T_{ac}^{FS}/(2M)$. Interestingly, it seems that the large constellation sizes of M would result in lower energy consumption due to the smaller values of T_{ac}^{QA}. However, as we will show later, the total energy-consumption of an MQAM scheme is not necessarily a monotonically decreasing function of M. For this purpose, we obtain the transmit energy consumption $P_t^{QA} T_{ac}^{QA}$ with a similar argument as for MFSK. It is shown in [10, p. 226] and [11] that the average SER of a coherent MQAM is upper bounded by

$$P_s = \frac{4(M-1)}{3\bar{\gamma}^{QA} + 2(M-1)} \left(1 - \frac{1}{\sqrt{M}}\right), \tag{5.17}$$

where $\bar{\gamma}^{QA} = \Omega E_t^{QA}/(L_d N_0)$ denotes the average received SNR with the energy per symbol E_t^{QA}. As a result,

$$E_t^{QA} \triangleq P_t^{QA} T_s^{QA} = \frac{2(M-1)}{3} \left[2\left(1 - \frac{1}{\sqrt{M}}\right)\frac{1}{P_s} - 1\right]\frac{L_d N_0}{\Omega}. \tag{5.18}$$

Thus, the output energy-consumption of transmitting N-bit during the active-mode period is computed as

$$P_t^{QA} T_{ac}^{QA} = \frac{T_{ac}^{QA}}{T_s^{QA}} E_t^{QA} = \frac{2(M-1)}{3} \left[2\left(1 - \frac{1}{\sqrt{M}}\right)\frac{1}{P_s} - 1\right]\frac{L_d N_0}{\Omega} \frac{N}{\log_2 M},$$

5.3 Energy consumption of uncoded scheme

which is a monotonically increasing function of M for every value of P_s, d, and η. For the transmitter with the MQAM modulator,

$$P_{ct}^{QA} = P_{DAC}^{QA} + P_{Sy}^{QA} + P_{Mix}^{QA} + P_{Filt}^{QA} + P_{Amp}^{QA}, \tag{5.19}$$

where P_{DAC}^{QA} and P_{Mix}^{QA} denote the power consumption of the DAC and mixer, respectively. It is shown in [4] that $P_{Amp}^{QA} = \alpha^{QA} P_t^{QA}$ with

$$\alpha^{QA} = \frac{\xi}{\vartheta} - 1, \quad \xi = 3\frac{\sqrt{M}-1}{\sqrt{M}+1}, \quad \vartheta = 0.35.$$

In addition, the circuit power-consumption of the receiver with the coherent MQAM is obtained as

$$P_{cr}^{QA} = P_{LNA}^{QA} + P_{Mix}^{QA} + P_{Sy}^{QA} + P_{Filr}^{QA} + P_{IFA}^{QA} + P_{ADC}^{QA}. \tag{5.20}$$

With a similar argument as for MFSK, we assume that the circuit power consumption during transient-mode period T_{tr}^{QA} is governed by the frequency synthesizer. As a result, the total energy consumption of an uncoded coherent MQAM for transmitting N-bit in each period T_N is obtained as

$$E_N^{QA} = (1+\alpha^{QA})\frac{2(M-1)}{3}\left[2\left(1-\frac{1}{\sqrt{M}}\right)\frac{1}{P_s} - 1\right]\frac{L_d N_0}{\Omega}\frac{N}{\log_2 M}$$

$$+(P_c^{QA} - P_{Amp}^{QA})\frac{N}{2B\log_2 M} + 2P_{Sy}^{QA}T_{tr}^{QA}. \tag{5.21}$$

Although there is no constraint on the maximum size M for MQAM, to make a fair comparison to the MFSK scheme we use the same M_{max} as MFSK. Taking this into account, the optimization problem is to determine the optimum $M \in [4, M_{max}]$ subject to $d > 0$ and $\eta_{min} \leq \eta \leq \eta_{max}$, such that E_N^{QA} can be minimized. It is seen that the first term in (5.21) is a monotonically increasing function of M for every value of P_s, d, and η, while the second term is a monotonically decreasing function of M, which is independent of d and η. For the above optimization and for a given P_s, we have the two following scenarios based on the distance d:

Case 1: For large values of d where the first term in (5.21) is dominant, the objective function E_N^{QA} is a monotonically increasing function of M and is minimized at $M = 4$, equivalent to the 4-QAM scheme.

Case 2: Assuming that d is small enough, one possible case that may happen is when the total energy-consumption E_N^{QA} for small sizes of M is governed by the second term in (5.21). For this situation, either the objective function behaves as a monotonically decreasing function of M for every value of M, or for a large constellation size M, the first term would be dominant, meaning that E_N^{QA} increases when M grows. In the former scenario, the optimum M is achieved at $\hat{M} = M_{max}$, whereas in the latter scenario, there exists a minimum value for E_N^{QA}, where the optimum M for this point is obtained by the intersection between the first and second terms in (5.21). Since P_s scales as $o(1)$ and

Table 5.1. System evaluation parameters [11, 12]

$\Omega = 1$	$P_{Sy} = 10$ mw	$P_{ADC} = 7$ mw
$B = 62.5$ KHz	$P_{ED} = 3$ mw	$P_{Mix} = 7$ mw
$M_l = 40$ dB	$P_{Filt} = 2.5$ mw	$P_{IFA} = 3$ mw
$L_1 = 30$ dB	$P_{Filr} = 2.5$ mw	$P_{LNA} = 9$ mw
$N_0 = -180$ dBm	$P_{DAC} = 7$ mw	

ignoring the term $P_{Sy}^{QA} T_{tr}^{QA}$ to simplify our analysis, the optimum M which minimizes (5.21) is obtained from the following equation:

$$M - 1 - \sqrt{M} + \frac{1}{\sqrt{M}} \approx \frac{\phi(d,\eta)}{1 + \alpha^{QA}}, \qquad (5.22)$$

where

$$\phi(d,\eta) \triangleq \frac{P_c^{QA} - P_{Amp}^{QA}}{2B} \frac{3 P_s \Omega}{4 L_d N_0},$$

and the fact that α^{QA} is a function of M.

To gain more insight to the above optimization problem, we use a specific numerical example with the simulation parameters summarized in Table 5.1.

We assume $P_s = 10^{-3}$, $4 \leq M \leq 64$ and $2.5 \leq \eta \leq 6$. Figure 5.2 illustrates the total energy consumption of MQAM versus M for different values of d. It is seen that E_N^{QA} exhibits different trends depending on the distance d and the path loss exponent η. For instance, for large values of d and η, E_N^{QA} is an increasing function of M, where the optimum value of M is achieved at $\hat{M} = 4$ as expected. This is because, in this case, the first term in (5.21) corresponding to the RF signal energy-consumption dominates E_N^{QA}. Table 5.2 details the optimum values of M that minimize E_N^{QA} for some values of $1 \leq d \leq 200$ m and $2.5 \leq \eta \leq 6$. We use these results to compare the energy efficiency of the optimized MQAM with the other schemes in the subsequent sections.

5.3.3 Offset-QPSK

For performance comparison, we choose the conventional OQPSK modulation which is used as a reference in the IEEE 802.15.4/ZigBee protocols. We also follow the same differential OQPSK structure mentioned in [6, p. 50] to eliminate the need for a coherent phase reference at the receiver node. For this configuration, the channel bandwidth and the data rate are determined by $B \approx 1/T_s^{OQ}$ and $R^{OQ} = 2/T_s^{OQ}$, respectively. As a result, the bandwidth efficiency of OQPSK is obtained as $B_{eff}^{OQ} \triangleq R^{OQ}/B = 2$ (b/s/Hz). Since we have 2 bits in each symbol period T_s^{OQ}, it is concluded that

$$T_{ac}^{OQ} = \frac{N}{2} T_s^{OQ} = \frac{N}{2B}. \qquad (5.23)$$

5.3 Energy consumption of uncoded scheme

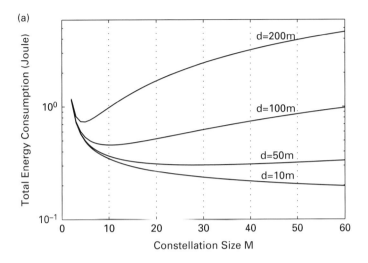

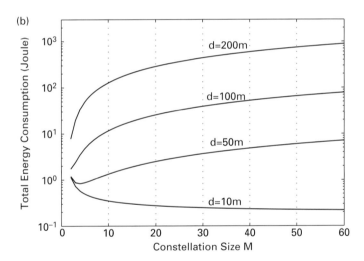

Figure 5.2 Total energy consumption E_N^{QA} vs. M over a Rayleigh fading channel with path loss for $P_s = 10^{-3}$, (a) $\eta = 2.5$, and (b) $\eta = 3.5$.

Compared to (5.5) and (5.16), we have $T_{ac}^{OQ} = (\log_2 M) T_{ac}^{QA}$ with $M \geq 4$, while for the optimized MFSK, $T_{ac}^{OQ} = \frac{1}{4} T_{ac}^{FS}$. More precisely, it is revealed that $T_{ac}^{QA} < T_{ac}^{OQ} < T_{ac}^{FS}$. To determine the transmit energy consumption of the differential OQPSK scheme, we derive E_t^{OQ} in terms of the average SER. It is shown in [11] and [13] that the average SER of the differential OQPSK is upper bounded by

$$P_s = \sqrt{\frac{1+\sqrt{2}}{2}} \frac{4}{(2-\sqrt{2})\bar{\gamma}^{OQ}+4}, \quad (5.24)$$

Table 5.2. Optimum constellation size M for the MQAM with $P_s = 10^{-3}$ and $2.5 \leq \eta \leq 6$

Distance (m)	$\eta = 2.5$	$\eta = 3$	$\eta = 4$	$\eta = 5$	$\eta = 6$
1	64	64	64	64	64
10	64	64	43	10	4
20	64	50	8	4	4
40	43	13	4	4	4
80	14	5	4	4	4
100	10	4	4	4	4
150	6	4	4	4	4
200	5	4	4	4	4

where $\bar{\gamma}^{OQ} = \Omega E_t^{OQ}/(L_d N_0)$. With a similar argument as for MFSK and MQAM, the energy consumption of transmitting N-bit during T_{ac}^{OQ} is computed as

$$P_t^{OQ} T_{ac}^{OQ} = \frac{T_{ac}^{OQ}}{T_s^{OQ}} E_t^{OQ} = \left[\frac{1}{2-\sqrt{2}} \left(\frac{4}{P_s}\sqrt{\frac{1+\sqrt{2}}{2}} - 4\right)\right] \frac{L_d N_0}{\Omega} \frac{N}{2}. \quad (5.25)$$

For the transmitter with the OQPSK modulator, $P_{ct}^{OQ} \approx P_{DAC}^{OQ} + P_{Sy}^{OQ} + P_{Mix}^{OQ} + P_{Filt}^{OQ} + P_{Amp}^{OQ}$, where we assume that the power consumption of the differential encoder is negligible, and $P_{Amp}^{OQ} = \alpha^{OQ} P_t^{OQ}$, with $\alpha^{OQ} = 0.33$. In addition, the circuit power consumption of the receiver with the differential detection OQPSK is obtained as $P_{cr}^{OQ} = P_{LNA}^{OQ} + P_{Mix}^{OQ} + P_{Sy}^{OQ} + P_{Filr}^{OQ} + P_{IF}^{OQ} + P_{ADC}^{OQ}$. As a result, the total energy consumption of a differential OQPSK system for transmitting N-bit in each period T_N is obtained as

$$E_N^{OQ} = (1 + \alpha^{OQ}) \left[\frac{1}{2-\sqrt{2}} \left(\frac{4}{P_s}\sqrt{\frac{1+\sqrt{2}}{2}} - 4\right)\right] \frac{L_d N_0}{\Omega} \frac{N}{2}$$

$$+ (P_c^{OQ} - P_{Amp}^{OQ}) \frac{N}{2B} + 2 P_{Sy}^{OQ} T_{tr}^{OQ}. \quad (5.26)$$

5.3.4 Numerical evaluations

In this section, we present some numerical evaluations using realistic parameters from the IEEE standards and state-of-the art technology to confirm the energy-efficiency analysis for the aforementioned uncoded modulation schemes. We assume that all the modulations operate in the carrier frequency $f_0 = 2.4$ GHz Industrial Scientist and Medical (ISM) unlicensed band used in the IEEE 802.15.4 standard [6]. According to the FCC 15.247 RSS-210 standard for United States/Canada, the maximum allowed antenna gain is 6 dBi [14]. In this work, we assume that $G_t = G_r = 5$ dBi. Thus for the $f_0 = 2.4$ GHz, L_1 (dB) $\triangleq 10 \log_{10}\left(\frac{(4\pi)^2}{G_t G_r \lambda^2}\right) \approx 30$ dB, where $\lambda \triangleq \frac{3 \times 10^8}{f_0} = 0.125$ m. We assume that in each period T_N, the data frame $N = 1024$ bytes (or equivalently

$N = 8192$ bits) is generated for transmission for all the modulations, where T_N is assumed to be 1.4 s. In addition, we assume that the path-loss exponent is in the range of 2.5 to 6. Note that $\eta = 2$ is regarded as a reference state for the propagation in free space and is unattainable in practice. Also, $\eta = 4$ is for relatively lossy environments, and for indoor environments, the path-loss exponent can reach values in the range of 4 to 6. We use the system parameters summarized in Table 5.1 for simulations. It is concluded from

$$\frac{M_{max}}{\log_2 M_{max}} = \frac{\zeta B}{N}(T_N - T_{tr}^{FS}) \qquad (5.27)$$

that $M_{max} \approx 64$ (or equivalently $b_{max} \approx 6$) for NC-MFSK. Since there is no constraint on the maximum M in MQAM, we choose $4 \leq M \leq 64$ for MQAM to be consistent with MFSK.

Figure 5.3 compares the energy efficiency of various modulation schemes versus M for $P_s = 10^{-3}$, $\eta = 3.5$, and different values of d. It is revealed from Figure 5.3(a) that for $M < 35$, NC-MFSK is more energy efficient than MQAM, differential OQPSK and coherent MFSK for $d = 10$ m and $\eta = 3.5$, while when M grows, 64-QAM outperforms the other schemes for $d = 10$ m. The latter result is well supported by Case 2 in the MQAM optimization discussed in Section 5.3.2. However, NC-MFSK for a small-sized M benefits from the advantage of less complexity and cost in implementation than 64-QAM. Furthermore, as shown in Figure 5.3-b, the total energy consumption of both MFSK and MQAM for large d increases logarithmically with M, which verifies the optimization solutions for the NC-MFSK and Case 1 for the MQAM in Section 5.3.2. Also, it is seen that NC-MFSK exhibits better energy efficiency than the other schemes when d increases.

The optimized modulations for different transmission distance d and $2.5 \leq \eta \leq 6$ are listed in Table 5.3. For these results, we use the optimized MQAM detailed in Table 5.2 and the fact that for NC-MFSK, $\hat{M} = 2$ is the optimum value that minimizes E_N^{FS} for every d and η. From Table 5.3, it is found that although 64-QAM outperforms NC-BFSK for very short-range communications, it should be noted that using MQAM with a large constellation size M increases the complexity of the system. In particular, when we know that MQAM uses the coherent detection at the receiver. In other words, there exists a trade-off between the complexity and the energy efficiency in using MQAM for small values of d.

Up to now, we have investigated the energy efficiency of the sinusoidal carrier-based modulations under the assumption of a Rayleigh fading channel with path loss. It is also of interest to evaluate the energy efficiency of the aforementioned modulation schemes operating over the more general Rician model, which includes the LOS path. For this purpose, let us assume that the instantaneous channel coefficient corresponding to symbol i is $G_i = h_i/L_d$, where h_i is assumed to be Rician distributed with pdf

$$f_{h_i}(r) = \frac{r}{\sigma^2} e^{-\frac{r^2 + A^2}{2\sigma^2}} I_0\left(\frac{rA}{\sigma^2}\right), \; r \geq 0,$$

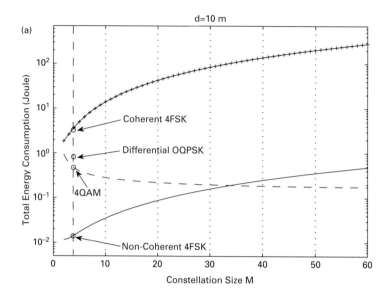
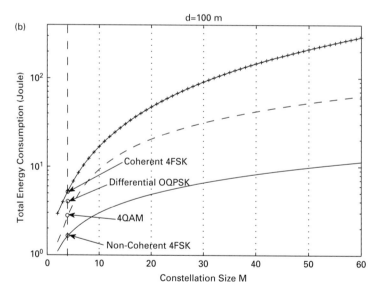

Figure 5.3 Total energy consumption of transmitting N-bit vs. M for MFSK, MQAM, and differential OQPSK over a Rayleigh fading channel with path loss and $P_s = 10^{-3}$, (a) $d = 10$ m, and (b) $d = 100$ m.

where A denotes the peak amplitude of the dominant signal, and $2\sigma^2 \triangleq \Omega$ is the average power of non-LOS multi-path components. For this model, $\mathbb{E}[|h_i|^2] = A^2 + 2\sigma^2 = 2\sigma^2(1 + K)$, where $K(\text{dB}) \triangleq 10\log(A^2/2\sigma^2)$ is the Rician factor. The value of K is a measure of the severity of the fading. For instance, $K(\text{dB}) \to -\infty$ implies Rayleigh fading and $K(\text{dB}) \to \infty$ represents AWGN channel. Table 5.4 summarizes the

Table 5.3. Energy-efficient modulation scheme for $P_s = 10^{-3}$ and $2.5 \leq \eta \leq 6$

Distance (m)	$\eta = 2.5$	$\eta = 3$	$\eta = 4$	$\eta = 5$	$\eta = 6$
1	64QAM	64QAM	64QAM	64QAM	64QAM
10	64QAM	64QAM	BFSK	BFSK	BFSK
20	64QAM	BFSK	BFSK	BFSK	BFSK
40	BFSK	BFSK	BFSK	BFSK	BFSK
80	BFSK	BFSK	BFSK	BFSK	BFSK
100	BFSK	BFSK	BFSK	BFSK	BFSK
150	BFSK	BFSK	BFSK	BFSK	BFSK
200	BFSK	BFSK	BFSK	BFSK	BFSK

Table 5.4. Total energy consumption (in Joule) of uncoded FSK, QAM, and OQPSK over a Rician fading channel with path loss for $P_s = 10^{-3}$ and $\eta = 3.5$

		$K = 1$ dB			$K = 15$ dB		
	M	OQPSK	NC-MFSK	MQAM	OQPSK	NC-MFSK	MQAM
d=10 m	4	1.1241	0.0173	0.5621	1.1241	0.0171	0.5620
	16		0.0769	0.2819		0.0765	0.2810
	64		0.6558	0.1924		0.6545	0.1874
d=100 m	4	1.2236	0.5835	0.8873	1.1310	0.0175	0.5627
	16		1.4920	3.2049		0.0767	0.2843
	64		4.6199	16.1010		0.6547	0.2002

energy-efficiency results of the previous modulations over a Rician fading channel with path loss for $P_s = 10^{-3}$ and $\eta = 3.5$. It is seen from Table 5.4 that NC-MFSK with a small-sized M has less total energy consumption than the other schemes in a Rician fading channel with path loss.

The above results make uncoded NC-MFSK with a small M attractive for using in energy-constrained wireless networks, since this modulation already has the advantage of less complexity and cost in implementation than MQAM, differential OQPSK, and coherent MFSK, and has less total energy consumption, in particular for the applications with lower data rates. The sacrifice, however, is the bandwidth efficiency of NC-MFSK (when M increases), which is a critical factor in band-limited networks. However, for the wireless network applications operating in unlicensed bands where large bandwidth is available, NC-MFSK can surpass the spectrum constraint in wireless systems.

5.4 Energy-consumption analysis of LT coded modulation

As has been mentioned in the previous section, among various sinusoidal carrier-based modulation techniques, uncoded FSK has been found to provide a good compromise between high data rate, simple radio architecture, low power consumption, and requirements on linearity of the modulation scheme. For energy-optimal designs, the impact of channel coding on the energy efficiency of the proposed wireless system must be considered for different distance d as well. It is well known that channel coding is a classical approach used to improve the link reliability along with the transmitter energy saving due to providing the coding gain. However, the energy saving comes at the cost of extra energy spent in transmitting the redundant bits in codewords as well as the additional energy consumption in the process of encoding/decoding. For a specific transmission distance d, if these extra energy consumptions outweigh the transmit energy saving due to the channel coding, the coded system would not be energy efficient compared with an uncoded system. In the subsequent sections, we will argue the above issue and determine at what distance use of specific channel coding becomes energy efficient compared to uncoded systems. In particular, we will show later that the rateless coded modulation surpasses this distance constraint in the proposed wireless system.

5.4.1 Energy efficiency of coded system

To get more insight into how channel coding affects the circuit and RF signal energy consumptions in the system, we modify the energy concepts in Section 5.3, in particular, the total energy consumption expression of uncoded FSK derived in (5.12) based on the coding gain, code rate, and the computation energy. We further present the tradeoff between the LT code rate and the coding gain required to achieve a certain Bit Error Rate (BER), and the effect of this tradeoff on the total energy consumption of LT coded MFSK for different transmission distances. For convenience, we drop the superscript "FS" from our notation. Also, for simplicity of notation, we use the superscripts "BC", "CC," and "LT" for BCH, convolutional and LT codes, respectively. We use the subscript "c" to distinguish coding parameters from the uncoded ones.

For an arbitrary channel coding scheme, each k-bit message, $B_j \in M_N$, is encoded into the codeword C_j with block length n and code rate $R_c \triangleq k/n$. In this case, the number of transmitted bits in T_N is increased from an N-bit uncoded message to an $N/R_c = (N/k)n$ bits coded one. To compute the energy consumption of the coded scheme, we use the fact that channel coding reduces the required average SNR value to achieve a given BER (i.e. the same BER as the uncoded one). Taking this into account, the wireless system benefits from the transmission energy saving of coded modulation specified by $E_{t,c} = E_t/\Upsilon_c$, where Υ_c is the coding gain of the utilized coded MFSK. Denoting $\bar{\gamma} = \Omega E_t/(L_d N_0)$ and $\bar{\gamma}_c = \Omega E_{t,c}/(L_d N_0)$ as the average SNR of uncoded and coded schemes, respectively, the coding gain (expressed in dB) is defined as the difference between the values of $\bar{\gamma}$ and $\bar{\gamma}_c$ required to achieve a certain BER. It should be noted that the cost of this energy saving is the bandwidth expansion B/R_c. In order to keep the bandwidth of the coded system the same as that of the uncoded case, we must keep the information transmission rate constant, i.e. the symbol duration T_s of uncoded and coded MFSK must be the same. However, the active-mode duration increases from $T_{ac} = (N/b)T_s$ in the uncoded system to

$$T_{ac,c} = \frac{N}{bR_c}T_s = \frac{T_{ac}}{R_c} = \frac{MN}{BR_c \log_2 M} \qquad (5.28)$$

5.4 Energy-consumption analysis of LT coded modulation

for the coded case. Thus, one would assume that the total time T_N increases to T_N/R_c for the coded scenario. It is worth mentioning that the active-mode period in the coded case is upper bounded by $(T_N/R_c) - T_{tr}$. As a result, the maximum constellation size M, denoted by $M_{max} \triangleq 2^{b_{max}}$, for the coded MFSK is calculated by

$$\frac{2^{b_{max}}}{b_{max}} = \frac{BR_c}{N}\left(\frac{T_N}{R_c} - T_{tr}\right)$$

which is approximately the same as that of the uncoded case.

We denote by E_{enc} and E_{dec} the computation energy of the encoder and decoder for each information bit, respectively. Thus, the total computation energy cost of the coding components for N/R_c bits is obtained as $N\frac{E_{enc}+E_{dec}}{R_c}$. Substituting (5.8) in $E_{t,c} = E_t/\Upsilon_c$, and using (5.10), (5.11), and (5.28), the total energy consumption of transmitting N/R_c bits in each period T_N/R_C for an arbitrarily coded MFSK, and a given P_b is obtained as

$$E_{N,c} = (1+\alpha)\left[\left(1 - \left(1 - \frac{2(M-1)}{M}P_b\right)^{\frac{1}{M-1}}\right)^{-1} - 2\right]\frac{L_d N_0}{\Omega \Upsilon_c}\frac{N}{R_c \log_2 M}$$

$$+ (P_c - P_{Amp})\frac{MN}{BR_c \log_2 M} + 2P_{Sy}T_{tr} + N\frac{E_{enc}+E_{dec}}{R_c}, \quad (5.29)$$

where we use the fact that the relationship between the average Symbol Error Rate (SER) P_s and the average Bit Error Rate (BER) P_b of MFSK is given by $P_s = \frac{2(M-1)}{M}P_b$ [3, p. 262].

To make a fair comparison between the uncoded and coded modulation, we use the same BER and drop the subscript "c" for P_b in the coded case. Thus, the optimization goal is to minimize the objective function $E_{N,c}$ over modulation and coding parameters. This is achieved by finding the optimum constellation size M under the constraint $2 \leq M \leq M_{max}$ for a specific channel coding scheme, and then minimizing $E_{N,c}$ over the coding parameters.

For the above optimization problem, we consider two scenarios: i) fixed-rate codes (e.g. BCH and convolutional codes), and ii) variable-rate codes (e.g. LT codes). To find the optimum M for a given fixed-rate code, we prove that (5.29) is a monotonically increasing function of M for every value of d and η. Since B and N are fixed and R_c is independent of M, it is concluded that the second term in (5.29) is a monotonically increasing function of M. Also, with a similar argument as for uncoded MFSK in Section 5.3.1,

$$\left(\left[1 - \left(1 - \frac{2(M-1)}{M}P_b\right)^{\frac{1}{M-1}}\right]^{-1} - 2\right)\frac{1}{\log_2 M} \approx \left(\frac{M}{2P_b} - 2\right)\frac{1}{\log_2 M}. \quad (5.30)$$

On the other hand, it is shown in [3] that Υ_c for MFSK is a decreasing function of M. Thus, it is concluded from (5.30) that the first term in (5.29) is also a monotonically increasing function of M. As a result, the minimum total energy consumption $E_{N,c}$ for a given fixed-rate code is achieved at $\hat{M} = 2$.

In the next section, we evaluate the above optimization problem for the LT codes using some simulation studies on the probability mass function of the LT code rate and coding gain. We show that the LT code parameters depend strongly on the constellation size M and exhibit different trends over fixed-rate codes. In addition, we present some beneficial uses of LT codes over block and convolutional codes in managing the energy consumption for different channel realizations.

5.4.2 Energy optimality of LT codes

LT codes are the first class of Fountain codes which are usually specified by the number of input bits, k, and the output-node degree distribution, $O(x)$. Without loss of generality and for ease of our analysis, we assume that a finite k-bit message, $B_1 \triangleq (m_1, m_2, \ldots, m_k) \in M_N$, is encoded to the codeword $C_1 \triangleq (a_1, a_2, \ldots, a_n)$. Each single-coded bit, a_i, is generated based on the encoding protocol proposed in [1]: i) randomly choose a degree $1 \leq D \leq k$ from a priori known degree distribution $O(x)$; ii) using a uniform distribution, randomly choose D distinct input bits, and calculate the encoded bit a_i as the XOR-sum of these D bits. The above encoding process defines a *bipartite graph* connecting encoded nodes to input nodes. The LT encoding process is extremely simple and has very low energy consumption. Unlike block and convolutional codes, in which the codeword block length is fixed, for the above LT code, n is a variable parameter, resulting in a random variable LT code rate $R_c^{LT} \triangleq k/n$. More precisely, $a_n \in C_1$ is the last bit generated at the output of LT encoder before receiving the acknowledgement signal from the receiver indicating the termination of a successful decoding process. This inherent property of LT codes means they can vary their codeword block lengths to match a wide range of channel conditions.

To describe the output-node degree distribution used in this work, let $\mu_i, i = 1, \ldots, k$, denote the probability that an output node has degree i. Following the notation of [15], the output-node degree distribution of an LT code has the polynomial form $O(x) \triangleq \sum_{i=1}^{k} \mu_i x^i$ with the property that $O(1) = \sum_{i=1}^{k} \mu_i = 1$. Typically, optimizing the output-node degree distribution for a specific wireless channel model is a crucial task in designing LT codes. In fact, for wireless fading channels the "optimal" $O(x)$ is still an open problem. In this work, we use the following output-node degree distribution, which was optimized for a Binary Symmetric Channel (BSC) using a hard-decision decoder [16, 17]:

$$O(x) = 0.00466x + 0.55545x^2 + 0.09743x^3 + 0.17506x^5 + 0.03774x^8$$
$$+ 0.08202x^{14} + 0.01775x^{33} + 0.02989x^{100}. \tag{5.31}$$

The LT decoder at the receiver side can recover the original k-bit message, B_1, with high probability after receiving any $(1+\epsilon)k$ bits in its buffer, where ϵ depends upon the LT code design [15]. For this recovery process, the LT decoder needs to correctly reconstruct the bipartite graph of an LT code. One practical approach suitable for the proposed system model is that the LT encoder and decoder use identical pseudo-random generators with a common seed value, which may reduce the complexity further. In this work, we assume that the receiver recovers k-bit message B_1 using a simple hard-decision "*ternary message passing*" decoder in a nearly identical manner to the "*Algorithm E*" decoder in [19] for low-density parity-check (LDPC) codes [18]. Also, the degree distribution, $O(x)$, in (5.31) was optimized for a ternary decoder in a BSC and we are aware of no better $O(x)$ for the ternary decoder in Rayleigh fading channels.

Unlike fixed-rate codes in which the active-mode duration of coded MFSK is fixed, for the LT coded MFSK, we have a non-fixed value for

$$T_{ac,c}^{LT} = \frac{N}{bR_c^{LT}} T_s = \frac{MN}{BR_c^{LT}\log_2 M}. \tag{5.32}$$

An interesting point raised from (5.32) is that $T_{ac,c}^{LT}$ is a function of the random variable R_c^{LT}, which results in an inherent adaptive duty-cycling for power management in each channel condition without any channel state information fed back from the receiver to the transmitter. Recalling from (5.29), we have

$$E_{N,c}^{LT} = (1+\alpha)\left[\left(1-\left(1-\frac{2(M-1)}{M}P_b\right)^{\frac{1}{M-1}}\right)^{-1} - 2\right]\frac{L_d N_0}{\Omega \Upsilon_c^{LT}} \frac{N}{R_c^{LT}\log_2 M}$$

$$+(P_c - P_{Amp})\frac{MN}{BR_c^{LT}\log_2 M} + 2P_{Sy}T_{tr} + N\frac{E_{enc}^{LT} + E_{dec}^{LT}}{R_c^{LT}}, \tag{5.33}$$

where the main goal is to minimize $E_{N,c}^{LT}$ in each distance d, in terms of M and the coding parameters. Toward this goal, we first compute the LT code rate and the corresponding LT coding gain. When using finite-k LT codes, we are treating the channel as static over one block length. In this case, the rate of the LT code for any block can be chosen to achieve the desired performance for that block. More precisely, for the particular block we are concerned with, the receiver could evaluate the channel instantaneous SNR, and then determine how many code bits it needs to collect in order to achieve its given BER target, thus essentially dynamically selecting its rate. For the next block, the receiver would once again evaluate the (new) instantaneous channel SNR and adjust its rate accordingly to collect more or fewer bits.

It should be noted that there is no (currently known) explicit equation governing the relationship between the instantaneous SNR and the required number of decoding bits for an LT code. In this work, we determine the necessary number of decoding bits in each case through simulation, as described below:

i) We decide upon a "*target BER*" for the decoded bits – e.g. the decoded message needed an average BER of 10^{-4} or better.
ii) Based on the assumption that instantaneous SNR is constant over at least one block we perform the following steps for a large range of possible instantaneous SNR:
  - Determine through numerical simulation the decoded BER using the LT code rate $R_c^{LT} = 1$.
  - If the decoded BER is greater than the target BER, then we reduce the LT code rate R_c^{LT} and try again.
  - Repeat the above step until the decoded BER is less than the target BER.

iii) At this point, for the given LT degree distribution, decoder, and SNR we can identify the highest LT code rate that will yield the target BER or better.

Armed with this information (computed ahead of time), a receiver can determine the appropriate rate at which to operate and hence determine how many bits it must collect for a given instantaneous SNR. This is also based on the common assumption that the receiver is capable of estimating the instantaneous SNR.

Based on the above arguments and for any given average SNR, the LT code rate is described by either a probability mass function (pmf) or a probability density function (pdf) denoted by $P_R(\ell)$. Because it is difficult to get a closed-form expression of $P_R(\ell)$, we use a discretized numerical method to calculate the pmf $P_R(\ell) \triangleq \Pr\{R_c^{LT} = \ell\}$, $0 \leq \ell \leq 1$, for different values of M.

Table 5.5 illustrates the average LT code rates and the corresponding coding gains of LT coded MFSK using $O(x)$ in (5.31), for $M = 2, 4$, and 8 and given $P_b = 10^{-3}$. The average rate for a certain average SNR is obtained by integrating the pmf over the rates from 0 to 1. It is observed that the LT code is able to provide a huge coding gain, Υ_c^{LT}, given $P_b = 10^{-3}$, but this gain comes at the expense of a very low average code rate, which means many additional code bits need to be sent. This results in higher energy consumption per information bit. In contrast to the fixed-rate codes in which the coding gain decreases when M grows, the LT coding gains display different trends in terms of M as illustrated in Table 5.5. Thus, in contrast with fixed-rate codes, $E_{N,c}^{LT}$ is not necessarily a monotonically increasing function of M. In the next section, we evaluate numerically $E_{N,c}^{LT}$ in terms of the optimized modulation and coding parameters compared to the uncoded and the fixed-rate codes.

An interesting point extracted from Table 5.5 is the flexibility of the LT code to adjust its rate (and its corresponding coding gain) to suit instantaneous channel conditions. For instance, in the case of favorable channel conditions, the LT coded MFSK is able to achieve $R_c^{LT} \approx 1$ with $\Upsilon_c^{LT} \approx 0$ dB, which is similar to the case of uncoded MFSK, i.e. $n = k$. The effect of LT code rate flexibility on the total energy consumption is also observed in the simulation results in the subsequent section.

5.5 Numerical results

5.5.1 Experimental setup

In this section, we present some numerical evaluations using the system parameters summarized in Table 5.1. In addition, it is assumed that $\eta = 3.5$. The results in Table 5.5 are also used to compare the energy efficiency of uncoded and coded MFSK schemes. To make a fair comparison between the energy consumption of different communication schemes, the bandwidth and the BER are assumed to be the same for all the coded schemes. In order to estimate the computation energy of the channel coding, we use the ARM7TDMI core, which is the industry's most widely used 32-bit embedded RISC microprocessor for an accurate power simulation [20]. For the energy and power

5.5 Numerical results

Table 5.5. Average LT code rate and coding gain of the LT coded MFSK over a Rayleigh fading channel for $P_b = 10^{-3}$

	M = 2			M = 4			M = 8	
$\frac{E_b}{N_0}$ (dB)	Average Code rate	Coding gain (dB)	$\frac{E_b}{N_0}$ (dB)	Average Code rate	Coding gain (dB)		Average Code rate	Coding gain (dB)
5	0.2560	25	0	0.0028	33.87		0.0012	36.46
6	0.3174	24	2	0.0140	31.87		0.0024	34.46
7	0.3819	23	4	0.0460	29.87		0.0095	32.46
8	0.4475	22	6	0.1100	27.87		0.0330	30.46
9	0.5120	21	8	0.2100	25.87		0.0870	28.46
10	0.5738	20	10	0.3300	23.87		0.1800	26.46
11	0.6315	19	12	0.4600	21.87		0.3000	24.46
12	0.6840	18	14	0.5900	19.87		0.4400	22.46
13	0.7307	17	16	0.7000	17.87		0.5700	20.46
14	0.7716	16	18	0.7800	15.87		0.6800	18.46
15	0.8067	15	20	0.8500	13.87		0.7700	16.46
16	0.8365	14	22	0.8900	11.87		0.8400	14.46
17	0.8614	13	24	0.9200	9.87		0.8800	12.46
18	0.8821	12	26	0.9400	7.87		0.9100	10.46
19	0.8991	11	28	0.9500	5.87		0.9300	8.46
20	0.9130	10	30	0.9600	3.87		0.9500	6.46
22	0.9333	8	32	0.9600	1.87		0.9500	4.46
24	0.9466	6	34	0.9600	−0.13		0.9600	2.46
26	0.9551	4	36	0.9700	−2.13		0.9600	0.46
28	0.9606	2	38	0.9700	−4.13		0.9700	−1.54
30	0.9640	0	40	0.9700	−6.13		0.9700	−3.54

calculations, the relations $E_{operation} = \frac{n_c}{n_o} \times E_{Hz}$ and $E_{block} = n_o \times E_{operation}$ are used, where

n_o : number of operations per block.

n_c : number of clock cycles on n_o operations.

E_{Hz} : total dynamic power-consumption per Hertz (W/Hz) of a single calculation cycle.

In addition, for the purpose of comparative evaluation, we use some classical BCH(n, k, t) codes with t-error correction capability, and convolutional codes (n, k, L) with the constraint length L. We use hard-decision prior to decoding for the BCH and convolutional codes. The main reason for using the hard-decision here is to make a fair comparison with the LT code, since this code involved a hard-decision in the ternary decoder.

5.5.2 Optimal configuration

As a starting point, we obtain the coding gain of some practical BCH and convolutional codes with MFSK for different constellation size M and given $P_b = (10^{-3}, 10^{-4})$ in

Table 5.6. Coding gain (dB) of BCH and convolutional coded MFSK over a Rayleigh fading channel for $P_b = (10^{-3}, 10^{-4})$

BCH Code(n,k,t)	R_c^{BC}	M=2	M=4	M=8	M=16	M=32
BCH (7, 4, 1)	0.571	(2.5, 2.8)	(0.3, 0.4)	(0.1, 0.2)	(0.0, 0.0)	(0.0, 0.0)
BCH (15, 11, 1)	0.733	(1.4, 1.6)	(0.2, 0.3)	(0.0, 0.0)	(0.0, 0.0)	(0.0, 0.0)
BCH (15, 7, 2)	0.467	(2.4, 3.3)	(2.0, 2.3)	(0.8, 1.0)	(0.3, 0.4)	(0.0, 0.0)
BCH (15, 5, 3)	0.333	(4.1, 4.6)	(2.7, 2.9)	(2.0, 2.1)	(1.5, 1.60)	(0.7, 0.8)
BCH (31, 26, 1)	0.839	(1.2, 1.5)	(0.2, 0.2)	(0.0, 0.0)	(0.0, 0.0)	(0.0, 0.0)
BCH (31, 21, 2)	0.677	(2.3, 2.9)	(1.7, 2.0)	(0.7, 0.8)	(0.2, 0.2)	(0.0, 0.0)
BCH (31, 16, 3)	0.516	(2.9, 3.1)	(2.1, 2.2)	(1.5, 1.6)	(1.3, 1.4)	(0.6, 0.7)
BCH (31, 11, 5)	0.355	(4.1, 4.4)	(3.5, 4.2)	(2.2, 2.3)	(2.0, 2.1)	(1.8, 2.0)
BCH (31, 6, 7)	0.194	(5.4, 5.9)	(4.3, 4.8)	(3.5, 3.8)	(3.2, 3.3)	(2.7, 2.8)
Convolutional Code	R_c^{CC}	M=2	M=4	M=8	M=16	M=32
trel(6, [53 75])	0.500	(3.8, 4.6)	(2.7, 3.1)	(2.1, 2.3)	(1.8, 2.0)	(1.4, 1.5)
trel(7, [133 171])	0.500	(4.0, 4.7)	(3.0, 3.5)	(2.2, 2.4)	(1.8, 2.0)	(1.5, 1.6)
trel(7, [133 165 171])	0.333	(5.7, 6.4)	(4.8, 5.1)	(3.7, 3.9)	(3.1, 3.3)	(2.7, 2.8)
trel([4 3], [4 5 17; 7 4 2])	0.667	(2.2, 2.6)	(1.5, 1.7)	(0.9, 1.1)	(0.6, 0.6)	(0.5, 0.5)
trel([5 4], [23 35 0; 0 5 13])	0.667	(2.9, 3.5)	(1.9, 2.4)	(1.4, 1.8)	(1.1, 1.2)	(0.8, 0.9)

Table 5.6. We observe from Table 5.6 that there is a tradeoff between coding gain and decoder complexity for the BCH codes. In fact, achieving a higher coding gain for a given M, requires a more complex decoding process, (i.e. higher t) with more circuit power consumption. In addition, it is seen that the coding gain of the BCH code is a monotonically decreasing function of M, as expected. For a given M, the convolutional codes with lower rates and higher constraint lengths achieve greater coding gains. In addition, it is observed that the coding gain of the convolutional coded MFSK is a monotonically decreasing function of M. By comparing the results in Table 5.5 with those in Table 5.6 for BCH and convolutional codes, one observes that LT codes outperform the other coding schemes in energy saving at comparable rates.

Figure 5.4 shows the total energy consumption versus distance d for the optimized BCH, convolutional and LT coded MFSK schemes, compared to the optimized uncoded MFSK for the case $P_b = 10^{-3}$. The optimization is performed over M and the parameters of the coding scheme. Simulation results show that for d less than the threshold level $d_T \approx 40$ m, the total energy consumption of optimized uncoded MFSK is less than that of the coded MFSK schemes. However, the energy gap between the LT coded and the uncoded MFSK is negligible compared to the other coded schemes, as expected. For $d > d_T$, the LT coded MFSK scheme is more energy efficient than the uncoded and other coded MFSK schemes. Also, it is observed that the energy gap between the LT and the convolutional coded MFSK increases when d grows. This result comes from the high coding gain capability of LT codes, which confirms our analysis in Section 5.4.2. The threshold level d_T (for the LT code) or d_T' (for the BCH and the convolutional codes) is obtained when the total energy consumptions of coded and uncoded systems become equal. For instance, using $L_d = M_l d^\eta L_1$, and the equality between (5.12) and (5.29) for uncoded and convolutional coded MFSK, we have

$$d'_T = \left[\hat{\Upsilon}_c^{CC} \frac{a_2(1 - \hat{R}_c^{CC}) + N(E_{enc}^{CC} + E_{dec}^{CC})}{a_1(\hat{\Upsilon}_c^{CC} \hat{R}_c^{CC} - 1)}\right]^{\frac{1}{\eta}},$$

where

$$a_1 \triangleq (1+\alpha)\left[\left(1 - \left(1 - \frac{2(M-1)}{M}P_b\right)^{\frac{1}{M-1}}\right)^{-1} - 2\right]\frac{L_1 M_l}{\log_2 M}\frac{NN_0}{\Omega},$$

$$a_2 \triangleq (P_c - P_{Amp})\frac{MN}{B\log_2 M},$$

$$(\hat{\Upsilon}_c^{CC}, \hat{R}_c^{CC}) = \arg\min_{\Upsilon_c^{CC}, R_c^{CC}} E_{N,c}^{CC}.$$

It should be noted that the above threshold level imposes a constraint on the design of the physical layer of some energy-constrained wireless networking applications, in particular mobile devices. To obtain more insight into this issue, let us assume that the location of the transmitter node is changed every $T_d \gg T_c$ time unit, where T_c is the channel coherence time. For the moment, let us assume that the transmitter aims to choose either a *fixed-rate coded* or an uncoded MFSK based on the distance between the transmitter and the receiver nodes. According to the results in Figure 5.4, it is revealed that using fixed-rate channel coding is not energy efficient for the short-distance transmission (i.e. $d < d'_T$), while for $d > d'_T$, convolutional coded MFSK is more energy efficient than other schemes. For this configuration, the transmitter must have the capability of an adaptive coding scheme for each distance d. However, as discussed previously,

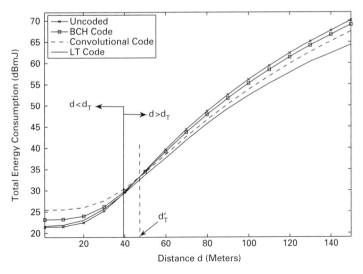

Figure 5.4 Total energy consumption of optimized coded and uncoded MFSK versus d for $P_b = 10^{-3}$.

the LT codes can adjust their rates for each channel condition and have (with a good approximation) minimum energy consumption for every distance d. This indicates that LT codes can surpass the above distance constraint in low-power wireless networks with dynamic position nodes over Rayleigh fading channels. This characteristic of LT codes results in reducing the complexity of the network design as well. Of interest are the strong benefits of using LT coded MFSK compared with the coded modulation schemes in [4]. In contrast to classical fixed-rate codes used in [4], the LT codes can vary their block lengths to adapt to any channel condition in each distance d. The simplicity and flexibility advantages of LT codes with an MFSK scheme make them the preferable choice for energy-constrained wireless networks, in particular for the systems with mobile nodes.

5.6 Conclusion

In the first part of this chapter, we analyzed the energy efficiency of some popular modulation schemes to find the distance-based green modulation in an energy-constrained wireless network over Rayleigh and Rician flat-fading channels with path loss. It has been demonstrated that among various sinusoidal carrier-based modulations, the optimized NC-MFSK is the most energy-efficient scheme in short-range wireless networks with sleep-mode processes for each value of the path-loss exponent, where the optimization is performed over the modulation parameters. In addition, the NC-MFSK with a small M is attractive for use in low-power wireless networking applications, since this modulation already has the advantage of less complexity and cost in implementation than MQAM, differential OQPSK, and coherent MFSK, and has less total energy consumption. Furthermore, the MFSK scheme has a faster start-up time than other schemes. Moreover, since for energy-constrained wireless networks, data rates are usually low, using M-ary NC-FSK schemes with a small M is desirable. The sacrifice, however, is the bandwidth efficiency of NC-MFSK when M increases. However, for the unlicensed band applications where large bandwidth is available, a loss in the bandwidth efficiency can be tolerated. In this case, the optimized NC-MFSK is attractive for use in the band-limited indoor wireless networks.

In the second part, we analyzed the energy efficiency of the LT coded MFSK in a proactive wireless network over Rayleigh fading channels with path loss. It has been shown that the energy efficiency of LT codes is similar to that of the uncoded MFSK scheme for $d < d_T$, while for $d > d_T$, the LT coded MFSK outperforms other uncoded and coded schemes, from the energy-efficiency point of view. This result follows from the flexibility of the LT code to adjust its rate and the corresponding LT coding gain to suit instantaneous channel conditions for any transmission distance d. This rate flexibility offers strong benefits in using LT codes in practical low-power wireless networks with dynamic distance and position nodes. In such systems and for every value of distance d, LT codes can adjust their rates to achieve a certain BER with a low energy consumption. The importance of our scheme is that it avoids some of the problems inherent in adaptive coding or incremental redundancy (IR) systems (channel feedback, large buffers, or multiple decodings), as well as the coding design challenge for fixed-rate codes used in wireless systems with mobile nodes. The simplicity and flexibility advantages of LT codes mean that the LT code with MFSK modulation can be considered as a *green modulation/coding* (GMC) scheme in energy-constrained wireless networks with dynamic nodes.

Acknowledgment

The authors would like to thank Dr. J. David Brown for his help in the LT code design section.

References

[1] M. Luby, "LT codes," in *Proc. of 43rd Annual IEEE Symposium on Foundations of Computer Science (FOCS)*, 2002, pp. 271–280.

[2] H. Karl and A. Willig, *Protocols and Architectures for Wireless Sensor Networks*. 1st ed., John Wiley and Sons Inc., 2005.

[3] J. G. Proakis, *Digital Communications*. 4th ed., New York: McGraw-Hill, 2001.

[4] S. Cui, A. J. Goldsmith, and A. Bahai, "Energy-constrained modulation optimization," *IEEE Trans. on Wireless Commun.*, vol. 4, no. 5, pp. 2349–2360, Sept. 2005.

[5] Hind Chebbo *et al.*, "Proposal for Partial PHY and MAC including Emergency Management in IEEE802.15.6," May 2009. [Online] Available at IEEE 802.15 WPAN TG6 in Body Area Network (BAN), www.ieee802.org/15/pub/TG6.html

[6] IEEE Standards, "Part 15.4: Wireless Medium Access control (MAC) and Physical Layer (PHY) Specifications for Low-Rate Wireless Personal Area Networks (WPANs)," in *IEEE 802.15.4 Standards*, Sept. 2006.

[7] F. Xiong, *Digital Modulation Techniques*. 2nd ed., Artech House, Inc., 2006.

[8] Q. Tang, *et al.*, "Battery power efficiency of PPM and FSK in wireless sensor networks," *IEEE Trans. on Wireless Commun*, vol. 6, no. 4, pp. 1308–1319, April 2007.

[9] R. E. Ziemer and R. L. Peterson, *Digital Communications and Spread Spectrum Systems*. New York: Macmillan, 1985.

[10] M. K. Simon and M.-S. Alouini, *Digital Communication over Fading Channels*. 2nd ed., New York: Wiley Interscience, 2005.

[11] J. Abouei, K. N. Plataniotis, and S. Pasupathy, "Green modulation in proactive wireless sensor networks," Technical report, University of Toronto, ECE Dept., Sept. 2009. [Online]. Available: www.dsp.utoronto.ca/~abouei/

[12] J. Abouei, K. N. Plataniotis, and S. Pasupathy, "Green modulations in energy-constrained wireless sensor networks," *IET Communications*, vol. 5, no. 2, pp. 240–251, Jan. 2011.

[13] M.-S. Simon, "Multiple-bit differential detection of offset QPSK," *IEEE Trans. on Commun.*, vol. 51, pp. 1004–1011, June 2003.

[14] "Range extension for IEEE 802.15.4 and ZigBee applications," *FreeScale Semiconductor, Application Note*, Feb. 2007.

[15] A. Shokrollahi, "Raptor codes," *IEEE Trans. on Inform. Theory*, vol. 52, no. 6, pp. 2551–2567, June 2006.

[16] J. Abouei, *et al.*, "On the energy efficiency of LT codes in proactive wireless sensor networks," *IEEE Trans. on Signal Processing*, vol. 59, no. 3, pp. 1116–1127, March 2011.

[17] J. Abouei, *et al.*, "On the energy efficiency of LT codes in proactive wireless sensor networks," in *Proc. of IEEE Biennial Symposium on Communications (QBSC'10)*, Kingston, Canada, May 2010, pp. 114–117.

[18] J. D. Brown, "Adaptive Demodulation Using Rateless Erasure Codes", Ph.D. Thesis, University of Toronto, 2008.

[19] T. J. Richardson and R. L. Urbanke, "The capacity of low density parity-check codes under message-passing decoding," *IEEE Trans. on Inform. Theory*, vol. 47, no. 2, pp. 599–618, Feb. 2001.

[20] "ARM7TDMI Technical Reference Manual," Tech. Rep.. [Online]. Available: http://infocenter.arm.com/help/topic/com.arm.doc.ddi0210c/

6 Cooperative techniques for energy-efficient wireless communications

Osama Amin, Sara Bavarian, and Lutz Lampe

6.1 Introduction

Cooperative communication techniques are envisioned as an integral part of next-generation wireless networks. Cooperative communication is based on extending the interactions between different communications nodes to obtain ubiquitous network access with the required quality of service (QoS). The virtual antenna array created by collaborating distributed communication nodes provides the network with a spatial diversity merit without the need to equip the nodes with multi-antenna transceivers. In addition to combating fading through spatial diversity, cooperative communication is a powerful technique to increase spectral efficiency, reduce energy consumption, and extend the network coverage with a lower cost than traditional networks [1, 2].

All the aforementioned advantages of cooperative communication encouraged the inclusion of relaying techniques in the International Mobile Telecommunication (IMT)-advanced fourth-generation (4G) standards IEEE 802.16m and Long Term Evolution-Advanced (LTE-A). These standards consider the two-hop relaying technique in their design, which is one of the well-studied cooperative transmission schemes. Multi-hop relaying is included in IEEE 802.16j, and other cooperative schemes, such as relay selection cooperative communication, are expected to be included in future generations of wireless communication networks to improve connectivity, achieve higher data rates, and reduce energy consumption compared to current networks [1, 3].

Energy-aware design is one of the main targets for the next generation of communication networks. This design helps both energy-constrained wireless devices and base stations to save energy and effectively work toward green communication solutions. "Green" communication is imperative considering the rising carbon footprint of mobile phones and networks, which are expected to reach 58% of the information and communication technology (ICT) industry contribution by 2020 (349 Megatons of CO_2) [4]. Therefore, energy efficiency (EE) is considered to be an important measure in designing future communications networks [5, 6].

In this chapter, we provide an overview of the use of cooperative techniques for EE communications. To this end, we first review EE metrics pertinent to measuring network performance and designing its parameters. Then the ability of relay networks for energy-efficient communication is investigated and compared with non-cooperative systems. Next, relay network parameters are optimized to achieve better EE and fulfill the energy-saving expectations of next-generation wireless communication networks.

Finally, we discuss the role of cooperative base stations in designing energy-efficient communications protocols.

6.2 Energy-efficiency metrics for wireless networks

An EE metric should capture all parameters pertinent to the energy consumption for communication, while considering the amount of delivered data possibly under QoS requirements. A widely used measure for EE is defined as [7]–[10]

$$EE = \frac{\text{Total amount of energy consumed}}{\text{Total amount of data delivered}} \text{ Joule/bit}. \quad (6.1)$$

The inverse of (6.1), i.e. $EE_i = 1/EE$ bit/Joule, has been adopted in [9], [11]–[13]. We note that to maximize EE, the measure EE from (6.1) and its inverse EE_i should be minimized and maximized, respectively.

The total energy consumption for a given transmission link consists of transmitted and transceiver system energy [7, 8]. The transceiver system has two types of circuit, namely analog and digital circuits. Analog circuits include a digital to analog converter, a power amplifier, a mixer, an intermediate frequency amplifier, a low-noise amplifier, an analog to digital converter, active filters and a frequency synthesizer. Digital circuits operate at baseband frequency and perform digital signal processing such as source and error correction coding, pulse modulation, digital modulation, and decoding processes. The energy consumption for a certain communication link can be expressed as

$$E = P_{\text{ont}}T_{\text{on}} + P_{\text{onr}}T_{\text{on}} + P_{\text{tr}}T_{\text{tr}} + P_{\text{sm}}T_{\text{sm}} + P_{\text{dc}}T_{\text{on}}, \quad (6.2)$$

where the total time duration is divided between the communication active mode T_{on}, transient mode T_{tr}, and sleep mode T_{sm}. During T_{sp}, the power consumption P_{sm}, is dominated by the leaking current of the switching transistors. This term is often neglected, i.e. P_{sm} can be set to zero [7, 8]. The transient-mode time arises mainly from the frequency synthesizer settling time, and the settling time for other devices such as the mixer and power amplifier (PA) can be neglected [7, 8]. During T_{on}, power is consumed at digital circuits (P_{dc}) and analog circuits at the transmitter (P_{ont}) and receiver side (P_{onr}). The analog-circuit power terms can be expressed as

$$P_{\text{ont}} = P_{\text{t}} + P_{\text{amp}} + P_{\text{ct}} = (1+\alpha)P_{\text{t}} + P_{\text{ct}}, \quad (6.3)$$

$$P_{\text{onr}} = P_{\text{cr}}, \quad (6.4)$$

where P_{t} is the transmitted power, P_{amp} is the PA power consumption, $\alpha = \xi/\eta - 1$ with η and ξ being the drain efficiency and the peak-to-average ratio, respectively, and P_{ct} and P_{cr} are the RF circuit power consumption at the transmitter and the receiver, respectively. P_{cr}, P_{ct}, and P_{tr} are often expressed as constants for given transceiver systems [7, 8].

Although a unified model for power/energy consumption of digital components is not available, there is some analysis for specific systems in the literature [11], [14]–[16]. The power consumption for Viterbi decoders is estimated in [14] by

$$P_{\text{VD}} = K_0 N_s R, \qquad (6.5)$$

where R is the throughput, N_s is number of states, and K_0 is a constant depending on some parameters such as clock frequency, total effective capacitance in the computing circuitry, voltage power supply, minimum transistor channel length, and code rate. The decoding energy consumption, E_{dec}, for a binary BCH code of length v is formulated in [15] as

$$E_{\text{dec}} = (2vg + 2g^2)(E_{\text{add}} + E_{\text{mult}}), \qquad (6.6)$$

where g is the error-correcting capability. E_{add} and E_{mult} are the energy consumption for addition and multiplication, respectively, which depend on the integrated circuits fabrication technology [15]. The general model

$$E_c = c_1 c_3^{c_2 R} T_{\text{on}} \qquad (6.7)$$

for decoder energy consumption is suggested in [16], where the parameters c_1, c_2, and c_3 vary for different decoders.

The power consumption for digital signal processing is usually considered to be relatively small compared to that of analog circuits. Especially in radio base stations, the power consumption of signal processing circuits accounts for only about 10% of the total consumed power, while analog circuits consume about 65% of the total power [4, 5]. This often permits to neglect power consumption of digital signal processing circuitry, especially when the employed signal processing techniques are not complicated such as in uncoded systems, single user communication, and non-iterative decoding [16].

The denominator of the EE metric in (6.1), i.e. the total amount of delivered data, is sometimes expressed simply in terms of raw bit rate (R_r) as $L_r = R_r T_{\text{on}}$. Another definition for the delivered data, which takes the data-rate loss due to pilots, feedback, and control overhead into account, only considers the payload $L_p = R_p T_{\text{on}}$ as delivered data, where R_p is the bit rate after removing all overheads. It is defined as [11]

$$R_p = R_r(1 - \rho) - v, \qquad (6.8)$$

where ρ is the average pilot density in time and frequency and v is the amount of feedback bits to obtain the channel state information (CSI). A further refined measure, which considers only the correctly detected data, is given in terms of error probability, P_e, as $L = L_p(1 - P_e)$ [10, 17] and the successful transmission rate, R, defined as

$$R = R_p(1 - P_e). \qquad (6.9)$$

In addition to considering the effect of overhead on data rate, the energy consumption for feedback symbols is also added to the total energy budget of the communication system in (6.2) [18].

The EE metric for an uplink and downlink communication scenario transmitting a total amount of L_T bits for both links can be written as [19]

$$EE = \frac{1}{L_T}(P_U T_U + P_D T_D), \tag{6.10}$$

with P and T denoting power consumption and transmission time for one direction and the subscripts U and D identifying the uplink and downlink, respectively. L_T is the sum of the bits transmitted in both links, i.e. L_U and L_D, based on the previous definition of L_r, L_p, or L.

An interesting dimensionless EE metric is used in [15, 20]. It is defined as the ratio of the energy used for delivering useful data and the total energy consumption used in the communication network. Considering our decomposition of energy terms in (6.2), it could be written as

$$EE_r = \frac{[(1+\alpha)P_t + P_{ct} + P_{cr}]T_{on} L/L_r}{E}. \tag{6.11}$$

This metric captures the energy-consumption sources of communication nodes and the transmission channel reliability. However, it does not consider the effect of the transmission rate on EE.

As seen in (6.9), and (6.11), EE metrics can be expressed in terms of successfully delivered data based on error probability. Another measure often used for reliably achievable rate is channel capacity, as we will see further below. In either case, the metric formulation depends on the channel quality. This gives rise to the notions of instantaneous EE metrics for a given instant of channel quality and average EE metrics, for which the expectation is taken over the random channel state variable.

6.2.1 Instantaneous EE metrics

In the following discussion, two metric types are considered as implementation examples for instantaneous EE metrics. The first one is applied to systems operating at a target instantaneous error probability considered in e.g. [13, 19, 22, 23, 24]. In this approach, the transmitted power, P_t, in (6.3) is adjusted to achieve a target error probability, P_e, for a given instantaneous channel state. With L from (6.9) the EE metric can be written as[1]

$$EE = \frac{[(1+\alpha)P_t + P_{ct} + P_{cr}]T_{on}}{L}. \tag{6.12}$$

Example 1: MQAM Considering uncoded M-ary quadrature amplitude modulation (MQAM), where the number of bits per symbol is $b = \log_2 M$, and using the bound for

---

[1] In the following, for brevity and as mentioned above, we drop the energy terms for digital signal processing and transients in (6.2).

the MQAM bit error probability (P_b) for the additive white Gaussian noise (AWGN) channel in [7, 23], the instantaneous transmitted power, P_t, can be expressed as

$$P_t = \frac{2N_0 N_f B \left(2^b - 1\right)}{3Gh} \ln \frac{4\left(1 - \frac{1}{\sqrt{2^b}}\right)}{bP_b}, \qquad (6.13)$$

where h, N_f, N_0, B, and G are the channel fading power gain, the receiver noise figure, the thermal noise power spectral density, the signal bandwidth, and the path loss (possibly including antenna gains), respectively. In the case of coded MQAM with coding gain G_c, the transmitted power, P_t, is reduced by G_c [8].

Example 2: MFSK Similarly, for uncoded M-ary frequency-shift keying (MFSK) and with non-coherent detection, the instantaneous transmitted power is obtained as [8]

$$P_t = \frac{2N_0 N_f B}{Gh} \ln \left(\frac{2^{b-2}}{P_b}\right) \frac{1}{2^b}. \qquad (6.14)$$

The second approach for instantaneous EE metrics considers a channel capacity constraint. In this approach, P_t is adjusted to achieve a target channel capacity C, where C is assumed to play the role of the successfully detected data rate (R). The corresponding EE metric used in [13, 19] is

$$EE_i = \frac{C}{(1+\alpha)P_t + P_{ct} + P_{cr}}, \qquad (6.15)$$

where for the AWGN channel and idealized Gaussian modulation and with coherent detection, P_t is given by

$$P_t = \left(2^{(C/B)} - 1\right) \frac{N_0 N_f B}{Gh}. \qquad (6.16)$$

6.2.2 Average EE metrics

Another strategy, which is adopted in some research works such as [17, 20, 25, 26, 27, 29], considers averaging over channel states. The average EE can be written as the ratio of average consumed energy and the average number of received bits ($\overline{(\cdot)}$ indicates average over channel states):

$$\overline{EE} = \frac{\overline{E}}{\overline{R}T_{on}} = \frac{(1+\alpha)\overline{P}_t + P_{ct} + P_{cr}}{R_p(1 - \overline{P}_e)}. \qquad (6.17)$$

If the transmitted power is kept constant, the average EE metric simplifies to

$$\overline{EE} = \frac{(1+\alpha)P_t + P_{ct} + P_{cr}}{\overline{R}}. \qquad (6.18)$$

This is, for example, the case when the transmitted power is adjusted to achieve a target average received signal-to-noise ratio (SNR) $\overline{\gamma}_r$ according to [20, 30]

$$P_t = \frac{\overline{\gamma}_r N_0 N_f B}{G}. \qquad (6.19)$$

Similarly, the transmitted power can be evaluated at a target average error probability as in [7, 10, 22, 29, 31]. For example, for uncoded MQAM, with multiple (M_t) transmit antennas, an average bit-error probability $\overline{P}_b$, and transmission over a flat Rayleigh fading channel, the transmitted power is given by [7]

$$P_t = \frac{2}{3}\left(\frac{\overline{P}_b}{4}\right)^{-\frac{1}{M_t}}\left(\frac{2^b - 1}{b^{\frac{1}{M_t}}}\right)\left(\frac{M_t N_0 N_f B}{G}\right). \qquad (6.20)$$

Instead of error rate, channel capacity can be used and averaged with respect to channel conditions. In this case, the average EE is expressed in [12] as

$$\overline{EE}_i = \frac{\mathbb{E}\{R_p \delta(R_p \leq C)\}}{(1+\alpha)P_t + P_{ct} + P_{cr}}, \qquad (6.21)$$

where $\delta(.)$ is the indicator function. Using $C = B\log_2(1 + \frac{GP_t}{N_0 N_f B}h)$ and assuming a flat Rayleigh fading channel, (6.21) can be written as

$$\overline{EE}_i = \frac{R_p}{(1+\alpha)P_t + P_{ct} + P_{cr}}\exp\left(-\frac{(2^{R_p/B} - 1)N_0 N_f B}{GP_t}\right). \qquad (6.22)$$

If the transmitted power is adjusted to achieve a target instantaneous received SNR and thus a fixed error probability, i.e. $P_e = \overline{P}_e$, the average EE metric is

$$\overline{EE} = \frac{(1+\alpha)\gamma_r N_0 N_f B/G\mathbb{E}\{1/h\} + P_{ct} + P_{cr}}{R_p(1 - P_e)}. \qquad (6.23)$$

6.3 Energy-efficient cooperative networks

In cooperative networks, nodes share resources to effectively form distributed antenna arrays. In doing so, the spatial diversity and beamforming advantages known from multiple-antenna transmission can be realized. Furthermore, as in conventional multi-hop transmission, network connectivity is improved. The nodes in cooperative networks are usually referred to as source (S), relay (R), and destination (D) nodes, with the source communicating data to the destination through (possibly multiple) relay(s). The relay(s) exploits the broadcasting nature of the wireless channel and retransmits the source message to the destination. The two main retransmission techniques are known as amplify-and-forward (AF) and decode-and-forward (DF). We will next discuss EE metrics applied to different forms of cooperative communications such as single relay, multiple relay and multi-hop communication.

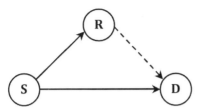

Figure 6.1 A single relay cooperative network.

6.3.1 Single relay cooperative network

We consider communication with a single relay as illustrated in Figure 6.1. A transmission from source to destination is organized in two time slots. In the first time slot, the source transmits and the relay and destination are listening (solid lines in Figure 6.1). In the second time slot, the relay retransmits a version of the received source signal to the destination (dashed line in Figure 6.1). In the following, we consider different retransmission mechanisms.

Selective decode-and-forward (SDF) is an adaptive DF technique in which the relay decides based on a certain criterion, such as SNR threshold or cyclic redundancy check (CRC), whether to retransmit a decoded and re-encoded version of the source signal or not [17, 29]. Assuming the use of a CRC at the relay and identical circuit-power-consumption for, respectively, transmission and reception in the source, relay, and destination, the EE of SDF is given similarly to [29] by

$$\overline{EE}_{\text{SDFc}} = \left[\frac{(1+\alpha)P_{t,S} + P_{ct} + 2P_{cr}}{\overline{R}_{\text{SDFc}}}\right] + \left[(1-\overline{P}_{p,\text{SR}})\frac{(1+\alpha)P_{t,R} + P_{ct} + P_{cr}}{\overline{R}_{\text{SDFc}}}\right], \quad (6.24)$$

where $P_{t,S}$ and $P_{t,R}$ are the transmit power at source and relay, respectively, and $\overline{P}_{p,\text{SR}}$ and $\overline{R}_{\text{SDFc}}$ are the average packet error rate (PER) of the S-R link and the average rate of the CRC-based SDF relay scheme, respectively. The first term in (6.24) represents the EE of the source transmission and the second term is the EE of the relay transmission phase. $\overline{R}_{\text{SDFc}}$ is the average rate for both the cooperative and non-cooperative cases, i.e.

$$\overline{R}_{\text{SDFc}} = \frac{R_p}{2}\left(\overline{P}_{p,\text{SR}}(1-\overline{P}_{p,\text{SD}}) + (1-\overline{P}_{p,\text{SR}})(1-\overline{P}_{p,\text{SRD}})\right), \quad (6.25)$$

where $\overline{P}_{p,\text{SD}}$ and $\overline{P}_{p,\text{SRD}}$ are the average PER for the S-D link and the S-R-D cooperative communication, respectively. Similarly, if SDF with an SNR threshold is used, we have

$$\overline{EE}_{\text{SDFs}} = \left[\frac{(1+\alpha)P_{t,S} + P_{ct} + 2P_{cr}}{\overline{R}_{\text{SDFs}}}\right] + \left[\Pr(\gamma_{\text{SR}} > \gamma_{\text{thr}})\frac{(1+\alpha)P_{t,R} + P_{ct} + P_{cr}}{\overline{R}_{\text{SDFs}}}\right], \quad (6.26)$$

where $\overline{R}_{\text{SDFs}}$ is the average transmission rate of SNR-based SDF, γ_{SR} is the SNR for the S-R link, and γ_{thr} is the S-R threshold SNR. Assuming that decoding at the relay is error free when $\gamma_{\text{SR}} > \gamma_{\text{thr}}$, then $\overline{R}_{\text{SDFs}}$ is given by (6.25) with $\Pr(\gamma_{\text{SR}} < \gamma_{\text{thr}})$ replacing $\overline{P}_{\text{p,SR}}$.

Incremental decode-and-forward (IDF) makes more frugal use of relay resources than SDF in that the relay only cooperates if the destination failed to detect the message after the first transmission phase [10, 17]. This requires a D-S and D-R feedback channel, but improves both energy and spectrum efficiency of relaying. The EE of IDF can be expressed as

$$\overline{EE}_{\text{IDF}} = \left(\frac{1}{\overline{R}_{\text{IDF}}}\right) \left(\left(1 - \overline{P}_{\text{p,SD}}\right) \left((1+\alpha)P_{\text{t,s}} + P_{\text{ct}} + 2P_{\text{cr}}\right) \right.$$
$$+ \overline{P}_{\text{p,SD}} \overline{P}_{\text{p,SR}} \left(P_{\text{t}}(1+\alpha) + P_{\text{ct}} + 2P_{\text{cr}}\right)$$
$$\left. + \overline{P}_{\text{p,SD}} \left(1 - \overline{P}_{\text{p,SR}}\right) \left((1+\alpha)P_{\text{t,s}} + 2P_{\text{ct}} + 3P_{\text{cr}}\right) \right). \qquad (6.27)$$

The first term in (6.27) is the EE when the destination succeeds in detecting the transmitted packet after the first transmission phase, while the second term is the EE when the relay decides not to participate in the second transmission phase although the destination could not detect the packet correctly. And the third term represents the scenario when the relay participates in the second transmission phase. $\overline{R}_{\text{IDF}}$ is the average transmission rate of the three previous scenarios, defined as

$$\overline{R}_{\text{p,IDF}} = \frac{R_{\text{p}}}{2}(1 - \overline{P}_{\text{p,SR}})\mathbb{E}\{\delta(\text{S-D error, S-R-D cooperation error free})\}$$
$$+ R_{\text{p}}(1 - \overline{P}_{\text{p,SD}}). \qquad (6.28)$$

In the previous examples of SDF and IDF, the durations of broadcasting and relaying phases are assumed to be equal. We note that if non-identical transmission phases are applied, the expressions in (6.24), (6.26), and (6.27) should be modified accordingly.

Hybrid automatic repeat request (HARQ) is an extension to IDF that allows data packet retransmission from both the source and the relay, if it detects the data packet successfully, until the destination signals successful detection or the number of transmission phases reaches its maximum allowable value (n) [27, 32]. There are several popular HARQ protocols applicable to relay transmission. HARQ type I (HARQ-TI) discards the erroneous data packets received in previous retransmissions and attempts detection using only the latest received packet. Chase-combining HARQ applies soft combining of several retransmissions in which the same data packet has been repeated. And incremental redundancy HARQ sends extra parity bits with new data packets and uses all received packets for decoding at the destination.

In the following discussion, the average EE of HARQ-TI is investigated for cooperative systems (CS). The EE metric captures the transmission attempts of the source, the activity of the relay in either transmission or detection and the attempts of the destination for detection. For this purpose, we define i and j as the number of source and relay transmissions, respectively, where $i = 1, 2, \ldots, n$ and $j = 1, 2, \ldots, i-1$. Under the assumption of equal transmit power of source and relay, the total power consumption after i transmissions of the source and j transmissions of the relay is expressed as [27]

$$P_{ij} = [(1+\alpha)P_t + P_{ct}](2i - j) + P_{cr}(2i + j). \tag{6.29}$$

Then, the average EE for the single relay HARQ-CS can be expressed as (see [27] for a similar expression)

$$\overline{EE}_{CS} = \frac{1}{\overline{R}_{HARQ}} \left[\sum_{i=1}^{n-1} \sum_{j=1}^{i-1} (q_{j-1} - q_j)(p_{i-1|i} - p_{i|j}) P_{ij} + \sum_{i=1}^{n-1} q_{i-1}(p_{i-1|i-1} - p_{i|i}) P_{ii} \right.$$
$$\left. + \sum_{j=1}^{n-1} (q_{j-1} - q_j) p_{n-1|j} P_{nj} + q_{n-1} p_{n-1|n-1} (P_{nn} - P_{cr}) \right], \tag{6.30}$$

where q_j is the probability of unsuccessful detection at the relay after j transmissions and $p_{i|j}$ is the probability of unsuccessful detection at the destination given that the relay detected the packet successfully at the jth transmission. Thus, the first term in (6.30) represents the average EE when the destination succeeds to detect the packet with the help of the relay and before n transmissions. The second term describes the successful detection before n transmissions and without the help of the relay. The third term stands for the failure of the destination to detect with the cooperation of the relay, before n transmissions. And the last term presents the scenario when both the relay and the destination fail to detect the packet correctly before n transmissions. The unsuccessful detection of the destination (with/without the cooperation of relay) at the nth transmission presents the outage case, which is expressed as [27]

$$P_{out,CS} = \sum_{j=1}^{n-1} (q_{j-1} - q_j) p_{n|j} + q_{n-1} p_{n|n}. \tag{6.31}$$

$\overline{R}_{HARQ}$ is the average rate of successful transmission and is written as

$$\overline{R}_{HARQ} = \sum_{i=1}^{n-1} \left(\frac{R_p}{i}\right) \sum_{j=1}^{i-1} (q_{j-1} - q_j)(p_{i-1|i} - p_{i|j}) + \sum_{i=1}^{n-1} \left(\frac{R_p}{i}\right) q_{i-1}(p_{i-1|i-1} - p_{i|i})$$
$$+ \left(\frac{R_p}{n}\right) \sum_{j=1}^{n-1} (q_{j-1} - q_j) p_{n-1|j} + \left(\frac{R_p}{n}\right) q_{n-1} p_{n-1|n-1}. \tag{6.32}$$

To evaluate the EE performance of the HARQ-CS, we compare it with the EE of the HARQ-TI non-cooperative system (NCS) [27]. In particular, we measure the EE gain of CS over NCS, which is defined as

$$\Upsilon = \frac{\overline{EE}_{NCS}}{\overline{EE}_{CS}}. \tag{6.33}$$

Table 6.1. Simulation parameters based on the 2.5 GHz radio in the industrial-scientific-medical band [8]

$P_{cr} = 112\,\text{mW}$	$P_{ct} = 98\,\text{mW}$	$\eta = 0.35$
$N_f = 10\,\text{dB}$	$N_0 = -171\,\text{dBm/Hz}$	$B = 10\,\text{kHz}$
$\xi = 1$ for MPSK, MFSK	$\xi = 3\dfrac{\sqrt{M}-1}{\sqrt{M}+1}$ for MQAM	Path-loss exponent $= 3.5$

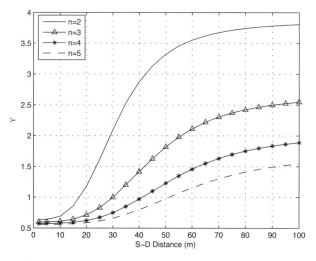

Figure 6.2 Comparison of EE performance (6.33) for a single relay cooperative network and direct transmission using HARQ-CS. System parameters from [27] and Table 6.1.

Figure 6.2 shows the relation of Υ versus the S-D distance for different values of n. The transmission is assumed to be through Rayleigh fading channels, where the fading gain is independent for each (re)transmission. The relay is located in the middle between source and destination, the transmitted power is computed to achieve an outage equal to 10^{-4} of (1.31). The other simulation parameters are taken from [27] and Table 6.1. We observe that, for the given parameters, NCS is preferable only for small S-D distances and CS achieves its best performance for $n = 2$. As n increases, the total transmitted power and circuit power consumption increase and thus the EE performance-gain over NCS decreases. The best n value changes when a different design criterion is adopted, such as optimizing the T_{on} duration [27]. In such cases, the delay in delivering the data should be considered in the system design.

Amplify-and-forward (AF) is a cooperative technique in which the relay transmits the amplified received signal from the source to the destination. Compared to DF relaying, the power consumption in the relay circuits is reduced. We therefore replace $(P_{ct} + P_{cr})$ by $P_{c,AF}$ for the relay-circuit power-consumption in the following. The instantaneous EE of AF is expressed in terms of the instantaneous error probability similar to (6.12) as

$$EE = \frac{\left[(1+\alpha)P_{t,S} + (1+\alpha)P_{t,R} + P_{c,AF} + P_{ct} + P_{cr}\right]T_{on}}{L_p(1-P_e)}. \tag{6.34}$$

6.3 Energy-efficient cooperative networks

If we adjust the source and relay power to achieve a target instantaneous bit-error rate, P_b, for MQAM transmission, then similarly to (6.13) we obtain

$$P_{t,R} = \frac{G_{SR}h_{SR}P_{t,S}\left[\left(\frac{3}{2}\right)G_{SD}h_{SD}P_{t,S} - N_0 N_f B(M-1)\ln\left(\frac{2^{b/2}-1}{b2^{b/2-1}P_b}\right)\right]}{G_{RD}h_{RD}\left[N_0 N_f B(M-1)\ln\left(\frac{2^{b/2}-1}{b2^{b/2-1}P_b}\right) - \left(\frac{3}{2}\right)G_{SR}P_{t,S}\right]}, \quad (6.35)$$

where $G_{SRD} = G_{SD}h_{SD} + G_{SR}h_{SR}$ and G_{XY} and h_{XY} are the path loss and the channel fading power gain for the X-Y link, X, Y $\in$ {S, R, D}, respectively. By further imposing equal transmitted power for source and relay, i.e. $P_{t,S} = P_{t,R} = P_t$, P_t is expressed in terms of P_b as

$$P_t = \frac{(2/3)N_0 N_f B(M-1)(G_{SR}h_{SR} + G_{RD}h_{RD})\ln\left(2\frac{2^{b/2}-1}{b2^{b/2}P_b}\right)}{(G_{SR}h_{SR} + G_{RD}h_{RD})G_{SD}h_{SD} + G_{SR}h_{SR}{}^2 G_{RD}h_{RD}}. \quad (6.36)$$

Alternatively, EE for AF can be expressed in terms of the instantaneous channel capacity similar to (6.18); that is

$$EE_i = \frac{C}{(1+\alpha)P_{t,S} + (1+\alpha)P_{t,R} + P_{c,AF} + P_{ct} + P_{cr}}. \quad (6.37)$$

The relay power, $P_{t,R}$, can be written in terms of the capacity C, source power $P_{t,S}$, and instantaneous channel conditions as

$$P_{t,R} = \frac{P_{t,S}G_{SR}h_{SR}\left(N_0 N_f B\left(2^{2C/B}-1\right) - P_{t,S}G_{SD}h_{SD}\right)}{G_{RD}h_{RD}\left(P_{t,S}G_{SR}h_{SR} + P_{t,S}G_{SD}h_{SD} - N_0 N_f B\left(2^{2C/B}-1\right)\right)}. \quad (6.38)$$

Under the assumption that $P_{t,S} = P_{t,R} = P_t$, we obtain

$$P_t = \frac{N_0 N_f B\left(2^{2C/B}-1\right)(G_{SR}h_{SR} + G_{RD}h_{RD})}{(G_{SR}h_{SR} + G_{RD}h_{RD})G_{SD}h_{SD} + G_{SR}h_{SR}G_{RD}h_{RD}}. \quad (6.39)$$

Incremental amplify-and-forward (IAF) is an improved AF technique that asks the relay to cooperate only if the received SNR at the destination is less than a certain threshold (γ_{thr}). Here, average EE is considered:

$$\overline{EE}_{IAF} = \left[\frac{(1+\alpha)P_{t,S} + P_{ct} + P_{cr}}{\overline{R}_{IAF}}\right]$$
$$+ \left[\Pr(\gamma_{SD} < \gamma_{thr})\frac{(1+\alpha)P_{t,R} + P_{c,AF}}{\overline{R}_{IAF}}\right], \quad (6.40)$$

where $\overline{R}_{IAF}$ is the average data rate of IAF, whose expression is somewhat complicated as error probabilities need to be conditioned on the event $\gamma_{SD} < \gamma_{thr}$.

6.3.2 Multi-relay cooperative network

The availability of multiple relays in the cooperation increases the degree of freedom in network design and improves the cooperation benefits. Selecting the best relay has proved its merit over all-relay working systems, where best-relay techniques achieve better spectral and energy efficiency [33]. In addition to that, selecting the best relay releases other relays to participate in another cooperative transmission and thus improves the deployment efficiency for all communication scenarios in the network [2]. Best-relay cooperative networks work in two phases similar to the single-relay network considered before. In the first transmission phase, the source broadcasts the data to the relay nodes and the destination. In the second transmission phase, only one selected (the "best") relay forwards the message to the destination. In the following, we consider relay selection based on EE metrics.

For the sake of illustration, we discuss EE multi-relay systems based on the opportunistic decode and forward (ODF) technique [28]. In this multi-relaying technique, the destination uses the channel state information and chooses either to receive the data directly only from the source or from one of the available relays, which apply DF and coded cooperation. The selection criterion is based on EE as

$$EE_i = \max_m \left(EE_{i,m} \right), \tag{6.41}$$

where m is the relay number and $m = 0$ means direct transmission. $EE_{i,0}$ is expressed as (6.15), while the EE of the mth relay with ODF is written as

$$EE_{i,m} = \frac{\min\left(tC_{\text{SR}},\ tC_{\text{SD}} + (1-t)C_{\text{RD}}\right)}{(1+\alpha)tP_{\text{t,S}} + (1+\alpha)(1-t)(P_{\text{t,R}}) + P_{\text{ct}} + (1+t)P_{\text{cr}}} \tag{6.42}$$

where t is the fraction of the transmission time used for the broadcasting phase. This dynamic time allocation feature can be used to further improve the EE performance of the cooperative network.

To investigate the effect of the relay-selection system in improving the EE, the EE gain of the ODF over NCS is studied versus the number of available relays in Figure 6.3. The relays communicate through Rayleigh fading channels with identical time allocation for the two transmission phases. All S-R and R-D channels experience the same path loss, with the S-R and R-D distances being q-times and $(1-q)$ times the S-D distance (500 m), respectively. The transmission power for all nodes is assumed to be equal to 50 dBm, and the other system parameters are taken from Table 6.1.

It can be seen from Figure 6.3 that using relay selection increased the EE gain of CS by more than three times when compared with NCS increases by more than 3 times. Near-to-the-source relay locations are somewhat preferable over their counterparts near to the destination due to improved decodability at the relays. The best relay's location is found to be in the middle region of the S-D distance, and the EE gain is significantly degraded at other locations. This observation can be related to the symmetrical time allocation between the two phases. Therefore proper adjustment of the parameter t will help in terms of EE at relay locations close to the source or destination.

6.3 Energy-efficient cooperative networks 137

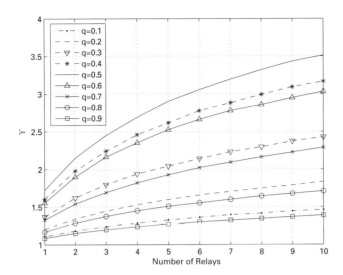

Figure 6.3 Comparison of EE performance for a relay-selection ODF and direct transmission versus number of available relays and different relay locations.

Figure 6.4 Multi-hop cooperative network.

To further improve the EE performance of relay selection, more than one relay can be selected and combined with cooperative beamforming to send the data to the destination [18].

6.3.3 Multi-hop cooperative network

Multi-hop cooperative relaying is a powerful technique to extend the network coverage and combat the attenuation resulting from path loss by breaking the low-SNR direct link into two or more high-SNR links, as shown in Figure 6.4. Multi-hop transmission can be performed through different relaying strategies such as fixed-rate relaying (FRR), adaptive-rate relaying (ARR), and incremental redundancy relaying (IRR) [20, 21].

In FRR, N relays help the source to transmit data to the destination through $N+1$ transmission phases (hops). In the first transmission phase, only the nearest relay to the source receives the transmitted data and forwards it using the DF technique to the next relay node. This relaying scenario repeats at each relaying node until the destination receives the transmitted data after $N+1$ hops. At each hop, FRR uses a fixed rate that is equal to the minimum achievable rate for all relaying links. On the other hand, ARR permits sending the data at different rates according to each link's reliability and by using the CSI of the link (hop). As a result, ARR can achieve a higher end-to-end transmission rate than FRR. Different from ARR and FRR, IRR. It assumes that more than one relay can receive the data after each transmission and hence the relays and destination can receive a combination of signals from the source and/or the relay nodes. IRR can achieve a higher end-to-end transmission rate than both FRR and ARR.

The aim of this study is to investigate when the multi-hop cooperative network improves EE over that of the NCS (single-hop) and what number of relays can give the best improvement in terms of EE. For simplicity, the simple FRR system with $N+1$ hops and the average EE

$$\overline{EE} = \frac{(1+\alpha)P_\text{t} + P_\text{ct} + P_\text{cr}}{R_N \left(1 - \overline{P}_\text{e}\right)} (N+1) \tag{6.43}$$

based on (6.18) are considered, where R_N is the rate for each hop transmission in the N-relay scenario and the transmitted power is computed from (6.19). $\overline{P}_\text{e}$ is the end-to-end error rate of the multi-hop system.

Figure 6.5 shows EE versus the S-D distance for different numbers of hops. We assume that the hop distances are equal so that all hops are subjected to the same path loss. Also, each link is assumed to be a non-fading channel and the transmit power is adjusted such that received SNR is 15 dB, and the other system parameters are taken from Table 6.1. For a fair comparison between different multi-hop scenarios, we fix the end-to-end transmission time for the same amount of delivered data by adjusting the transmission bit-rate as $R_N = (N+1)R_0$, where R_0 is the bit-rate of single-hop system (NCS). As expected, we observe from Figure 6.5 that direct (1-hop) transmission is preferable for short S-D distance. For large distances, the EE performance of single-hop degrades sharply and multi-hop relaying becomes the method of choice. Each hopping scenario achieves the best EE performance for some S-D distance range. This is also the

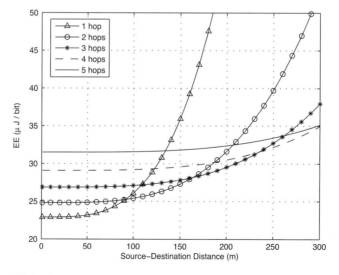

Figure 6.5 EE for single and multi-hop cooperative networks.

case for ARR. In IRR, we expect EE performance to improve with number of hops until the additional power consumption due to the listening of multiple nodes, i.e. $(N+1)P_{\text{cr}}$, becomes notable.

6.4 Optimizing the EE performance of cooperative networks

Cooperative networks have been optimized according to capacity or error rate. In this section, the EE performance of cooperative networks is improved by managing the available resources. Throughout this section, the single-relay scenario is considered and modulation level and transmitted power are chosen as optimization parameters.

6.4.1 Modulation constellation size

The EE performance of the cooperative system depends on several parameters such as, bandwidth, error probability, and constellation size ($M = 2^b$). In the following study, we assume a fixed bandwidth and target error rate and choose the modulation level (b) that minimizes EE. We consider the EE metric AF relaying from (6.34) and write [8]

$$T_{\text{on}} = \frac{L}{bB} \tag{6.44}$$

to express EE of AF as

$$EE = \frac{(1+\alpha)P_{\text{t,S}} + (1+\alpha)P_{\text{t,R}} + P_{\text{c,AF}} + P_{\text{ct}} + P_{\text{cr}}}{bB(1-P_{\text{e}})}. \tag{6.45}$$

We note that AF relaying uses only one modulation size for the end-to-end transmission, while DF systems permit the relay to change the modulation size in its relaying phase to the destination. Therefore, the EE of DF systems is a two-dimensional function with two different modulation sizes, while the EE of AF relaying is a one-dimensional function of modulation size.

The transmitted power is computed under a target instantaneous error rate from (6.36). Considering (6.45), we note that a small modulation level is preferred for a lower transmit power for a given error rate, but it suffers from a low data rate. Using a large modulation level increases the transmitted data rate, but also the transmit power increases for a given error probability. This trade-off is illustrated in the relation between the EE and constellation size as shown in Figure 1.6 for different target error rates. An S-D distance of 100 m with the relay located half-way in-between source and destination, and Rayleigh fading channels are assumed. The other system parameters are taken from Table 1.1

Figure 6.6 shows that the optimal constellation size increases as the target error probability relaxes. For example, for $P_{\text{b}} = 10^{-6}$ the optimal modulation level is 4 bits/symbol and as the P_{b} decreases to 10^{-3} and 10^{-1}, more bits are allowed for transmission, increasing the optimal modulation level to 6 bits/symbol and 8 bits/symbol, respectively.

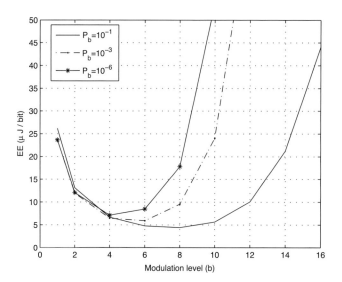

Figure 6.6 The average EE (1.45) as a function of constellation size for different target error rates. Rayleigh fading channels, an S-D distance of 100 m with the relay in the middle between S and D, and the parameters from Table 1.1 are assumed.

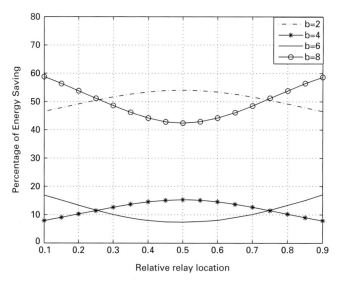

Figure 6.7 Energy savings when using adaptive compared to fixed constellation sizes $M = 2^b$ for $b = 2$; 4; 6; 8 and $P_e = 10^{-3}$ versus relative relay location between source and destination. System parameters are as for Figure 1.6.

Optimizing the constellation size is a discrete optimization problem and can be solved using full enumeration considering the few possibilities. In Figure 6.7, we demonstrate the energy savings due to adaptation of the constellation size. The figure shows the relative energy savings over systems with fixed constellations of size $M = 2^b$ as a function of the relative location of the relay on a line between the source and destination.

The error probability constraint is assumed to be 10^{-3} and other simulation parameters are as for the results in Figure 6.6. Choosing the optimal constellation leads to significant energy savings especially compared to fixed small and large constellation sizes, where we can have up to 60% reduction of the energy consumption. The energy savings become less pronounced in the middle region of constellation sizes which is consistent with the results shown in Figure 6.6.

6.4.2 Power allocation

In the previous section, we investigated the energy saving gained from optimizing the constellation size to minimize EE with a target error rate. In this part, we investigate the EE improvement that can be achieved from power allocation in cooperative networks.

Depending on the adopted EE metric, a suitable optimization problem could be formulated as

$$\min_{P_{t,S},\ P_{t,R}} EE$$
$$\text{subject to} \quad P_{t,S},\ P_{t,R} \leq P_{t,\max} \quad (6.46)$$
$$C \geq C_{\min},$$

where $P_{t,\max}$ is the maximum power constraint and $C_{\min}$ is the minimum acceptable capacity (i.e. achievable rate). Alternatively, the formulation

$$\min_{P_{t,S},\ P_{t,R}} EE$$
$$\text{subject to} \quad P_{t,S},\ P_{t,R} \leq P_{t,\max} \quad (6.47)$$
$$\bar{P}_e \leq \bar{P}_{e,\max},$$

where $\bar{P}_{e,\max}$ is the target error rate, which can be translated to an equivalent successful data rate constraint as $\bar{R} \geq \bar{R}_{\min} = R_p(1 - \bar{P}_{e,\max})$.

Reference [29] considers a version of the EE metric of CRC-based SDF described in (6.24) for the optimization problem (6.46), with the simplification of no peak-power constraint. Then, assuming Rayleigh fading channels, we obtain

$$\min_{P_{t,S},\ P_{t,R}} \frac{(1+\alpha)}{R_r}\left(P_{t,S} + \left(1 - \frac{KN_0 N_f R_r}{G_{SR} P_{t,S}}\right) P_{t,R}\right)$$
$$+ \frac{1}{R_r}\left(\left(2 - \frac{KN_0 N_f R_r}{G_{SR} P_{t,S}}\right) P_{ct} + \left(3 - \frac{KN_0 N_f R_r}{G_{SR} P_{t,S}}\right) P_{cr}\right) \quad (6.48)$$
$$\text{subject to} \quad \left(1 - \frac{KN_0 N_f R_r}{G_{SR} P_{t,S}}\right)\left(\frac{3N_0^2 N_f^2 R_r^2}{16 G_{SR} G_{RD} P_{t,S} P_{t,R}}\right) + \frac{KN_0^2 N_f^2 R_r^2}{4 G_{SR} G_{SD} P_{t,S}^2} \leq \bar{P}_{e,\max},$$

where K is a constant that depends on a high-SNR approximation of the block-error rate [29]. The adaptive scheme optimizes the power allocation of both the source and the relay for a given relay location and average target error rate. A good approximation to the optimal power allocation can be expressed as [29]

$$P_{t,S} = \sqrt{\frac{3N_0^2 N_f^2 R_r^2}{16 G_{SR} G_{RD} \bar{P}_{e,\max}}} + \sqrt{\frac{K N_0^2 N_f^2 R_r^2}{4 G_{SR} G_{SD} \bar{P}_{e,\max}}} \qquad (6.49)$$

and

$$P_{t,R} = \frac{\left(1 - \frac{K N_0 N_f R_r}{P_{t,S} G_{SR}}\right) 3 N_0^2 N_f^2 R_r^2}{16 G_{SR} G_{RD} P_{t,S} \left(\bar{P}_{e,\max} - \frac{K N_0^2 N_f^2 R_r^2}{4 G_{SR} G_{SD} P_{t,S}^2}\right)}. \qquad (6.50)$$

To measure the effectiveness of this power allocation, the EEs of the adaptive power allocation (APA) and the equal power allocation (EPA) cooperative systems are compared in Figure 1.8. It shows the relative energy savings when using APA instead of EPA for different target error rate constraints versus the relative relay location for an S-D distance of 500 m. The simulation parameters are again taken from Table 1.1. We observe that the SDF system does not save energy due to power allocation when the relay is near to the source. This is because in this case the relay acts as a second transmit antenna of the source and APA will essentially be identical to EPA. As the S-R distance increases, the energy savings start to become noticeable and reach up to 20% for the selected parameters at the middle relay location. The maximum energy savings occur when the relay is close to the destination. In this scenario, the relay is assigned little power compared to the source, rendering APA a quasi multiple receive antenna system. Choosing small

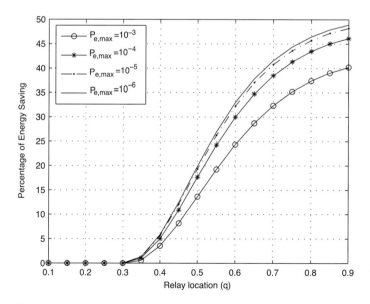

Figure 6.8 Energy saving of APA compared to EPA versus relative relay location and for target error-rates.

error probability constraint permits the APA system to save more power, but this saving saturates after some values as shown in Figure 1.8 for near-to-the-destination relay location.

6.5 Energy efficiency in cooperative base stations

Cooperation between the base stations (BSs), generally referred to as coordinated multipoint (CoMP), is a promising concept to increase the spectral efficiency and improve coverage in cellular systems, and it is considered a key element in the LTE-Advanced roadmap [35, 36]. CoMP is different from the relay cooperation discussed so far, because the cooperating BSs are usually connected through a backhaul infrastructure that can be physically implemented by direct optical-fiber links, or multi-hop virtual connections using different technologies. Improving the EE of network operation is another potential benefit of CoMP that has recently attracted considerable interest [37]. This section discusses the ongoing research on energy-efficient CoMP protocols.

The basic premise behind CoMP is to coordinate the operation of BSs either to reduce the harmful effect of intercell interference or to exploit the spatial diversity offered through interfering signals transmitted or received by neighboring BSs (macrodiversity). Figure 6.9 shows a mobile terminal in a CoMP system maintaining simultaneous connections to multiple controlling BSs. Joint processing of signals transmitted or received by cooperating BSs (full cooperation) presents a network of spatially spread virtual antenna arrays. Primary studies explored the benefits of macrodiversity to increase the system-wide capacity [38]–[40] and to improve the QoS, particularly for cell edge users, and proposed innovative algorithms for joint BS processing [41]–[45]. Recent research is mostly focused on low-complexity CoMP schemes such as coordinated scheduling

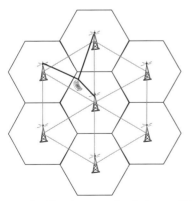

—— Backhaul connection between BSs
—— Wireless links o cooperating BSs

Figure 6.9   Illustration of CoMP.

and beamforming techniques, and explores practical issues facing future commercial application, including standardization process, performance evaluation, system requirements, and field trials [35]–[49].

The CoMP concept is also considered in the "green" network operation of cellular systems [5]. CoMP has the ability to substantially increase the average throughput in cellular systems. However, it also increases BS energy needs due to extra overheads, including additional pilots, increased backhaul traffic, and sophisticated signal processing. The tradeoff between these two trends controls the EE of CoMP schemes. Channel propagation characteristics, network density, and cooperation cluster size (the number of cooperating BSs) are some of the other factors to be considered in studying the EE of CoMP networks.

A simple energy-consumption model for CoMP is proposed in [11]

$$P_{BS} = a \cdot P_{tx} + b \cdot P_{sp} + c \cdot P_{bh}, \tag{6.51}$$

where P_{BS}, P_{tx}, P_{sp}, and P_{bh} denote the average power consumption, radiated power, signal processing power, and backhauling power per BS, respectively. The coefficients a, b, and c are used to capture the effects of miscellaneous factors such as amplifier or feeder losses, cooling, or battery backup. P_{tx} can be determined according to the power control protocol and path-loss channel characteristics. Assuming minimum mean square error (MMSE) signal detection (uplink) and MMSE linear beamforming (downlink), the rate of increase in signal processing power is modeled by

$$P_{sp} = p_{sp}(0.87 + 0.1 N_c + 0.03 N_c^2), \tag{6.52}$$

where p_{sp} is the base value for signal processing power and N_c is the cooperation size. P_{bh} is estimated by a linear model $P_{bh} = r \cdot c_{bh}$, where r is a coefficient and c_{bh} is the average backhaul requirement per BS. Using MMSE upper bounds for uplink and downlink data transmission and reception, and subtracting the additional pilot and control traffic, the authors in [11] estimate the EE of BSs in CoMP networks. Figure 6.10 from [11] shows EE gains of 10% to 20% due to CoMP for small and large cell sites cluster size 3 (cluster size refers to the number of cooperating BSs). Larger cluster sizes decrease the EE gains because of the requirements for additional backhauling and overhead.

An interesting approach is to take advantage of the extended coverage benefit of CoMP in designing dynamic policies for the energy-efficient operation of cellular networks [37]. BSs consume the largest share of the total energy consumption (60–80 percent) in cellular systems. Even in periods of low activity, a BS consumes more than 90% of its peak energy. Therefore, there is significant room for energy savings by turning off redundant BSs during periods of low usage and transferring their loads to neighboring cells. The key technical challenge is how to implement this energy-saving approach without reducing the QoS to mobile users. Coordinated transmission and reception through CoMP can compensate the negative effect of inactive BSs by extending the coverage area of active BSs in the neighboring cells, as depicted in Figure 6.11.

6.5 Energy efficiency in cooperative base stations

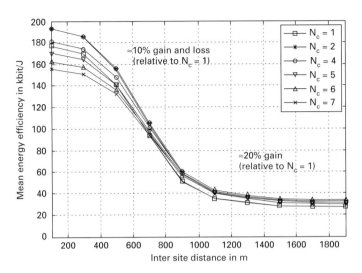

Figure 6.10 Bit per Joule efficiency of CoMP schemes for different inter-site distance (from [11], © [2010] IEEE).

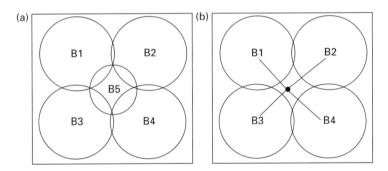

Figure 6.11 Reducing the number of active BSs in low-activity periods using coverage extension in CoMP, according to [37].

Cooperation between different operators (particularly in dense urban centers) can also substantially improve the EE of cellular systems. The practical implementation of such cooperation schemes faces many challenges including complex network operation, cross-operator authentication, and billing. Evaluating cross-operator cooperation using a game-theoretic approach is an interesting area of research that is beneficial in designing energy-efficient operation policies [37]. There are, however, concerns as to whether such agreements between the operators can reduce competition and hurt the customers.

Designing EE power control techniques for CoMP is another potential approach to improve energy efficiency. The existing power control mechanisms are designed for traditional cellular systems where one BS connects to a mobile user [50]. In the CoMP paradigm, multiple BSs maintain simultaneous links to a user to take advantage of the macrodiversity either through interference suppression or coherent joint detection. Cell-edge users benefit the most from macrodiversity and their error-rate performance improves significantly. For example, a mobile user located at the halfway point between two cooperating BSs enjoys the diversity order of twice the order of one in a non-cooperating system and hence, can use less power to achieve equal throughput. Such system-wide power reduction, in turn, leads to reduced inter-cell interference in CoMP networks. A simple power control scheme for cooperating BS is presented in [51], and it is shown through simulations how it can reduce the interference level in cellular systems. More research is required in this area to devise optimal power control techniques for CoMP and examine their effect on system-wide energy-efficiency measures.

In summary, the CoMP offers a promising solution for future cellular systems, with the potential to increase average spectral efficiency and cell-edge data rates, and it can also be beneficial in improving system-wide EE. There still remain many practical challenges that need to be addressed through further research in order to realize the potential benefits of the CoMP schemes. Some of these challenges include low-latency and high-bandwidth backhaul to support increased traffic, efficient clustering and scheduling policies, synchronization, integration into future standards, and designing energy-saving schemes to reduce the operating costs and to address environmental concerns.

6.6   Conclusion

In this chapter, we have considered cooperative communication as an energy-efficient candidate for the next-generation communication networks. For this purpose, we outlined the EE metrics that are used to measure and evaluate EE performance. We have also discussed how these metrics include both classical metrics, such as channel capacity or error probability, and the total cost of energy consumption, i.e. the transmitted energy and energy consumption in circuits of different communication nodes. According to EE analysis, relay networks achieve EE improvements (compared to non-cooperative schemes) through exploiting the broadcasting and scalability features of cooperative communication. Although having more relays improves the reliability of a communication link, there is an optimal number of relays for each distance range to ensure EE in multi-hop scenarios. By taking advantage of the increased degrees-of-freedom represented in transmission techniques, forms, and free-system parameters, adaptive cooperative communication systems can be designed to achieve higher EE than non-adaptive systems. Relay location is an important factor that can significantly affect the energy requirements of a communication link. Therefore, optimizing the location of relay(s) jointly with other system-design parameters is highly encouraged in adaptive systems to guarantee the maximum possible EE improvement and avoid bad relay(s) locations with negligible EE improvements. Cooperation between the base stations of cellular systems, referred to as CoMP, is another area where cooperative techniques have the potential to improve EE measures. We have discussed the state-of-the-art research in this area that present and evaluate energy-saving CoMP schemes to offset the extra power consumption of backhaul transmission and sophisticated signal processing required in CoMP systems. Although the research on green cooperative communications is still ongoing, with many open technical and practical issues to be resolved to facilitate commercial applications, it is certainly a key promising area to improve the EE of future communication networks.

References

[1] F. H. P. Fitzek and M. D. Katz, *Cooperation in wireless networks: principles and applications; real egoistic behavior is to cooperate!*, New York: Springer-Verlag, 2006.

[2] Y. Chen et al., "Fundamental trade-offs on green wireless networks," *IEEE Commun. Mag.*, vol. 49, no. 6, pp. 30–37, June 2011.

[3] K. Loa et al., "IMT-advanced relay standards," *IEEE Commun. Mag.*, vol. 48, no. 8, pp. 40–48, Aug. 2010.

[4] O. Blume, D. Zeller and U. Barth, "Approaches to energy efficient wireless access networks," in *Proc. of 4th Int. Symp. Communications, control and signal processing* (ISCCSP'10), March 2010.

[5] L. M. Correia et al., "Challenges and enabling technologies for energy aware mobile radio networks," *IEEE Commun. Mag.*, vol. 48, no. 11, pp. 66–72, Nov. 2010.

[6] K. Pentikousis, "In search of energy-efficient mobile networking," *IEEE Commun. Mag.*, vol. 48, no. 1, pp. 95–103, Jan. 2010.

[7] S. Cui, A. J. Goldsmith, and A. Bahai, "Energy-efficiency of MIMO and cooperative MIMO techniques in sensor networks," *IEEE J. Select. Areas Commun.*, vol. 22, no. 6, pp. 1089–1098, Aug. 2004.

[8] S. Cui, A. J. Goldsmith, and A. Bahai, "Energy-constrained modulation optimization," *IEEE Trans. Wireless Commun.*, vol. 4, no. 5, pp. 2349–2360, Sept. 2005.

[9] T. Chen, H. Kim, and Y. Yang, "Energy-efficient metrics for green wireless communication schemes," in *Proc. of Int. Conf. Wireless Communication and Signal Processing*, (WSCP'10) Oct. 2010.

[10] S. Wang and J. Nei, "Energy efficiency optimization of cooperative communication in wireless sensor networks," *EURASIP J. Wireless Commun. Netw.*, vol. 2010, Apr. 2010.

[11] A. J. Fehske, P. Marsch, and G. Fettweis, "Bit per Joule efficiency of cooperating base stations in cellular networks," in *Proc. of IEEE Globecom Workshop Green Communication*, Dec. 2010.

[12] S. Zhang, Y. Chen, and S. Xu, "Joint bandwidth-power allocation for energy efficient transmission in multi-user systems," in *Proc. of IEEE Globecom Workshop Green Communication*, Dec. 2010.

[13] G. Miao et al., "Energy-efficient design in wireless OFDMA," in *Proc. of IEEE Int. Conf. Communication*, (ICC'08), pp. 3307–3312, May 2008.

[14] S. Cui, A. J. Goldsmith, and A. Bahai, "Power estimation for Viterbi decoders," in *Wireless Systems Lab, Stanford Univ., Stanford, CA, Technical Report*, 2003.

[15] Y. Sankarasubramaniam, I. F. Akyildiz, and S. W. McLaughlin, "Energy efficiency based packet size optimization in wireless sensor networks," in *Proc. of IEEE 1st Int. Workshop on Sensor Network Protocols and Applications*, 2003.

[16] P. Rost and G. Fettweis, "On the transmission-computation-energy trade-off in wireless and fixed networks," in *Proc. of IEEE Globecom Workshop on Green Communication*, Dec. 2010.

[17] Q. Chen and M. C. Gursoy, "Energy efficiency analysis in amplify-and-forward and decode-and-forward cooperative networks," in *Proc. of IEEE Wireless Communications and Networking Conf.*, (WCNC'10), Apr. 2010.

[18] R. Madan *et al.*, "Energy-efficient cooperative relaying over fading channels with simple relay selection," *IEEE Trans. Wireless Commun.*, vol. 7, no. 8, pp. 3013–3025, Aug. 2008.

[19] W. Yang *et al.*, "Energy-efficient relay selection and optimal relay location in cooperative cellular networks with asymmetric traffic," *The Journal of China Universities of Posts and Telecommunications*, vol. 17, no. 6, pp. 80–88, Dec. 2010.

[20] Z. Shelby *et al.*, "Energy optimization in multi-hop wireless embedded and sensor networks," *International Journal on Wireless Information and Networks*, vol. 12, no. 1, pp. 11–21, Jan. 2005.

[21] O. Oyman, J. Laneman, and S. Sandhu, "Multihop relaying for broadband wireless mesh networks: from theory to practice," *IEEE Commun. Mag.*, vol. 45, no. 11, pp. 116–122, Nov. 2010.

[22] S. K. Jayaweera, "Virtual MIMO-based cooperative communication for energy-constrained wireless sensor networks," *IEEE Trans. Wireless Commun.*, vol. 5, no. 5, pp. 984–989, May 2006.

[23] C. Schurgers, O. Aberthorne, and M. Srivastava, "Modulation scaling for energy aware communication systems," in *Proc. of Int. Symp. on Low Power Electronics and Design*, pp. 96–99, Aug. 2001.

[24] C.-L. Wang, Y.-W. Huang, and Y.-C. Huang, "An energy-efficient cooperative SIMO transmission scheme for wireless sensor networks," in *Proc. IEEE Int. Conf. on Communication* (ICC'09), pp. 1–5, June 2009.

[25] N. Krishnan and B. Natarajan, "Energy efficiency of cooperative SIMO schemes - amplify forward and decode forward," in *Proc. of Int. Conf. Computer Communications and Networks*, (ICCCN'09), pp. 1–5, Aug. 2009.

[26] G. G. D. O. Brante, M. T. Kakitani, and R. D. Souza, "On the energy efficiency of some cooperative and non-cooperative transmission schemes in WSNs," in *Proc. 45th Ann. Conf. Information Sciences and Systems* (CISS'11), pp. 1–6, Mar. 2011.

[27] I. Stanojev *et al.*, "Energy efficiency of non-collaborative and collaborative hybrid-ARQ protocols," *IEEE Trans. Wireless Commun.*, vol. 8, no. 1, pp. 326–335, Jan. 2009.

[28] D. Gunduz and E. Erkip, "Opportunistic cooperation by dynamic resource allocation," *IEEE Trans. Wireless Commun.*, vol. 6, no. 4, pp. 1446–1454, Apr. 2007.

[29] L. Simić, S. M. Berber, and K. W. Sowerby, "Partner choice and power allocation for energy efficient cooperation in wireless sensor networks," in *Proc. IEEE Int. Conf. Communication* (ICC'08), pp. 4255–4260, May 2008.

[30] L. Simić, S. M. Berber, and K. W. Sowerby, "Energy-efficiency of cooperative diversity techniques in wireless sensor networks," in *Proc. IEEE Int. Symp. Personal, Indoor and Mobile Radio Communication* (PIMRC'07) pp. 1–5, 2007.

[31] H. Naqvi, S. Berber, and Z. Salcic, "Energy efficiency of collaborative communication with imperfect frequency synchronization in wireless sensor networks," *International Journal of Multimedia and Ubiquitous Engineering*, vol. 5, no. 4, Oct. 2010.

[32] B. Zhao and M. C. Valenti, "Practical relay networks: a generalization of hybrid-ARQ," in *IEEE J. Select. Areas Commun.*, vol. 23, no. 1, pp. 7–18, Jan. 2005.

[33] A. Bletsas *et al.*, "A simple cooperative diversity method based on network path selection," *IEEE J. Select. Areas Commun.*, vol. 24, no. 3, pp. 659–672, Mar. 2006.

[34] Z. Huang, T. Yamazato, and M. Katayama, "Modulation scaling for energy aware communication systems," in *Proc. of Int. Conf. Intelligent Sensors, Sensor Networks and Information Processing* (ISSNIP'08), pp. 96–99, Dec. 2008.

[35] R. Irmer *et al.*, "Coordinated multipoint: concepts, performance, and field trial results," *IEEE Commun. Mag.*, vol. 49, no. 2, pp. 102–111, Feb. 2011.

[36] M. Sawahashi *et al.*, "Coordinated multipoint transmission/reception techniques for LTE-advanced [Coordinated and Distributed MIMO]," *IEEE Wireless Commun.*, vol. 17, no. 3, pp. 26–34, June 2010.

[37] E. Oh *et al.*, "Toward dynamic energy-efficient operation of cellular network infrastructure," *IEEE Commun. Mag.*, vol. 49, no. 6, pp. 56–61, June 2011.

[38] S. V. Hanly and P. Whiting, "Information-theoretic capacity of multi-receiver networks," *Telecommun. Syst.*, vol. 1, no. 1, pp. 1–42, 1993.

[39] A. D. Wyner, "Shannon-theoretic approach to a Gaussian cellular multiple-access channel," *IEEE Trans. Info. Theory*, vol. 40, no. 6, pp. 1713–1727, Nov. 1994.

[40] O. Somekh and S. Shamai, "Shannon-theoretic approach to a Gaussian cellular multiple-access channel with fading," *IEEE Trans. Info. Theory*, vol. 46, no. 4, pp. 1401–1425, Jul. 2000.

[41] Z. J. Haas and C.-P. Li, "The multiply-detected macrodiversity scheme for wireless cellular systems," *IEEE Trans. Veh. Technol.*, vol. 47, no. 2, pp. 506–530, May 1998.

[42] L. Welburn, J. K. Cavers, and K. W. Sowerby, "A computational paradigm for space-time multiuser detection," *IEEE Trans. Commun.*, vol. 52, no. 9, pp. 1595–1604, Sept. 2004.

[43] M. C. Valenti and B. D. Woerner, "Iterative multiuser detection, macrodiversity combining, and decoding for the TDMA cellular uplink," *IEEE J. Select. Areas Commun.*, vol. 19, no. 8, pp. 1570–1583, Aug. 2001.

[44] E. Aktas, J. Evans, and S. Hanly, "Distributed base station processing in the uplink of cellular networks," in *Proc. of IEEE Int. Conf. Communications*, (ICC '06.), vol. 4, pp. 1641–1646, June 2006.

[45] S. Bavarian and J. K. Cavers, "Reduced-complexity belief propagation for system-wide MUD in the uplink of cellular networks," *IEEE J. Select. Areas Commun.*, vol. 26, no. 3, pp. 541–549, Apr. 2008.

[46] V. Jungnickel *et al.*, "Interference aware scheduling in the multiuser MIMO-OFDM downlink," *IEEE Commun. Mag.*, vol. 47, no. 6, pp. 56–66, June 2009.

[47] R. Irmer *et al.*, "Multisite field trial for LTE and advanced concepts," *IEEE Commun. Mag.*, vol. 47, no. 2, pp. 92–98, Feb. 2009.

[48] P. Marsch, "Coordinated multi-point under a constrained backhaul and imperfect channel knowledge," Ph.D. thesis, 2010.

[49] A. Muller and P. Frank, "Cooperative interference prediction for enhanced link adaptation in the 3GPP LTE uplink, in *Proc. of IEEE Vehicular Technology Conf.*, (VTC - Spring'10), pp. 1–5, 2010.

[50] M. Chiang *et al.*, "Power control in wireless cellular systems," now Publishers Inc., 2008. [Online]. Available: www.princeton.edu/~chiangm/powercontrol.pdf

[51] S. Bavarian and J. K. Cavers, "Total power control for cooperative base stations uplink," in *Proc. of IEEE Vehicular Technology Conf.* (VTC- Spring'09), pp. 1–5, April 2009.

7 Effect of cooperation and network coding on energy efficiency of wireless transmissions

Nof Abuzainab and Anthony Ephremides

7.1 Introduction

There is a growing interest in studying and improving the energy efficiency of wireless transmissions to reduce CO_2 emissions and to combat climate change. One promising technique to reduce the energy consumption of wireless transmissions is the use of cooperation between the nodes of the network, as cooperation has been proven to achieve performance improvements in wireless networks [1]–[3]. Cooperation can be achieved by adding relays that have better link qualities with the destinations than the source node, and hence can assist the source in transmitting the information to the target destinations. Another form of cooperation is user cooperation. User cooperation works whenever a source node is multicasting packets to multiple destinations: the destinations that first receive the data successfully from the source can assist the source in transmitting the data to the remaining destinations. This form of cooperation is motivated by the fact that some destinations may have better channel quality than the source node due to the nature of wireless channels. Hence, it is anticipated that this method will decrease the total energy consumed by the network to deliver the required data.

Relay cooperation is expensive since new resources (i.e. the relays) are added to the network. In user cooperation, on the other hand, no extra resources are added, and hence it is less expensive than relay cooperation. However, obtaining better performance is not always guaranteed in user cooperation since the users that act as relays may not always have better link quality with the remaining users than the source. Hence, it is essential to study the cases in which user cooperation can achieve performance improvement compared to when no cooperation is used and design techniques that decide whether the source or the users should transmit based on the channel quality between the nodes in the network.

Further at the physical layer, some techniques have been designed to enhance the forwarding capabilities of the relaying node and to achieve more performance improvements when either user or relay cooperation is used. One of the main physical layer techniques for cooperation is the use of space-time codes and in particular Alamouti coding [4]. Alamouti coding is favorable because it allows the source node and the relaying node to transmit simultaneously without causing interference to each other. However, perfect synchronization between the source and the relaying node should be maintained to achieve the desired performance. Cooperative techniques that have used Alamouti coding were studied in [5]–[7]. In [5], amplify-and-forward (AF) and decode-and-forward

(DF) protocols based on Alamouti coding were proposed in a network consisting of a source, a relay, and a destination. The protocols were evaluated in terms of the outage probability. In [6], the symbol error probability (SEP) was studied in a network consisting of a single source, a destination, and two AF relays that use Alamouti coding. Then, the optimal power allocation that minimizes the SEP was computed. Further, recent studies have considered a joint physical and network layer framework for cooperative communications. In [7], automatic repeat request (ARQ) protocols were proposed for a two-user cooperative diversity system using Alamouti coding and the performance of these protocols was compared based on the system throughput.

More recent work has focused on finding which cooperative schemes are energy efficient. In [8], a wireless fading network consisting of a single source, a single destination, and N relays was considered. The trade-off between decreasing the overhead of obtaining the channel state information (CSI) by using fewer relays and decreasing the energy consumption was shown. In [9], an energy-efficient cooperative scheme was proposed in a wireless sensor network where the cooperating nodes employ Alamouti coding, and it was shown that under certain distance ranges between the nodes, the energy of the cooperative scheme is reduced compared with non-cooperative schemes.

At the network layer, network coding, originally proposed in [10], is an emerging communication concept that has achieved high improvements in terms of throughput and energy efficiency in wireless networks, especially in multicasting [11]–[13], and thus it is important to examine its effect on the network performance and to incorporate it in the design of energy-efficient systems. Further, network coding has recently been incorporated into cooperation schemes to improve network performance. In [14], a cognitive and cooperative scheme was proposed in a network composed of a single source, a relay, and a destination in which the relay transmits during the source's silent periods and uses random network coding (RNC) on the packets it receives. It was shown that the stable throughput at the source using this scheme is increased compared to ARQ-based protocols. In [14], a network composed of a source, a relay, and destination was considered and it was shown that cooperation using network coding at the relay increases the source's stable throughput.

The objective of this chapter is to present different physical and network layer cooperative techniques for wireless fading transmissions and to evaluate their energy efficiency. Two models for wireless transmissions are considered. The first model considers transmissions over a wireless link, and a relay is used to assist the source node to deliver its data to the destination node. The second model considers multicast transmissions in which the source node is multicasting its data to two destinations. In this case, user cooperation is utilized, i.e. the destination node that first receives the data successfully can assist the source in transmitting the data to the remaining destination. In both cases, a packetized system is considered in which packets arrive at the source with a certain rate and are then stored in a queue. The energy efficiency of each cooperative technique is determined by computing the minimum energy consumed per successfully delivered packet. Also, the maximum stable throughput achieved at the source is computed. Further, this chapter investigates whether using Alamouti coding at the physical layer or using network coding at the network layer can achieve better performance.

The rest of the chapter is organized as follows: Section 7.2 presents the system model and explains the cooperative techniques used for the case of transmission over a wireless link. Section 7.3 presents the system model and the cooperative techniques used for the case of multicast wireless transmission. Section 7.4 defines the energy cost and the method to minimize it. Section 7.5 evaluates the maximum stable throughput achieved for every cooperative protocol. Section 7.6 presents some performance evaluation results. Finally, conclusions are drawn in Section 7.7.

7.2 Relay cooperation in single link wireless transmissions

7.2.1 System model

This part considers transmissions over a wireless Rayleigh fading link in which a source node is required to deliver packets to the destination node in the form of packets. Each packet is composed of N symbols (N is fixed for all packets). Packets arrive at the source according to a Bernoulli process with rate λ. A relay node is present that assists the source node in delivering the packets to the destination node, and it is assumed that the channel quality between the relay and the destination is better than the channel quality between the source and the destination. The relay employs a store-and-forward (SF) protocol, i.e. the relay should successfully decode the packet before transmitting it to the destination. Half-duplex transmissions are assumed, i.e. a node cannot transmit and receive at the same time. Time is slotted, and in each time slot the source node can transmit a packet to the destination node. The values of the packet size and the time slot duration are fixed prior to transmission. Also, in every time slot, transmission occurs over a block fading channel, i.e. the channel gain does not change within a time slot duration. Additive white Gaussian noise (AWGN) is present at both the relay and the destination.

Due to fading and the presence of noise at the destination, the packet can be received only with a certain probability. This probability is given by the probability that the signal-to-noise ratio (SNR) of the received packet exceeds the required threshold at the receiving node (i.e. the destination or the relay). The required threshold depends on the communication parameters such as the transmission rate, the target probability of error, the modulation and coding scheme used for communication; we do not analyze this dependence but rather assume a given probability of success for a given threshold. Figure 7.1 shows the system model where h_1, h_2, and h_3 are the channel gains between the source and the destination, between the source and the relay, and between the relay and the destination, respectively.

Hence to transmit the packets reliably, the transmitting node (i.e. the source or the relay) uses either one of the following methods. The first is a simple automatic repeat request (ARQ), i.e. the source selects the packet at the head of the queue and keeps transmitting the same packet in the subsequent time slots until it is received successfully by the destination node, which then sends an acknowledgment packet. The second is random network coding (RNC), i.e. the source selects a predetermined number of packets (this number is fixed during all transmissions) from the head of the queue and transmits

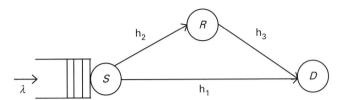

Figure 7.1   Transmission over a wireless link.

in every time slot a random linear combination of the selected packets. Then, the source keeps transmitting random linear combinations of the same group of packets until the destination can successfully decode the packets. Successful decoding occurs when the destination has received enough linearly independent combinations of the current transmitted group of packets. Once the destination decodes the group of packets successfully, it sends an acknowledgement packet. In both cases of ARQ and RNC, perfect channel feedback is assumed, i.e. acknowledgments are received instantaneously and they are error-free.

Although there are other more effective and sophisticated methods used for reliable data delivery than simple ARQ (such as HARQ), the focus in this part is to investigate the effect of random network coding on the energy efficiency of the transmission and hence simple ARQ is used as a base line for comparison with RNC. Also, each of the source and the relay can transmit with a power value over the interval $[0, P_{max}]$, where P_{max} is the maximum transmission power.

Now we explain the cooperation protocols used for the above setting.

7.2.2   Cooperation protocols

To transmit the packets reliably from the source to both destinations D_1 and D_2, we consider the following cooperation protocols.

- *Plain relaying (PR) using ARQ*: In this case, the source transmits each packet using ARQ until either the destination or the relay receives the packet. If the destination receives the packet successfully, the transmission is completed, and the source starts transmitting the next packet. If the relay successfully receives the packet before the destination, the relay transmits the packet using ARQ to the destination until the destination receives the packet successfully. Using this scheme, the received SNR values are given as follows:
i) from the source to the destination:

$$SNR_{SD} = \frac{|h_1|^2}{N_0} P_1, \qquad (7.1)$$

ii) from the source to the relay:

$$SNR_{SR} = \frac{|h_2|^2}{N_0} P_1, \qquad (7.2)$$

Figure 7.2 The structure of the original packet and the encoded packet, respectively.

iii) from the relay to the destination:

$$SNR_{RD} = \frac{|h_3|^2}{N_0} P_2, \quad (7.3)$$

where N_0 is the power spectral density of the AWGN at both the relay and the destination. The variables P_1 and P_2 are the values of the transmission power of the source and the relay, respectively.

- *Relaying with Alamouti coding (AC) using ARQ*: The first stage of this protocol is similar to plain relaying, i.e. the source transmits the packet using ARQ until either the destination or the relay receives the packet. If the relay receives the packet successfully before the destination, it forms an encoded packet by applying Alamouti coding to every pair of consecutive symbols of the original packet. The structures of the original packet and the Alamouti-coded version of the packet are shown in Figure 7.2, where x_i is the ith symbol of the original packet and x_i^* is the complex conjugate of x_i. Note that in this case we assume that the symbols of the packet are mapped to a complex constellation.

After the relay forms the encoded packet, both the source and the relay transmit in the next time slot where the source transmits the original packet, and the relay transmits the encoded packet until the destination receives the packet successfully.

In this stage, although both the sender and the relay transmit simultaneously, they do not interfere with each other. This is because Alamouti coding [4] constructs a packet that is orthogonal to the original packet.

Assuming channel estimation is performed at the receiver, the decoding process of the transmitted signals using Alamouti coding is similar to maximum ratio combining (MRC) as proved in [4]. Hence the signal-to-noise ratio at the destination in the cooperation phase is

$$SNR_{AC} = \frac{|h_1|^2 P_1 + |h_2|^2 P_2}{N_0}. \quad (7.4)$$

The expressions for SNR_{SD} and SNR_{SR} in the non-cooperative phase are the same as given by equations (7.1) and (7.2).

- *Plain relaying with random network coding*: In this case, the source transmits random linear combinations of every group of L packets (L is determined prior to transmission) until either the destination or the relay successfully decodes the L packets.

If the destination successfully decodes the L packets before the relay, transmission is considered to be successful, and the source starts transmitting the next group of L packets. If the relay successfully decodes the L packets before the destination, it starts transmitting the L packets to the destination using RNC until the destination successfully decodes the L packets. The destination uses the previously successfully received random linear combinations directly from the source along with the new ones generated by the relay to perform its decoding. Using this scheme, the received SNR expressions are identical to the case of plain relaying using ARQ but this time they apply to the coded packets.

- *Relaying using Alamouti coding with pseudo random network coding*: When using Alamouti coding in the cooperation phase, the relay should transmit the Alamouti-coded version of the packet transmitted by the source. Therefore, using Alamouti coding in conventional random network coding is not feasible because the source and the relaying node independently select a different random linear combination in every time slot. Hence, in order to be able to use Alamouti coding with random network coding, we suggest the following scheme: the source starts transmitting random linear combinations of every group of L packets until either the relay or the destination decodes the L packets successfully. If the relay decodes the L packets before the destination, then in every subsequent time slot, the source forms a new random linear combination and sends the coefficients to the relaying node in order to form the same linear combination. Then, the relay forms the Alamouti-coded version of the network-coded packet and subsequently the source and the relay transmit simultaneously to the destination. This process is repeated in every time slot until the destination node decodes the L packets successfully.

  Although this scheme has an additional overhead (because the source transmits the coefficients of the random linear combination to the relay), it is interesting to study its effect on the performance achieved.

  In this case, SNR_{SD} and SNR_{SR} in the non-cooperation phase, and SNR_{AC} in the cooperation phase are again given by equations (7.1), (7.2), and (7.4), respectively.

7.3 User cooperation in wireless multicast transmissions

7.3.1 System model

In this part, we consider a source node multicasting packets to two destinations, D_1 and D_2, over a single-hop wireless network. Packets arrive at the source according to a Bernoulli process with rate λ. Again, the channels between each pair of nodes are assumed to be independent time-invariant Rayleigh block fading, and additive white Gaussian noise is present at each of the destination nodes. Thus, the packet will be received successfully with a certain probability, which is given by the probability that the received SNR at the destination node exceeds the required threshold. Also, half-duplex transmissions are assumed, i.e. each node cannot transmit and receive at the same time. Figure 7.3 shows the system model where the variables h_1 and h_2 are the

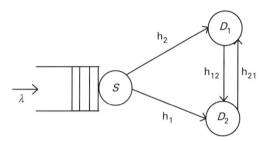

Figure 7.3 Multicast transmission.

channel gains between the source and destinations D_1 and D_2, respectively. The variable h_{ij} ($i \neq j, i, j \in \{1, 2\}$) is the channel gain between destination D_i and D_j.

In this part, user cooperation is considered, i.e. the destination that first receives the information successfully will act as a relay and assists the source in transmitting the information to the remaining destination.

As in the previous part to transmit the packets reliably, the transmitting node will use either simple ARQ or RNC. Perfect channel feedback is assumed, i.e. acknowledgments are received instantaneously and error free. Also, the source node and destinations D_1 and D_2 transmit with power values P_s, P_1, and P_2, respectively, where P_s, P_1, and P_2 $\in [0, P_{max}]$.

Now we explain the cooperation protocols used for this setting.

7.3.2 Cooperation protocols

To transmit the packets reliably from the source to both destinations D_1 and D_2, we consider the following cooperation protocols.

- *Plain relaying using ARQ*: The source transmits each packet until either of the destinations D_1 or D_2 receives the packet. If both destinations receive the packet in the same time slot, transmission is successful, and the source starts transmitting the next packet. If only one of the destinations receives the packet successfully, this destination transmits the packet using ARQ to the remaining destination. Using this scheme, the received SNR values are given as follows:
  i) from the source to destination D_1:

$$SNR_1 = \frac{|h_1|^2}{N_0} P_s, \quad (7.5)$$

ii) from the source to destination D_2:

$$SNR_2 = \frac{|h_2|^2}{N_0} P_s, \quad (7.6)$$

iii) from destination D_i to destination D_j ($i, j \in \{1, 2\}, i \neq j$):

$$SNR_{ij} = \frac{|h_{ij}|^2}{N_0} P_i, \quad (7.7)$$

where N_0 is the power spectral density of the AWGN at both destinations D_1 and D_2.
- *Relaying with Alamouti coding (AC) using ARQ*: The source transmits the packet using ARQ until either of the destinations receives the packet. If both destinations receive the packet in the same time slot, transmission is successful, and the source starts transmitting a new packet. If only one of the destinations receives the packet successfully, it forms an encoded packet by applying Alamouti coding to every pair of consecutive symbols of the original packet. Then, both the source and this destination transmit in the next time slot where the source transmits the original packet, and the destination transmits the encoded packet until the remaining destination receives the packet successfully.

The received SNR at destinations D_1 and D_2 when the source is transmitting in the non-cooperative phase are the same as the expressions given by equations (7.5) and (7.6).

During the cooperation phase, the SNR at destination D_1 is

$$SNR_{AC} = \frac{|h_1|^2 P_s + |h_{21}|^2 P_2}{N_0}. \tag{7.8}$$

The SNR at destination D_2 in the cooperation phase is

$$SNR_{AC} = \frac{|h_2|^2 P_s + |h_{12}|^2 P_1}{N_0}. \tag{7.9}$$

- *Plain relaying with random network coding*: In this case, the source transmits random linear combinations of every group of L packets (L is determined prior to transmission) until either of the destinations decodes the L packets. If both destinations successfully decode the L packets in the same time slot, transmission is successful and the source starts transmitting the next group of L packets. If only one of the destinations successfully decodes the L packets, it starts transmitting the L packets to the remaining destination using RNC until the remaining destination successfully decodes the L packets. The remaining destination retains the coded packets that were received successfully from the source's transmissions. The received SNR expressions are identical to the case of plain relaying using ARQ but this time they apply to the coded packets.
- *Relaying using Alamouti coding with pseudo-random network coding*: Similar to the case of wireless unicast transmission, we propose a scheme that combines Alamouti coding with random network coding. The scheme works as follows: The source starts transmitting random linear combinations of every group of L packets until one of the destinations decodes the L packets successfully. If both destinations decode the L packets in the same time slot, transmission is successful, and the source starts transmitting the next group of L packets. If only one of the destinations decodes the L packets successfully, in every subsequent time slot, the source forms a new random linear combination and sends the coefficients to this destination node to form the same linear combination. Then, the destination forms the Alamouti-coded version of the packet, and subsequently, the source and the transmitting destination node

transmit simultaneously to the remaining destination. This process is repeated until the remaining destination decodes the L packets successfully. In this case, the SNR expressions are the same as in the case of relaying using Alamouti coding with ARQ.

The following section defines the energy cost used to evaluate each of the cooperation protocols. It also presents the method of minimizing the energy cost for each of the considered protocols.

7.4 Energy-cost minimization

For both models considered (i.e. unicast transmissions and multicast transmissions), the energy cost is defined as the expected energy spent per successfully delivered packet. Analytic expressions can be easily obtained for the case when ARQ is used as a transmission scheme. This follows from the fact that the times that each of the nodes receives the packet successfully from the source's transmission are independent and geometrically distributed. However, for the case when RNC is used, the times that the nodes decode the current delivered group of packets from the source's transmissions are correlated. This is because the source is transmitting the same linear combinations to both of the receiving nodes, and hence the numbers of linearly independent packets received by each of the receiving nodes from the source are correlated. Thus, evaluating the energy cost from the joint distribution of the times that each of the nodes decodes the L packets successfully from the source's transmission is more complicated than in the case of ARQ.

Thus, the computation of the energy cost for the case when RNC is used is done through a Markov-chain model that keeps track of the number of linearly independent coded packets received by each of the receiving nodes, as well as the linearly independent packets received by both of them.

After evaluating the cost functions for the different cooperation protocols, the objective is to find the optimum transmission power values of the nodes for each of the cooperation strategies that minimize their corresponding cost and to find the channel conditions under which each performs best.

However, since the cost functions have complicated structures and in the case of RNC does not have a closed-form expression, numerical calculation is used to determine the values of the powers that minimize the cost functions. This is achieved by generating "dense" sets of power values over the interval $[0, P_{max}]$. Then for every power value, the cost function for every cooperation scheme is computed. Finally, the power values that correspond to the lowest cost are selected.

7.5 Stable throughput computation

The objective of this part is to investigate the effect of minimizing energy on the achieved maximum stable throughput. Hence, the optimal power values that minimize the energy cost are used to compute the maximum stable throughput achieved for each of the cooperation protocols. In this setting, since there is only one source, and given that the service and arrival processes are jointly stationary, stability is achieved [16] if and only if

$$\lambda_s < \mu_s, \quad (7.10)$$

where λ_s is the arrival rate of the source and μ_s is the service rate.

The service rate is given by the reciprocal of the expected completion time of the successful transmission of the current delivered packet when ARQ is used as a transmission scheme and is given by the ratio of the network coding parameter, L, over the completion time of the successful transmission of the current delivered L packets.

7.6 Performance evaluation

7.6.1 Relay cooperation

In this section, we present some numerical results that illustrate the effect of varying the channel conditions on the performance of each of the cooperation protocols for the case of single link wireless transmissions. Since there are three different channels in the network, to limit the number of variables in the analysis, we vary the channel quality between the source and the destination while fixing the quality of the remaining channels. Thus, the channel quality between the source and the destination is varied by varying the variance s_1 of the Rayleigh fading distribution between the source and the destination. The variance is varied between 10 and 100 (it can be shown that the higher the value, the better the channel quality) while the values of the variances s_2 and s_3 of the fading distributions corresponding to the channels between the source and the relay, and between the relay and the destinations, respectively, are kept fixed at 100, and the optimal cost for each cooperation protocol is computed for every considered value of s_1. The optimal costs for every cooperative protocol as a function of the variance s_1 are shown in Figure 7.4. Also, the optimal power values obtained are used to compute the service rate for every cooperation protocol. The results are shown in Figure 7.5.

Figures 7.4–7.5 first show that using Alamouti coding with ARQ achieves a higher service rate and consumes less energy per successfully transmitted packet compared to the case when ARQ is used with plain relaying; this is because when Alamouti coding is used in the cooperation phase, the probability that the destination receives the packet successfully (from the simultaneous transmission of the source and the relay) is higher than the case when plain relaying is used with ARQ, since in the case of plain relaying, the relay will only be transmitting the packet (during the cooperation phase) to the destination.

Also, Figures 7.4–7.5 show that as the network coding parameter, L, increases, the service rate increases and using RNC becomes more energy efficient. This is because when the relay receives the packet successfully from the source's transmission using ARQ, it should start transmitting the packet again to the destination. While in the case of RNC, even though the destination may not have successfully decoded the L packets when the relay successfully decodes the L packets, it may have successfully received linearly

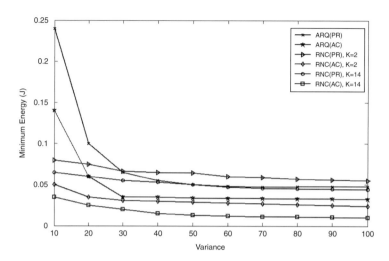

Figure 7.4 Optimal cost for each cooperation scheme as a function of the variance s_1.

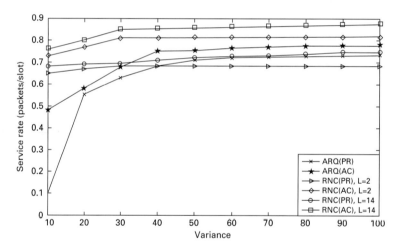

Figure 7.5 Service rate (in packets/slot) for each cooperation scheme as a function of the variance s_1.

independent packets from source's transmission, and thus the relay should not retransmit the L packets again but only enough random linear combinations of the currently delivered L packets until the destination successfully decodes the L packets. This results in decreasing the total number of time slots required per successfully delivered packet, and the performance of RNC becomes better even than the case when ARQ is used with Alamouti coding. Further, more energy reductions are observed when Alamouti coding is used with random network coding.

Hence under ARQ, relaying using Alamouti coding reduces the energy expenditure and gives a higher stable throughput at the source than in the case when plain relaying is used. Also, by properly selecting the value of the network coding parameter L, cooperation using RNC achieves the highest stable throughput at the source and the lowest energy expenditure scheme among the communication protocols.

7.6.2 User cooperation

In this section, we will investigate the effect of the channel conditions on each of the cooperation protocols for the case of user cooperation in wireless multicast. Hence, the channel qualities between the destinations D_1 and D_2 are varied simultaneously while fixing the quality of the remaining channels.

The channel qualities between the destinations D_1 and D_2 are varied by simultaneously varying the values of the variances s_{12} and s_{21} for the Rayleigh distribution of the channel between destination D_1 and destination D_2 and the channel between destination D_2 and D_1. The values of the variances are varied between 10 and 150, while the variance values of the Rayleigh distribution of the channel between the source and destinations D_1 and D_2 are kept fixed at 50, and the minimum energy for each cooperation protocol is computed for every considered value of s_{12} and s_{21}. Also, to determine when user cooperation achieves better performance, we consider the case when no cooperation is used, i.e. the source is the only transmitting node in the network. Then, the optimal costs for every transmission scheme as a function of the value of the variances are computed and are shown in Figure 7.6. Also, the optimal power values obtained are used to compute the service rate for every protocol. The results are shown in Figure 7.7. In Figure 7.7, NC refers to the scheme when no cooperation is used.

Figures 7.6–7.7 show that using Alamouti coding with ARQ achieves higher service rate and consumes less energy per successfully transmitted packet compared to the other strategies. As the channel quality between the destinations becomes higher and as the network coding parameter, L, increases, the service rate increases and using RNC becomes

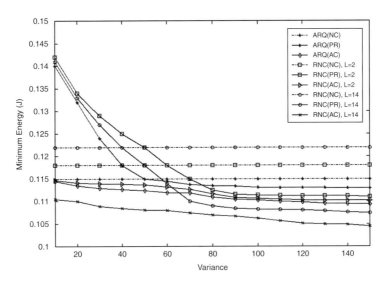

Figure 7.6 Optimal cost for each cooperation scheme as a function of s_{12}/s_{21}.

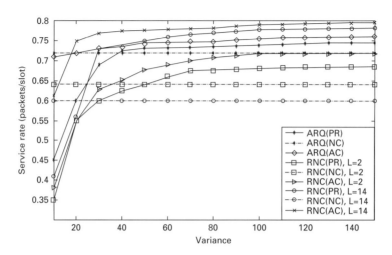

Figure 7.7 Service rate (in packets/slot) for each cooperation scheme as a function of s_{12}/s_{21}.

more energy efficient even than the case when ARQ is used with Alamouti coding. Also, in the case of wireless multicast, for certain values of the coding parameter, L, random network coding combined with Alamouti coding achieves the best performance.

7.7 Conclusion

This chapter focuses on joint physical and network layer cooperative techniques applied on two scenarios. The first is cooperative relaying transmission over a wireless link. The second is user cooperation in wireless multicast over a single-hop network. In both cases, channels are assumed to be time-invariant. The cooperative techniques are evaluated by finding the optimal power values that minimize the energy consumed per successfully delivered packet, and then using the optimal power value to find the maximum stable throughput.

Also, this chapter investigates the effect of random network coding at the network layer and the effect of Alamouti coding at the physical layer on the performance of the cooperative techniques. The results show that by properly choosing the coding parameter, the random network coding-based cooperative technique achieves better performance than ARQ-based cooperative technique even when it is enhanced with Alamouti coding. Also, further improvements in the performance are achieved when random network coding is used in combination with Alamouti coding.

Further, the chapter confirms that user cooperation does not always achieve performance improvement, and that the performance of user cooperation depends on the channel quality between the different nodes in the network.

This work is supported by MURI Grant W911NF-08-1-0238, by ONR Grant N000141110127 and by NSF Grant CNS1147730.

References

[1] A. Sendonaris, E. Erkip, and B. Aazhang, "User cooperative diversity-part I: system description," *IEEE Transactions on Communications*, vol. 51, no. 11, pp. 1927–1938, 2003.

[2] J. Laneman, D. Tse, and G. Wornell, "Cooperative diversity in wireless networks: efficient protocols and outage behavior," *IEEE Transactions on Information Theory*, vol. 50, no. 12, pp. 3062–3080, 2004.

[3] J. Laneman and G. Wornell, "Distributed space-time coded protocols for exploiting cooperative diversity in wireless networks," *IEEE Transactions on Information Theory*, vol. 49, no. 10, pp. 2415–2425, 2003.

[4] S. M Alamouti, "A simple transmit diversity technique for wireless communications," *IEEE Journal on Selected Areas in Communications*, 1998.

[5] C. Hucher, G. R. Rekaya, and J. Belfiore, "AF and DF protocols based on Alamouti ST code," *IEEE International Symposium on Information Theory*, pp. 1526–1530, 2007.

[6] T. Duong *et al.*, "On the symbol error probability of distributed-Alamouti scheme," *Journal of Communications*, vol. 4, no. 7, 2009.

[7] C. Zhang, W. Wang, and G. Wei, "Design of ARQ protocols for two-user cooperative diversity systems in wireless networks," *Journal of Computer Communications*, vol. 32, no. 6, pp. 1111–1117, 2009.

[8] R. Madan *et al.*, "Energy-efficient cooperative relaying over fading channels with simple relay selection," *IEEE Transactions on Wireless Communications*, vol. 7, no. 8, 2008.

[9] Z. Zhou *et al.*, "Energy-efficient cooperative communication based on power control and selective relay in wireless sensor networks," *IEEE transactions on Wireless Communications*, vol. 7, no. 8, pp. 3066–3078, 2008.

[10] R. Ahlswede *et al.*, "Network information flow," *IEEE Transactions on Information Theory*, vol. 46, no. 4, 2000.

[11] T. Ho *et al.*, "On randomized network coding," in *Proc. of 41st Annu. Allerton Conf. Communication, Control, and Computing*, Monticello, IL, Oct. 2003.

[12] T. Ho *et al.*, "A random linear network coding approach to multicast," *IEEE Trans. Inf. Theory*, vol. 52, no. 10, 2006.

[13] X. Tao, C. Zhang, and J. Lu, "Network coding for energy efficient wireless multimedia transmission in ad hoc network," in *Proc. of International Conference on Communication Technology*, pp. 1–5, 2006.

[14] A. Fanous and A. Ephremides, "Network-level cooperative protocols for wireless multicasting: stable throughput analysis and use of network coding," in *Proc. IEEE Information Theory Workshop (ITW)*, Dublin, August 30–September 3, 2010.

[15] P. Fan *et al.*, "Reliable relay assisted wireless multicast using network coding," *IEEE Journal on Selected Areas of Communications*, vol. 27, pp. 749–762, 2009.

[16] R. Loynes, "The stability of a queue with non-independent inter-arrival and service times," *Mathematical Proceedings of the Cambridge Philosophical Society*, vol. 58, pp. 497–520, 1962.

Part III

Base station power-management techniques for green radio networks

8 Opportunistic spectrum and load management for green radio networks

Oliver Holland, Christian Facchini, A. Hamid Aghvami,
Orlando Cabral, and Fernando Velez

8.1 Introduction

Historically, the radio spectrum has been managed in a rather rigid fashion where systems have been constrained to very specific bands in order to avoid interference and maintain the spectrum's viability. This regime is extremely inefficient, because at any one time many systems are not being used thereby leaving their associated spectrum also unused. Alternative spectrum management, where systems not designated for a particular band may nevertheless use it if it is available, would greatly increase spectrum usage efficiency and capacity.

Communications traffic has also historically been managed in a somewhat inefficient manner, whereby traffic load has usually only been carried on a specific band as directed by the "owner" of the user/device carrying the traffic. Improved traffic-load management techniques, where the traffic can be shared among bands and systems, would also increase efficiency or capacity. Although the end-user may sometimes have a limited choice of which band to receive traffic on (e.g. via a Wi-Fi interface using an ISM/UNII band, or via a 3G mobile communications interface using a UMTS band), centralized control of that choice, in a timely fashion, can far better manage efficiency and capacity than the end-user operating alone.

Such opportunistic load and spectrum management between bands/systems is being made feasible by operators having an increasingly wide range of spectrum bands at their disposal, of very different frequencies and physical characteristics. Operators may typically operate a range of different systems on this range of spectrum bands. Even considering a single-operator case, already in most major cities around the world some operators are concurrently providing services in multiple bands, such as GSM 900 MHz and 1800 MHz, UMTS 2 GHz, and 2.4 GHz Wi-Fi services [1]. The introduction of IMT-Advanced bands in the short-to-medium term future will further increase the range of spectrum bands available to the operator: There are many such bands widely identified, some examples being 450–470 MHz, 790–862 MHz, and 2.3–2.4 GHz [2]. The greater use of unlicensed spectrum such as UNII bands (5.15–5.825 GHz), will also facilitate better spectrum availability, and hence more opportunistic use of spectrum or traffic load sharing between bands.

Other recent developments, from both regulatory and technical viewpoints, are also facilitating freer use of the spectrum, as well as the sharing of traffic loads among spectrum bands. Some examples here include innovative technical paradigms for spectrum access [3], novel regulatory ways of managing spectrum which allow spectrum sharing [4, 5], and pioneering communications network management techniques [6]. Such developments will in many cases consign the current status quo, where a particular communications technology is used only within a particular spectrum band or where traffic load is constrained to a designated system/band, to history. Alternatively, operators and collaborations of operators, as well as to some extent independent devices and systems, will be able to use and share their range of spectrum bands for whichever technologies they see fit. This is, of course, within constraints such as a maximum allowable transmission power. Technical developments such as software-defined radio and other forms of adaptive radio are also facilitating spectral freedom, leading to scenarios where operators and devices/systems can adapt radio access technologies (RATs) to requirements in specific bands, ultimately even being able to dynamically and autonomously custom-engineer RATs on-the-fly to optimally fit the used spectrum band and purpose.

There is clear evidence emerging that carbon emissions, and hence the associated energy consumption, must be reduced to save the planet. Carbon emissions associated with mobile communications can be quite significant; a major operator covering the UK, for example, emits over 200,000 tonnes of CO_2 per year, the biggest contributor to which is the 35 MW consumed in operating the network, where the majority of this (some 26 MW) is consumed by base stations (BSs) [7]. As in all areas of technology, it is necessary to reduce these emissions by as much as practically possible. Moreover, the means of reducing carbon emissions through power-consumption reduction can significantly reduce operational expenditure, which is beneficial to operators from the financial as well as the "corporate responsibility" perspective.

To reach such ends, this chapter investigates the novel concepts of opportunistic spectrum and load management among the range of spectrum bands available to the operator or a collaboration of operators, the objective of which is to reduce power consumption without impacting QoS. The work concentrates on improving the efficiency of radio access networks and particularly reducing the power consumption of BSs, given that this is where the majority of the power is consumed within the network. The context in which this work is placed is that where an operator or a collaboration of operators operates many systems and has multiple spectrum bands available, spaced significantly in the frequency domain. It is emphasized again that such a scenario, where there are many networks/bands covering the same area with systems operated by a single operator, is already routinely the case in many areas/countries and the proliferation of used systems and available bands will further increase in the future.

The rest of this chapter is organized as follows. In the next section, the concepts that are leveraged are explained. Section 8.3 investigates aspects of the performances of the proposed schemes, showing significant potential for power consumption reduction. Finally, Section 8.4 concludes this chapter.

8.2 Opportunistic spectrum and load management concepts

A range of opportunistic spectrum and load management techniques are proposed in this chapter to save energy for operators' systems. They comprise: (i) the opportunistic moving of traffic loads into particularly active bands from other bands, through the sharing of those particularly active bands and the associated radio network equipment, allowing radio network equipment operating in the other bands to be switched off or put into stand-by mode when possible; (ii) the opportunistic moving of traffic loads between bands, or opportunistic spectrum usage, to take advantage of more appropriate propagation bands and reduce necessary transmission power; and (iii) the sharing of spectrum to allow channel bandwidths to be increased or better "balanced," thus allowing the transmission power to be significantly decreased. It is noted that the schemes and assessment reported in this chapter are a considerable advancement on the work reported in [8].

8.2.1 Opportunistic load management to power down radio network equipment

The switching off (or entering into stand-by) of radio equipment through reallocating users or traffic loads to other bands at times of low load, illustrated in Figure 8.1, is extremely promising as it implies a guaranteed power saving through radio equipment being virtually "switched off at the mains." It is noted that for macro-cell BSs in particular, by far the biggest contribution to power consumption of the BS is it merely being switched on and in an operational state; the variation of power consumption with transmission power is relatively less significant, although such variation depends greatly on the exact manufacture of the BS hence can only be very broadly generalized. The opportunistic powering down of radio network equipment based on solutions such as presented in this chapter is a readily achievable way of reducing actual from-the-mains power consumption.

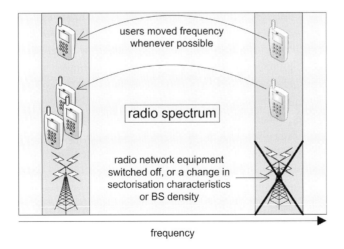

Figure 8.1 Opportunistic load management to power down radio network equipment.

There are two possibilities concerning the dynamic powering down of radio equipment considered in this chapter: (i) turning off cells entirely in one network or spectrum band at that time/location through traffic being sufficiently carried by a single network or spectrum band, and (ii) using the spare capacity of one network/band to cover the required drop in load of another network/band to enable that other network/band to operate in omnidirectional mode instead of tri-sectored mode. It should be noted, however, that the former scheme is also extended to consider cases where it is possible to not only carry all users in one band, but also to switch off every other cell within that remaining used band when traffic load is low enough. This can be implemented in cases where there is a sufficient surplus allowable (unused) transmission power at the BSs, as might be the case in urban scenarios where cells are tightly packed. Such a concept might also be implementable in conjunction with the opportunistic usage of better propagation spectrum to save power.

Given the assumption that each active radio chain consumes the same from-the-mains power, the latter sectorization removal (solution (ii), mentioned above) would save 50% of the overall BS power consumption at the two spectrum bands, for the case where traffic load is reallocated to enable two out of three radio chains to be switched off at the sectorized network/band while the one radio chain at the non-sectorized network/band remains switched on. For the case where both bands were operating in sectorized mode before the schemes were employed, the scheme would reduce six operational radio chains (three at each band) to four operational radio chains (one remaining operational at the band that users were reallocated from, and the three remaining operational at the other band), thus saving 33% power. Considering solution (i) mentioned above, switching off one of two networks/bands by collating all users into the other network/band saves 50% of the power consumption at the times when it can be implemented. The ability to switch off every other cell would save an additional 25% compared with the case where the networks were both operating in omnidirectional mode beforehand. Considering both schemes in operation together (i.e. solution (i) and solution (ii)), the maximum power saving that is implied is some 91.6%, reflecting the situation where both bands were operating in tri-sectored mode before implementation of the power-saving solution, meaning that six radio chains in total were active. If traffic considerations were to allow all of the load to be sufficiently carried by a single band operating in omnidirectional mode, the six active radio chains would be reduced to one, and if traffic load were further reduced and spectrum propagation or the network power budget allowed, it would be possible for only every other cell to be switched off while sufficiently carrying traffic in the remaining band. This gives a saving of eleven out of twelve radio chains, i.e. 91.6%. However, the opportunities where such significant savings are likely to be possible are very few and far between.

It is noted that these schemes might be employed on a macro-scale taking advantage of variations in loads at certain times of the day over large areas, or might be employed on a micro-scale also taking advantage of statistical variations in traffic loads in individual cells or small groups of cells. The latter of these solutions presents better overall power-saving performance, although of course it introduces network coordination challenges, for example, in terms of frequency reuse. In this context and others, we observe that

8.2 Opportunistic spectrum and load management concepts

for such radio network equipment dynamic powering down solutions to be possible, centralized management by the operator or a collaboration of operators of the range of networks/frequencies is required, as well as means of informing devices dynamically of the altered range of connectivity options. Solutions for conveying information about changed connectivity options have been under consideration/development by industry, academias and regulators for some time; one possibility is the cognitive pilot channel (CPC) concept [9]. The CPC, at least in the "in-band" form, would be very easy to implement. As regards centralized management, one viable solution aimed at precisely the kinds of spectrum/network management scenarios considered in this chapter, which also considers/maintains QoS in the resource management process, is the IEEE 1900.4 standard architecture [10].

A further consideration in the implementation of this concept is that of how to select the band at which to turn radio equipment off. In the powering down of cells, if one band supports the total traffic load (or total number of users) at both bands but the other band doesn't, then clearly users should be moved to the higher-capacity band such that equipment at the lower-capacity band can be switched off. Alternatively, if both bands can support the total number of users, then the band with the higher number of users already present might remain on, such that the number of users to be reallocated between the bands, hence complexity, is reduced. In the case of sectorization removal, if one band is operating in omnidirectional mode but the other is in sectorized mode, clearly users should always be moved from the sectorized band to the omnidirectional band such that sectorization in the sectorized band can be removed. If both bands were operating in sectorized mode, the consideration in deciding which band to switch to omnidirectional mode would, again, most likely be the minimization of the number of users moved between bands.

Another consideration in the decision of which band to remove users/links from, hence which band to switch off is, of course, the relative power consumptions of bands. It is noted that, in many cases, these bands may be at considerably different frequencies, in which case it would generally be preferable to switch off the higher-frequency band due to the higher power consumption of such a band. Such a decision must, again, be made by the centralized management entity on a case-by-case basis.

8.2.2 Opportunistic spectrum management to improve propagation characteristics

The opportunistic reallocation of links or users to lower-frequency spectrum bands at times when that spectrum becomes available, illustrated in Figure 8.2, decreases the necessary transmission power due to improved propagation. Alternatively, in a frequency reuse scenario, the opportunistic use of spectrum bands with more appropriate propagation characteristics based on the user density and hence the necessary cell density, as well as the local propagation environment, can reduce inter-cell interference because there is less power "leaking" into the cells that are operating on a co-channel basis. Note that in the former case, such a concept might be employed in tandem with the above-mentioned powering down of radio network equipment, through the higher frequency always being powered down and its users being reallocated to the lower frequency,

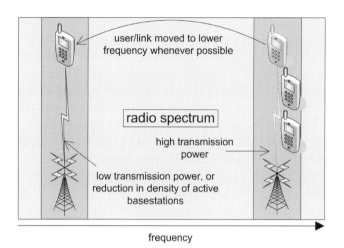

Figure 8.2 Opportunistic spectrum management to improve propagation characteristics.

thereby allowing the network at the high-frequency band being switched off while maintaining coverage.

Considering a GSM example, for a range of path-loss models, at least 6 dB less path loss occurs at a given distance for 900 MHz transmission compared with 1800 MHz transmission; for many path-loss models this value may be much higher. This simple and crude model, ignoring aspects such as antenna gain, implies that *at least* 75% less, and often much more than 75% less, transmission power can be used in the 900 MHz band compared with the 1800 MHz band. As an alternative solution, a network might be able to switch off half of its BSs when the traffic load is reduced by 50% or more, if it opportunistically operates at 900 MHz, as 900 MHz can travel *at least* twice as far as 1800 MHz before suffering an equivalent level of path loss, even according to the most conservative path-loss models. This latter solution knits nicely with the observation that if traffic is reduced by 50% or more in the 900 MHz band, in the small hours of the night for example, 50% or more of the 900 MHz band might be used by the other conventionally high-frequency system through dynamic spectrum reallocation. Of course, the reallocated high-frequency network will also be experiencing a 50%+ reduction in traffic load, hence will require only half of its prior cell density and so will be able to switch off half of its BSs while maintaining adequate coverage by opportunistically operating at the new lower-frequency band.

In frequency-reuse scenarios, it is noted that operators are conventionally tied to the use of each system at its particular spectrum band. Such default spectrum allocation, due to differences in the propagation characteristics of different bands, is inherently suboptimal. Optimality could be improved through the opportunistic use of more appropriate spectrum in which the power level in electromagnetic waves falls away with a more suitable exponent against distance, in view of the propagation characteristics in the local area, based on the necessary cell density, and hence the necessary propagation distance (whereby the necessary cell density is, of course, dependent on the active user density,

8.2 Opportunistic spectrum and load management concepts

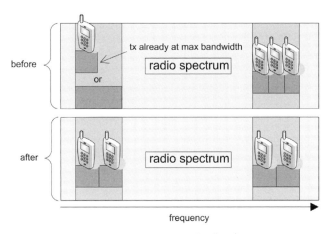

Figure 8.3 Opportunistic spectrum management to increase or balance channel bandwidth.

and hence the traffic density). Opportunistic selection of more appropriate spectrum in this way could significantly decrease inter-cell interference, there by reducing the necessary transmission power or, alternatively, improving capacity.

8.2.3 Power saving by channel bandwidth increase or better bandwidth balancing

Repartitioning of spectrum bands and/or the reallocation of users to increase or balance channel bandwidths, illustrated in Figure 8.3, significantly reduces the necessary transmission power. If there is an imbalance in the spectrum allocated in transmissions to users, far better power efficiency can be achieved by moving users among bands to address this imbalance. This is simply a consequence of the power-saving proportion against bandwidth increase factor being an increasing function of decreasing gradient—the reader is referred to the later Section 8.3.4 for an expression of this. If the bandwidths allocated in transmissions to users are made more equal by moving users/traffic loads from crowded bands to less crowded bands, the power saved by the increase in channel bandwidths for the users moved from the crowded band is generally far higher than the extra power expended by decreasing the channel bandwidths for the users in the non-crowded bands to which the other users are moved. Of course, if the traffic loads (in bits per second) to users are different, the spectrum allocated to each user should be proportional to its traffic load.

Such a concept might alternatively increase the average channel bandwidths among users per se. Refer to the depicted example in Figure 8.3, in which the bandwidth for one user in the case prior to reallocation of users between the bands is already at the maximum allowed bandwidth for the radio access technology being used, and there is still spare capacity in the band. As shown, the opportunistic reallocation of users between the bands could increase bandwidth by 50% for three out of four users, while the bandwidth would remain the same for the other user. As will be shown later in Section 8.3.4, such a bandwidth increase could give a significant decrease in the necessary transmission power for three of the four users.

It is noted that, statistically, if there is a large number of users on average in the cell there is less likely to be a significant variation in the number of users as a proportion of the average number. If there is a low number on average in the cell, this variation as a proportion might be quite significant, thereby more often leading to there being a significant difference in bandwidths for users among the bands. This observation tends to suggest that such schemes are far better suited to pico- or femto-cells, or other cases where there is likely to be a lower number of users being served per cell on average.

8.3 Assessment of power-saving potential

Using numerical simulations, here we put some numbers to the power-consumption savings that are achievable through some of the aforementioned concepts. Various approaches to the simulations are taken; these approaches are introduced at appropriate points throughout this section.

8.3.1 Example reflecting GSM networks

The performance for network configurations broadly reflecting GSM is first investigated. We choose GSM as one example of where our spectrum/load management solutions might be readily employed with particularly advantageous properties for power saving.

One of the most important factors affecting the performance of such solutions is of course the offered traffic load. Regarding our utilized traffic model, the average traffic load over a 24-hour period at each network/band is seen as varying according to a scaled (in both time and amplitude) and shifted sine cycle [11], [12], parameterized by the time of day of the busy hour load ϕ, the busy hour load itself in terms of average number of active users in the cell (given as $BusyLoad$), and the quiet hour load (given as $QuietLoad$). This average traffic load, at time of day t, is therefore given by

$$L(t) = \frac{BusyLoad + QuietLoad}{2} + \frac{BusyLoad - QuietLoad}{2} \cos(2\pi(t - \phi)/24). \tag{8.1}$$

Further to this, it is assumed that the number of active users in the cell is Poisson distributed, the expectation of which at any one time of day can be set using the aforementioned average traffic load at that time of day. Assuming this model, the probability of there being k number of users in the cell at time of day t is expressed as

$$P(k, t) = \frac{[L(t)]^k e^{-L(t)}}{k!}, \tag{8.2}$$

where $L(t)$ is given in (8.1).

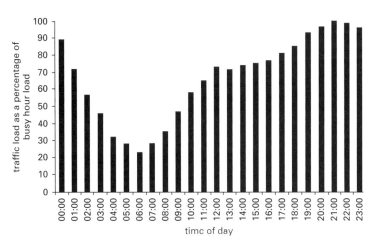

Figure 8.4 Hourly variation of traffic load as a percentage of busy hour load over a typical day for a mobile network operator in London, UK.

Our utilized traffic is parameterized to closely match real operator statistics in terms of average traffic load as a proportion of busy hour load. Given this, our sine-cycle representation achieves results in terms of power-saving performance that are very similar to those for real operator average traffic-loading statistics, although this comparison is omitted here for conciseness. These real operator statistics pertaining to the Vodafone 3G network in London, UK, are plotted in Figure 8.4, and are used later in this chapter. They were obtained via internal communication within the Mobile VCE Core 5 Green Radio research program.

We simulate an Erlang-based approach, which is appropriate in the GSM case where each active user consumes the same amount of resource. For our GSM-like example, we assume that 252 channels are available to the system in each band, and that the maximum acceptable blocking probability is 1%. Given this configuration, an omnidirectional cell can support 25 Erlangs per band, whereas the use of tri-sectored cells can support 38 Erlangs per band ([13], pp. 28–30). Our pertinent simulation parameters are summarized in Table 8.1.

Opportunistic load management to power down radio network equipment
First we investigate the opportunistic reallocation of traffic loads to power down radio network equipment. We assume that an operator or collaboration of operators has the aforementioned GSM-like systems operating at both 900 MHz and 1800 MHz, and, in order to assess performances against different network/band traffic loads, we assert that there can be different busy hour loads at each network/band, i.e. network/band 1 has the busy hour load $BusyLoad_1$ and network/band 2 has the busy hour load $BusyLoad_2$. $QuietLoad$s for both networks/frequencies are set at 25% of the respective $BusyLoad$s in order to mirror the real operator statistics presented in Figure 8.4.

Given this configuration, our numerical assessment cycles in outer loops through a 24-hour period in steps in time of one-tenth of an hour, and uses the value of the average

Table 8.1. Simulation configuration parameters (GSM example)

Parameter	Value
System configuration	Reflecting GSM [13]
Operating frequency, low-frequency network	900 MHz
Operating frequency, high-frequency network	1800 MHz
Channel path-loss model	Lenient path-loss, reflecting, e.g. rural scenario, implying the high-frequency network requires twice the power to operate
Number of channels available to the system	252 ([13], pp. 28–30)
Blocking probability	1% ([13], pp. 28–30)
Number of users supported per cell in omnidirectional mode	25 ([13], pp. 28–30)
Number of users supported per cell in tri-sectorized mode	38 ([13], pp. 28–30)
Busy hour load (users per cell)	Varied
Quiet hour load (users per cell)	25% of respective busy hour load

traffic load according to the sine-cycle representation (in (8.1)) at each time instance to parameterize the mean of the Poisson distribution (i.e. $L(t)$ in (8.2)), representing the statistical distribution of the number of users in the cell at that time instance. In inner loops, for each time instance, it then cycles through each possible value of k representing each possible number of users in the cell (thus populating the other parameter of the Poisson distribution), for each participating spectrum band in the process, and for each set of k's among the spectrum bands ascertains the power consumption that would be required given the dynamic spectrum access power-saving solution being applied. The actual power consumption for each such case is then given as this power consumption multiplied by the probability of it happening, which is of course the product of the Poisson probabilities $P(k,t)_{network1} \cdot P(k,t)_{network2}$ for the participating networks/frequencies. This result is then summed with equivalent results for all possible chosen values of k at each spectrum band to obtain the overall power consumption at that time instance. The same operation is performed over all time instances in the 24-hour period, and the average power-consumption is then taken among all time instances. This average power-consumption is then compared with the average power-consumption that would be required without the dynamic spectrum access power-saving solution taking place, as ascertained through the same process.

To illustrate the radio network equipment powering-down concept, Figure 8.5 gives an example of variation in traffic loads over a 24-hour period for two networks/frequencies, before and after applying the sectorization switching and cell powering-down solutions in tandem, using the sine-cycle traffic load representation. Furthermore, Figures 8.6 and 8.7 give the proportion of "from-the-mains" power that is saved by applying the schemes,

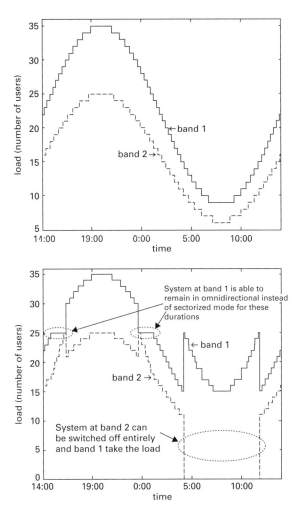

Figure 8.5 Example of loads before (upper plot) and after (lower plot) applying the sectorization switching and network equipment powering-down solutions in tandem (busy loads for spectrum band 1 and spectrum band 2 are 35 users and 25 users respectively, quiet loads are 25% of respective busy loads, supported number of users in omnidirectional mode is 25 users per band).

first for the cell powering-down solution only, then for both solutions, including sectorization switching. It is clear from these figures that very significant savings can be achieved: most significant savings are realized by switching off the system at one spectrum band and allocating its users to another spectrum band, particularly if networks are lightly loaded as might occur in urban areas at vacation times, for example. Additional significant improvements in efficiency can be achieved by the sectorization adjustment concept, especially if one network is heavily loaded and the other is lightly loaded as might occur, for example, after a network is newly deployed. Referring to Figure 8.7, the combined effect of these two concepts, for the most realistic traffic configurations (i.e. busy hour loads for the

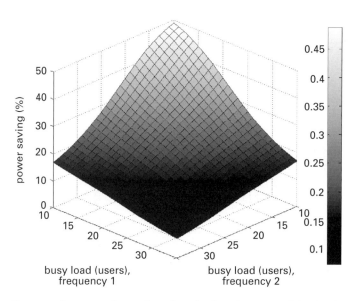

Figure 8.6 Power-saving proportion vs. busy hour loads at the two participating frequencies with the opportunistic cell powering-down solution only applied (GSM example).

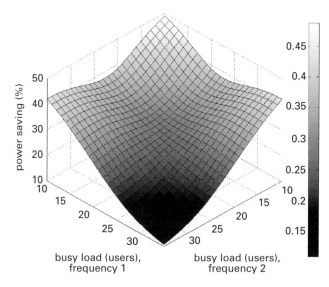

Figure 8.7 Power-saving proportion vs. busy hour loads at the two participating frequencies with the opportunistic cell powering-down solution and sectorization switching solutions both applied (GSM example).

two networks being close to their capacities), is typically of the order of 20–30% power saving. For less common traffic configurations these savings can be up to 50%.

While absorbing these results, it might be noted that the benefits of these solutions are greatly accentuated if there is a low correlation between traffic loads at the two

8.3 Assessment of power-saving potential 179

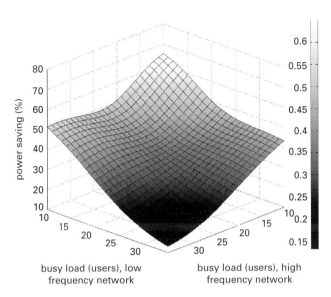

Figure 8.8 Power-saving proportions under the same configuration as Figure 8.7, where in this case the high-frequency network consumes twice the from-the-mains power as the low-frequency network (GSM example).

spectrum bands or networks. Indeed, our further results, which are not plotted in this chapter, where the traffic loads for the networks/frequencies in question are set to be not so highly correlated, show a significant performance increase of more than 50% better than the power savings in this chapter, particularly if the networks/bands are highly loaded. Other further results indicate that savings can be far more significant (of 60%, or greater) if there is a disparity in the power consumptions of the networks at the two spectrum bands, as might occur, for example, due to differences in hardware capabilities and physical characteristics at different spectrum bands. Figure 8.8 plots such results for the case where the higher-frequency network requires twice the "from-the-mains" power consumption of the lower-frequency network.

Opportunistic spectrum management to improve propagation characteristics
Next assessed is the concept of user/link reallocation for propagation improvement, under the same GSM example. We maintain precisely the same dynamic traffic configuration over the 24-hour period, where the high-frequency and low-frequency bands are assumed to be 1800 MHz and 900 MHz; as discussed in Section 8.2.2, this implies that links at the higher frequency require at least four times more transmission power than the lower frequency at an equivalent propagation distance, although this factor is generally far higher for more specific propagation models. Here we assess only the concept of opportunistically reallocating users/links to a lower frequency to reduce the necessary transmission power by bettering propagation; we don't investigate the concept of opportunistically using spectrum of more appropriate propagation distance as would be useful in frequency-reuse scenarios.

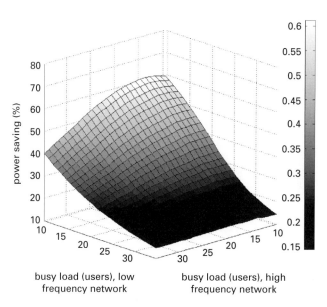

Figure 8.9 Transmission power-saving proportion through propagation improvement achievable by opportunistically allocating users/links to the lower-frequency band when possible (GSM example).

Results in Figure 8.9 are for the case where cells are omnidirectional only. These results show a significant transmission power-saving potential for the proposed scheme, of up to 55% or more in low load conditions for the network (at holiday times, for example), and a lesser power saving potential of some 25–35% in more normal conditions. It should be emphasized that these results are conservative in the sense that the chosen propagation model is the least conducive to good performance for the scheme: our further results, not depicted in this chapter, which use more realistic/severe propagation models, show savings far higher – in some cases in excess of 80%.

8.3.2 Example reflecting LTE networks

Next we investigate the performance of the proposed concept for networks reflecting LTE. As opposed to the GSM example, here we focus on single cells at each band and use a per-cell/sector system capacity limit to parameterize the traffic load that can be supported in omnidirectional mode. Various configuration parameters for this case are as given in Table 8.2.

Opportunistic load management to power down radio network equipment

Again, we start by investigating the opportunistic reallocation of traffic loads to power down radio network equipment. Results in Figure 8.10 are for the case where *BusyLoad* is varied and the *QuietLoad* is 25% of each respective *BusyLoad*, whereby the same simulation procedure is taken as in Section 8.3.1. Although here it is assumed that each active user is receiving a video traffic flow of rate 384 kbps, and the number of users

8.3 Assessment of power-saving potential

Table 8.2. Simulation configuration parameters (LTE and HSDPA examples)

Parameter	Value
Spectral-efficiency (LTE)	1.5 b/s/Hz [14]
Spectral-efficiency (HSDPA)	0.8 b/s/Hz [14]
Bandwidth per LTE band	20 MHz [14]
Bandwidth per HSDPA band	5 MHz
Ratio of loading at busy hour to capacity (used in LTE simulations)	60% [14]
Channel path-loss model (where applicable)	Weissberger
Weissberger foliage depth	3 m
Transmission center frequencies (for propagation improvement work under LTE example)	460 MHz, 820 MHz, 2.35 GHz (reflecting allocated IMT-Advanced bands) [2]
Video communications rate per user	384 kb/s
Data (FTP) communications OFF period duration	Exponentially distributed, mean 180 s [15]
Data (FTP) communications ON period duration	Pareto distributed file size, mean 2 MB [15], $\alpha = 1.5$ (unless otherwise stated). Pareto parameter k calculated from the mean and α, and ON duration calculated from each sampled file size assuming a fixed data rate of 1Mb/s per user
Data (HTTP) traffic reading time (OFF duration)	Exponentially distributed, mean 30 s [15]
Data (HTTP) traffic parsing time (OFF duration)	Exponentially distributed, mean 0.13 s [15]
Data (HTTP) traffic main object size (contributes to ON duration)	Truncated Log-normally distributed, $\sigma = 1.37$, $\mu = 8.35$, $min = 100$ B, $max = 2$ MB [15]
Data (HTTP) traffic embedded object size (contributes to ON duration)	Truncated Log-normally distributed, $\sigma = 2.36$, $\mu = 6.17$, $min = 50$ B, $max = 2$ MB [15]
Data (HTTP) traffic number of embedded objects per page (contributes to ON duration)	Truncated Pareto distributed, $\alpha = 1.1$, $k = 2$, $max = 55$ (k subtracted from each sampled value) [15]
Per-user rate in FTP/HTTP ON durations	1 Mbps (LTE), 64 kbps (HSDPA)

supported per cell is calculated according to the quoted spectral efficiency, bandwidth, and the appropriate ratio of loading at busy hour to available capacity in Table 8.2. The system capacity estimation procedure in [14] is used. These results consider cases where two, three, and four bands are participating in the process, where in all cases it is assumed that the *BusyLoad* and *QuietLoad* are the same for each participating band. From these results, it is clear that very significant savings can be achieved, of up to 50% or more if more than two bands are participating in the process, and more typically in the range of 20–50% if there are lesser bands participating or there is a greater network

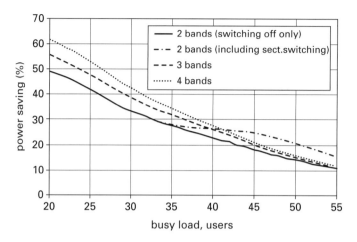

Figure 8.10 Power saving against busy hour load for network powering-down solutions (video traffic example in LTE).

loading. Moreover, among the two-band cases, it is noted that the sectorization switching solution considerably improves performance if the networks are heavily loaded, but gives little or no improvement if networks are lightly loaded. Other simulations that we have performed show an additional significant improvement in performance attained by the sectorization switching solution if there is a significant difference in traffic loads among the participating bands.

We have also performed simulations where each user receives an independent ON/OFF traffic flow, with ON and OFF durations mirroring FTP traffic. Configuration parameters again are as given in Table 8.2, whereby the number of users at time of day t, each of which receives an ON/OFF traffic flow, is represented by the same sine cycle as (8.1) (i.e. (8.2) is omitted). Results are again averaged over the 24-hour period. Parameters for ON/OFF durations from [15] are chosen, with the exception that a Pareto distribution is used for ON periods, in order to be able to represent power-tailed file sizes and associated traffic self-similarity. The mean file size is the same as configured in [15].

Results in Figure 8.11 again show a considerable saving, which decreases as the busy hour load increases. Moreover, these results show that as the *QuietLoad* is decreased as a proportion of the *BusyLoad*, further improvement in performance is possible, especially if the *BusyLoad* is high. Furthermore, it is shown in Figure 8.12 that if there is a difference in the from-the-mains power consumptions of networks, additional significant savings are possible, increasing from a peak saving of 50% to a peak saving of 80% if the power consumption difference factor is increased from 1 to 4.

Finally, simulations have also been performed for the alternative configuration where average traffic loads depicted in Figure 8.4 have been used instead of the sine-cycle configuration in (8.1), moreover, HTTP (web browsing) traffic flows have been simulated in addition to the aforementioned FTP flows, as parameterized in Table 8.2. Figure 8.13 plots power-saving results for the FTP and HTTP (web browsing) ON/OFF traffic models over the LTE configuration, where two bands are participating in the process and the

8.3 Assessment of power-saving potential

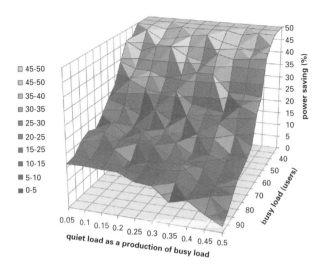

Figure 8.11 Power saving against busy hour load and quiet hour load, for network powering-down solutions (FTP ON/OFF traffic example in LTE).

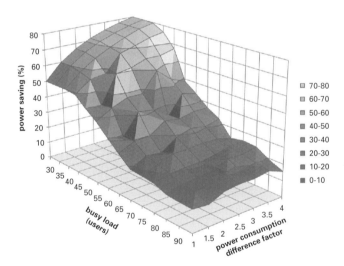

Figure 8.12 Power saving against busy hour load and power consumption difference, for network powering-down solutions (FTP ON/OFF traffic example in LTE).

assumption is that the network powering-down solution only is employed. Results again show a significant power-saving potential of up to 50% for low network loads. In the FTP case, power saving begins to reduce at a BusyLoad of ≈ 20 users, reaching as low as 10% at a BusyLoad of ≈ 50 users. In the HTTP case, power saving begins to reduce at a BusyLoad of ≈ 150 users, and hits 10% at a BusyLoad of ≈ 500 users. It is emphasized here that the per-user traffic load for the HTTP (web browsing) case is very light compared with FTP downloads.

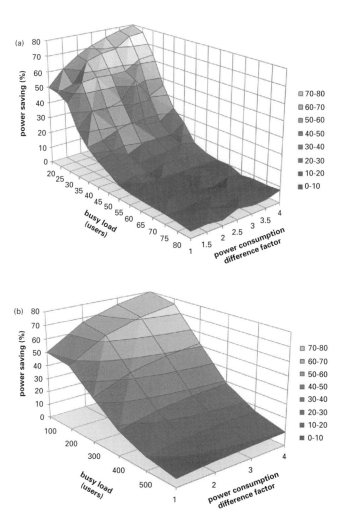

Figure 8.13 Power saving against busy hour load and power consumption difference, for network powering-down solutions in LTE: (a) FTP ON/OFF traffic, (b) HTTP ON/OFF traffic.

It is also clear from Figure 8.13 that power savings greatly increase as the difference factor in power consumption between the bands is increased.

Opportunistic spectrum management to improve propagation characteristics

Next we look at the opportunistic reallocation of traffic loads to improve propagation. Here we revert to the prior dynamic traffic configuration combining the operator statistics in Figure 8.4 with (8.2), under the assumption of the per-user video flow traffic in Table 8.2 under LTE. The Weissberger loss model is assumed as one example of a continuous function loss model that appropriately applies across an extremely wide range of spectrum bands. As results are applicable to any allowed propagation distance under the Weissberger model, propagation distance is not specified.

8.3 Assessment of power-saving potential

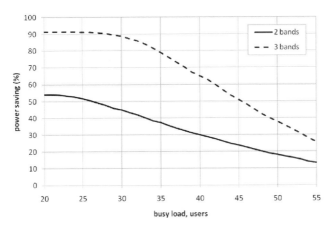

Figure 8.14 Power saving against busy hour load through opportunistic use of better propagation bands (video traffic example in LTE).

We investigate both the two-band and three-band opportunistic reallocation cases, where configuration parameters are as in Table 8.2 and it is assumed that all bands have the same $BusyLoad$ to simplify the representation of the results. For the two-band case (bands operating at 460 MHz and 820 MHz bands), results in Figure 8.14 indicate that there is transmission power-saving potential of some 20–50%, particularly if the network is lightly loaded (at holiday times, for example). For the three-band case, given that the upper band is of much higher frequency (poorer propagation), transmission power saving is far greater, of up to 90%.

8.3.3 Example reflecting HSDPA networks – combining opportunistic reallocation to power down radio network equipment and opportunistic reallocation to improve propagation

Here we attempt to combine the effect of opportunistic reallocation to power down radio network equipment and opportunistic reallocation to improve propagation, whereby we look at the overall from-the-mains power consumption as a result of these efforts, not merely the saving in transmission power consumption. To this end, it is necessary to understand the mapping between transmission power (which is, for the purpose of this section, analogous to traffic load), and overall from-the-mains power consumption for a modern base station. Given that the most modern such statistics we could obtain were for an HSDPA Release 7 base station, we hereby work with HSDPA Release 7.

For an anonymous manufacturer responding to a call for information, internal documentation within the Mobile VCE Core 5 Green Radio Research Program indicates from-the-mains power consumption for an HSDPA BS at 100% transmission power to be 857 W, and at 20% transmission power to be 561 W. It is widely observed that from-the-mains power consumption against transmission power broadly varies with an $m \cdot p + c$ relationship, comprising a fixed-term c that is independent of transmission power p, and a term that varies with transmission power, $m \cdot p$. Given this, the above numbers regress to give 487 W as the fixed part from-the-mains power consumption c, and the gradient of variation with transmission power m as 9.25 from-the-mains Watts per transmission Watt. These values are used throughout this section.

In ascertaining the necessary transmission power, we use the values in Table 3 of reference [16], which gives the total transmit power to support a maximally loaded BS. We set 80% of this available power as being scaled by the number of users present in the system and 20% as being allocated to pilot transmission. The work in [16] is for full HSDPA networks operating at 2 GHz (as per current HSDPA deployments), and at 5 GHz as argued as an interesting future (additional) deployment option in [16]. A 600 m cell radius is chosen by us, where again we assume the aforementioned FTP ON/OFF traffic model as described at the start of Section 8.3.2. Moreover, we again assume that the *BusyLoad* is the same for both bands, facilitating the depiction of results.

Results in Figure 8.15 show that there is significant transmission power-saving potential through the opportunistic reallocation scheme. Power saving initially increases to some 58% as the busy hour load is increased to 30; this is because it is always possible to reallocate users to power down radio equipment, so adding more users simply increases the number that are reallocated to better spectrum thereby saving additional transmission power. However, as the traffic load is further increased, power saving decreases and a difference begins to emerge in the performance of the solutions with and without opportunistic reallocation to save transmission power. It is noted that, especially if the networks are experiencing moderate load, the opportunistic reallocation of links to save transmission power saves up to an additional 10% compared with just opportunistically reallocating users/links to be able to power down radio equipment. This additional saving would be far higher if the value of m, the gradient of the from-the-socket power to transmission power relationship, were greater than the modest value of 9.25 assumed here. Moreover, further results not depicted in this chapter show a far greater power saving if there is a difference in the traffic loadings at the two bands, particularly if the high-frequency band is heavily loaded and the low-frequency band is lightly loaded.

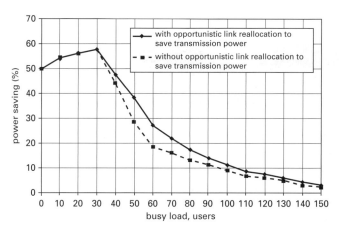

Figure 8.15 Power saving against busy hour load through opportunistic link reallocation to use better propagation bands (FTP ON/OFF traffic in HSDPA).

8.3.4 Power saving by channel bandwidth increase or better bandwidth balancing

Finally, through a simple capacity analysis, we consider the concept of power saving by increasing or better balancing channel bandwidths. Our assessment is independent of the type of deployed system. A simple manipulation of the Shannon capacity formula for a given initial bandwidth B indicates that increasing that bandwidth by a factor A under the same required (Shannon) capacity C in the before and after increase cases, leads to a power-saving proportion of

$$\text{Power-saving proportion} = \frac{(2^{C/(B \cdot A)} - 1)A}{2^{C/B} - 1}. \tag{8.3}$$

We simulate the concept of bandwidth "balancing" to save energy, whereby the case where users' bandwidths are automatically maximized in each band independently is compared with the case where active users (those currently in the "ON" state) can be moved between bands to make the bandwidths allocated to users more equal. We consider a two-band case where the busy load for each band varies independently between 5 and 50 users, and the simulation configuration is otherwise as used previously, under the FTP traffic model described in Table 8.2 as parameterized by the operator statistics in Figure 8.4.

Results in Figure 8.16 show a power saving of up to 15% under this concept, for the range of network loadings investigated. Moreover, power saving is more significant if the number of users per band is lower, or there is a difference in the average loading of each band. Both of these situations lead to it being more likely that there will be a significant

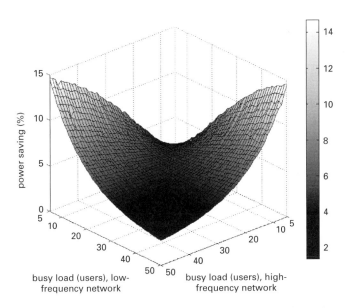

Figure 8.16   Power saving against busy hour load through user reallocation to better balance bandwidths (FTP ON/OFF traffic).

difference in the number of users in each band, which presents an opportunity for energy saving by moving users between bands to make users' bandwidths more equal.

8.4 Conclusion

Radio spectrum is a precious resource, which to date has been used with poor efficiency. Furthermore, traffic load is often managed in a suboptimal manner, whereby it is frequently constrained to being carried by a particular system and associated spectrum band as provided by the "owner" of the user and device creating that traffic. This has commonly been done for reasons such as simplicity and to facilitate/allow charging for carrying that traffic.

Spectrum usage and traffic management intransigence lead to inefficiency, in terms of the power consumption of the systems carrying traffic, and in terms of realizing the achievable capacity by the associated systems and spectrum bands. This chapter has discussed various concepts through which the dynamic adaptation of spectrum and traffic load allocations by an operator or collaboration of operators can reduce the power consumption for providing mobile and wireless services. Although these concepts are being worked on primarily to save power, it is noted that they can also apply to capacity improvement for the operator. Indeed, there is a trade-off between the power saving and capacity improvement advantages of such approaches.

The range of results in this chapter has shown real potential for power saving of up to 50% or more for the individual schemes, with greater power savings being possible if combinations of the schemes are appropriately applied together.

Acknowledgment

This work was supported by the Green Radio Core Research Program of the Virtual Centre of Excellence in Mobile & Personal Communications, Mobile VCE, www.mobilevce.com, the ICT-ACROPOLIS Network of Excellence, FP7 project number 257626, www.ict-acropolis.eu, the UBIQUIMESH and OPPORTUNISTIC-CR Portuguese projects, the PLANOPTI Marie Curie Reintegration Grant, and COST Actions IC0902, IC0905 "TERRA", and IC1004. The authors thank Adnan Aijaz and Paul Pangalos of King's College London, and Terence Dodgson of Roke Manor Research Ltd., part of the Chemring group, for various assistance and partaking in some very useful discussions.

References

[1] S. Buljore *et al.*, "Architecture and enablers for optimized radio resource usage in heterogeneous wireless access networks: the IEEE 1900.4 working group," *IEEE Communications*, vol. 47, no. 1, pp. 122–129, Jan. 2009.

[2] International Telecommunication Union (ITU), "Results of WRC-07". [Online]. Available: www.itu.int/en/ITU-R/space/Presentations/resultsWRC07.pdf, accessed Apr. 2011.

References

[3] A. Wyglinski, M. Nekovee, and T. Hou (Eds.), *Cognitive Radio Communications and Networks: Principle and Practice*. Elsevier, Nov. 2009, ISBN: 978-0-12-374715-0.

[4] Office of Communications (Ofcom), "Spectrum usage rights: technology and usage neutral access to the spectrum." [Online]. Available: www.ofcom.org.uk/consult/condocs/sur/, accessed Jun. 2011.

[5] Federal Communications Commission (FCC), Second Memorandum Opinion and Order, "In the matter of unlicensed operation in the TV broadcast bands; additional spectrum for unlicensed devices below 900 MHz and in the 3 GHz band," Sept. 2010. [Online]. Available: www.fcc.gov/Daily_Releases/Daily_Business/2010/db0924/FCC-10-174A1.pdf, accessed Apr. 2011.

[6] 3rd Generation Partnership Project (3GPP) TS 23.251, "Technical specification group services and system aspects; network sharing; architecture and functional description," version 9.4.0, Mar. 2011.

[7] Vodafone UK Corporate Responsibility Report 2009-10. [Online]. Available: www.vodafone.com/content/dam/vodafone/UK, accessed Apr. 2011.

[8] O. Holland *et al.*, "Opportunistic load and spectrum management for mobile communications energy efficiency," IEEE PIMRC2011, Toronto, Canada, Sept. 2011.

[9] J. Perez-Romero *et al.*, "A novel on-demand cognitive pilot channel enabling dynamic spectrum allocation," in *Proc. of IEEE DySPAN 2007*, Dublin, Ireland, Apr. 2007.

[10] IEEE Standard 1900.4, published Feb. 2009. [Online]. Available: www.ieee.org

[11] S. Thajchayapong and J. M. Peha, "Mobility patterns in microcellular wireless networks," *IEEE Transactions on Mobile Computing*, vol. 5, no. 1, Jan. 2006, pp. 52–63.

[12] L. Chiaraviglio, M. Mellia, and F. Neri, "Energy-aware backbone networks: a case study," in *Proc. of International Workshop on Green Communications (GreenComm '09)*, Dresden, Germany, Jun. 2009.

[13] J. Eberspcher *et al.*, *GSM Architecture, Protocols, and Services*. Wiley, 2009, ISBN: 978-0-470-03070-7.

[14] Intel Corporation White Paper, "WiMAX, 3G and LTE: a capacity analysis," 2010.

[15] 3rd Generation Partnership Project (3GPP) TR 25825, "Technical Specification Group Radio Access Network; Dual-Cell HSDPA Operation," version 1.0.0, May 2008.

[16] O. Cabral *et al.*, "Integrated common radio resource management with spectrum aggregation over non-contiguous frequency bands," *Springer Wireless Personal Communications Journal*, vol. 59, no. 3, pp. 499–523, Aug. 2011.

9 Energy-saving techniques in cellular wireless base stations

Tao Chen, Honggang Zhang, Yang Yang, and Kari Horneman

9.1 Introduction

The rapid growth of mobile communications comes with the prominent energy-consumption challenge. It has become so critical that, without being dealt with in advance, it will eventually prevent the sustainable growth of the mobile industry [1]. Conventional treatments on the energy-efficiency study largely focus on the component and equipment level. It is shown that novel architecture and advanced methods allow for significant improvement of the energy efficiency (EE) of wireless systems [2]. For this it is necessary to extend the study to the system/network level.

Network energy-saving techniques tune the parameters and protocols of networks for interference mitigation, resource optimization, and energy saving. It is a prerequisite to understand key energy-consumption problems in a network. Cellular wireless access networks have been identified as the main consumer of energy in the wireless industry, while statistics show that radio base stations (RBS) in such a network consume most of the energy [2]. Various approaches have been proposed to reduce the energy consumption of an RBS, for instance, passive cooling techniques, energy-efficient backhaul solutions, and distributed base station design by using a remote radio head (RRH). The most promising approaches target the energy-consumption reduction of the power amplifier (PA) in an RBS since the PA dominates the energy consumption of a cellular RBS [3]. Due to limitations on cost and technology, the power efficiency of a PA used in recently developed RBS is less than 50%. For a network energy-saving solution applied to cellular networks, it is necessary to reduce the impact of the PA on the EE of the RBS. Numerous solutions have been proposed to address this issue [4]. However, they are only applied for a single RBS. It would be beneficial to consider the energy-saving problem from the network perspective, which takes into account the interactions of different kinds of RBSs, for instance, macro-cell RBS, femto-cell RBS, and other short-range RBS.

The chapter will first overview the network energy-saving techniques proposed for 3rd Generation Partnership Project (3GPP) Long-Term Evolution (LTE) systems. The aim is to provide comprehensive understanding on current studies regarding RBS energy saving. Those techniques are divided into the time, frequency, and space domain. Next we propose the use of a layered structure for the energy saving of RBSs. The layered structure studied in this chapter is a two-layer structure consisting of LTE macro-cells and femto-cells. In this layered structure, femto-cells are assumed to be mainly deployed to boost capacity; however, they also improve the EE of macro-cell RBSs. We are looking at the LTE system because it is chosen as the next-generation mobile system by the International Telecommunications Union (ITU) Radio Communication Sector (ITU-R). The basic mechanism to use the layered structure for energy saving is to off-load the downlink traffic of macro-cells to femto-cells in a smart way so that network energy-saving techniques proposed for LTE can be applied in the macro-cell RBS. We derive an EE metric, propose an energy-aware handover algorithm, and study the energy-saving gain of the proposed architecture by simulation. Finally, the conclusion is drawn by providing future research directions on this topic.

9.2 Energy-consumption model of RBS

We start the chapter by describing a generic energy-consumption model of RBS for wireless access networks. As RBSs consume most of the energy in a wireless access network, identifying energy-consumption problems in an RBS provides clues for better energy-saving solutions for the whole system.

The energy-consumption reference model is defined in [5]. We only use the outdoor reference model here as the main difference between the outdoor and indoor models is the climate control. As shown in Figure 9.1, the reference model of an RBS is composed by an RBS equipment and support system infrastructure. An RBS equipment is a network component that serves one or more cells and interfaces the mobile station through air interface and a wireless network infrastructure. Transceivers and associated PAs are included in an RBS. The support system of an RBS includes a power supply, climate control for temperature control, a transmission module connected to the core network, and a battery backup. The power loss of the battery backup is not taken into account in this model.

The typical power-consumption figures of different RBSs give a rough idea of the EE of widely deployed cellular RBSs. For instance, a GSM system with 12 transceivers

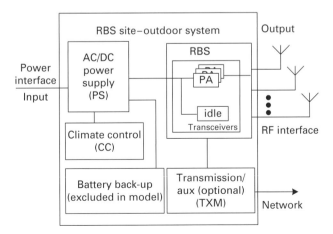

Figure 9.1 Energy-consumption reference model of RBS (adapted from [5]).

takes 3802 W of AC power supply as the feeding power [3]. For each transceiver, the input power is 200 W, of which 60 W is consumed at idle time and 140 W is used for transmission. The output power after a power amplifier is 40 W, and the final power sent over air is only 10 W. In this case we can easily see that the PAs consume 70% of the whole power, PA efficiency is 28%, and the overall efficiency of the RBS is only 3.1%. For a 3G RBS, PA efficiency can be 45% and the overall efficiency of a 3G RBS is less than 12%. PA efficiency is still the main problem for the overall EE of the RBS.

It is clear that the PA is the main concern in energy-efficient solutions for the RBS and wireless access networks. Physical constraints put a significant limit on PA efficiency. PA efficiency of 60% is now the target by advanced PA techniques. In addition to continuously improve PA efficiency, the alternative way is to switch off the PA when there is no traffic to transmit. Many network energy-saving proposals for the RBS fall into this category [4]. We will describe them in detail. But before describing the energy-saving solutions, it is necessary to know how and by which metrics the energy consumption and EE of the RBS and wireless access system/network are measured.

9.3   EE metric

The standard body European Telecommunications Standards Institute (ETSI) defined the metrics and methods to evaluate EE of RBSs in [5]. The current technical specification [5] covers three systems: GSM/enhanced data-rates for global evolution (EGDE), wideband code-division multiple access (WCDMA), and worldwide inter-operability for microwave access (WiMAX). The EE for LTE systems will be included in the next version.

The ETSI document [5] specifies the standardized energy-consumption measurement for RBS equipment and an RBS site. A site normally includes RBS equipment, rectifiers, climate control, power distribution loss between units, and other auxiliary equipment and cabinets. A reference model for an RBS site is shown in Figure 9.1. Note that the figure only shows the outdoor reference model.

RBSs defined in [5] are divided into concentrated RBSs and distributed RBSs. The difference is that a concentrated RBS puts all but the antenna element in the same location, while a distributed RBS uses the RRH close to the antenna element to reduce feeder loss.

The measurement takes power-consumption samples at different load conditions, i.e. busy hour load, medium-term load, and low-load. The duration in each load condition is denoted as t_{BH}, t_{Med}, t_{Low}, respectively. For a concentrated RBS, assuming the average power-consumption measured in each load condition to be P_{BH}, P_{Med}, P_{Low}, respectively, the average power-consumption (in watt) is defined as:

$$P_{equipment} = \frac{P_{BH}t_{BH} + P_{Med}t_{Med} + P_{Low}t_{Low}}{t_{BH} + t_{Med} + t_{Low}}. \tag{9.1}$$

For a distributed RBS, the power consumption for RRH and the elements in the central location are measured separately. Similar to (9.1), we can obtain the power consumption of RRH and central elements P_{RRH} and P_C. The total average power-consumption of a distributed RBS is

$$P_{equipment} = P_C + P_{RRH}. \tag{9.2}$$

The power consumption of a site takes into account the power consumed by the power supply, cooling, and other auxiliary equipment. Two correction factors are applied in the power consumption of the RBS equipment to obtain the power consumption of a site: the power supply correction factor, denoted as PSF, and the cooling factor, denoted as CF. Both of these two factors are unitless and the values are related to the specific system. The average power-consumption for concentrated RBS is defined as

$$P_{site} = PSF \cdot CF \cdot P_{equipment}. \tag{9.3}$$

For a distributed RBS site, the specific correction factors for central elements and RRH are applied separately, and the power consumption of the site is obtained from the sum.

The Flexi EDGE base station from Nokia Siemens Networks has been announced as the first one to be measured following the ETSI technical specification [6]. In a typical configuration with 3 sectors and 4 carriers per sector, the average power-consumption of a Flexi EDGE base station is 978 Watts. By using energy-saving features, it can be reduced to 562 Watts in some instances.

In the current version of the ETSI document [5], no EE metric is proposed for the RBS equipment. However, with the average power-consumption, we can use the metrics proposed by the energy-consumption rating (ECR) Initiative [7]. The basic form of ECR is

$$ECR = \frac{E}{T}, \tag{9.4}$$

where E is the energy consumption in Watts and T is the effective system throughput in bits per second. The effective system throughput counts the frame overhead from the physical and link layer.

Based on the measured power consumption of RBS, ETSI defines two network-level EE metrics for the GSM system [5]. The network-level EE considers not only energy consumed by the RBS site, but also the features and properties related to capacity and coverage of the network. In rural areas, the network is seldom fully loaded. The coverage area is used in the EE metric to reflect energy to achieve coverage. The EE metric for rural areas is defined as

$$PI_{rural} = \frac{A_{coverage}}{P_{site}}, \tag{9.5}$$

where $A_{coverage}$ is the RBS coverage area in km^2, and P_{site} is the average site power-consumption. The coverage area is calculated based on uplink and downlink system values.

In urban areas, the traffic demand is often larger than the capacity of the RBS. The capacity, instead of the coverage, is reflected in the EE metric. The metric for urban areas is defined as

$$PI_{urban} = \frac{N_{BH}}{P_{site}}, \qquad (9.6)$$

in which N_{BH} is the number of subscribers based on average busy hour traffic demand by subscribers and average RBS busy hour traffic.

Referring to the area spectral-efficiency [8], Richter et al. proposed the concept of area power-consumption for cellular networks [9] which is defined as

$$\rho = \frac{P_C}{A_C}, \qquad (9.7)$$

where P_C is the power consumption of a cell site, and A_C is the coverage area in km^2 of the cell. The area power consumption reflects the power consumption of cellular cells at different site density. Note that this metric is the inversion of (9.5).

As we can see, the aforementioned metrics are not directly related to the throughput performance of the system. The reason is that for the GSM and WCDMA systems the main service is voice, the performance of which is not measured by the data rate. However, in LTE, all services are carried out by packets. It provides a possibility to measure the performance of the system by throughput represented by the data rate. *Bit/joule* is expected to be the basic EE metric for the LTE system and beyond.

9.4 RBS energy-saving methods

This section summarizes LTE energy-saving studies performed in the 3GPP standardization body. We assume that readers have a basic knowledge of the LTE standard. The philosophy behind all proposed methods is the same: making the energy consumption in the RBS scale with the traffic load. Most methods introduced here aim at reducing the PA operating time. Other solutions are possible, for instance, by reducing the operating bandwidth so as to reduce the transmission power, or using the combination of different cell sizes to enable better energy consumption in a covered area. We divide those solutions into time, frequency, and spatial domain. Performance of different solutions is compared based on a realistic traffic pattern during a day. Note that hybrid approaches combining solutions from three domains are possible. The study in [4] showed that the hybrid approach provides better energy-saving performance.

We should emphasize that from the standard viewpoint those solutions can be divided into two categories: implementation-based or standard-impacted. The implementation-based solutions require no changes in the standard and therefore can be easily integrated into products. For standard-impacted solutions, it is possible that they will never be included in the standards.

9.4 RBS energy-saving methods 195

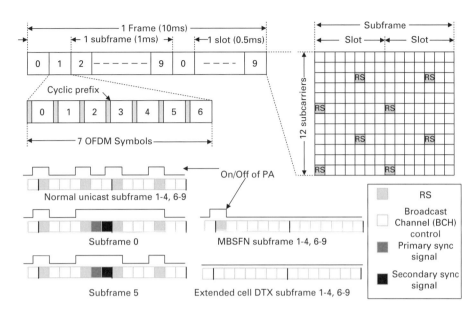

Figure 9.2 Reduced control overhead in frame for energy saving in LTE.

9.4.1 Time-domain approaches

The common idea of time-domain solutions is to shut down PAs whenever there is no traffic in the downlink. The energy saving is measured by the time fraction when PAs are off during a time period, normally measured by a frame. Before describing those solutions, it is worth introducing the frame structure of LTE. In LTE, the channel is structured by contiguous time frames. Each frame consists of 10 subframes with a fixed length of 1 ms. Each subframe in turn has two equal-sized time slots. According to the configuration, each time slot accommodates a number of orthogonal frequency-division multiplexing (OFDM) symbols. Normally that number is 7. The structure of a downlink subframe is illustrated in Figure 9.2, which includes symbols for control signals and data traffic. Among a variety of control signals, reference signals (RS) are regularly transmitted in each subframe, used to obtain downlink channel state information (CSI). If there is no or less downlink traffic, the frequency to transmit RSs can be reduced. This reduces the energy consumption of PAs. This is the basis for the time domain energy-saving solutions.

There are three ways to shut down PAs for energy saving. The most straightforward approach is to turn off PAs for signal-free downlink data symbols. This is an implementation-based approach with no change to the standard. The downside is that the number of RSs remains the same. A simple calculation from Figure 9.2 shows that for 47% of time in a frame, PAs are in the operating mode [10]. The second approach, as shown in Figure 9.2, uses multicast broadcast single frequency network (MBSFN) structure to reduce RSs. MBSFN is proposed to deliver services such as Mobile TV using the LTE infrastructure. In an MBSFN frame, 6 subframes out of 10 only need to transmit 1 RS, while in the normal case it is 4 RSs. The PA on time is therefore reduced to 28%

[10]. The feature of MBSFN has been included in LTE Release-8; therefore, there is no backward compatibility problem.

The most interesting solution is the extended cell discontinuous transmission (DTX) approach [11]. This approach further reduces RSs as compared to the MBSFN approach. As shown in Figure 9.2, if there is no downlink traffic, in the extended cell DTX mode, there is no need to have any transmission in 8 subframes. The PA on time is further reduced to 7.1%.

The extended cell DTX approach has some limitations. Firstly, it does not provide backward compatibility. Secondly, it only works when there is no downlink traffic. This traffic condition is very rare during a day. Thirdly, the reduction of RSs has an impact on the performance of user equipment (UE). In LTE, some control procedures are performed with the assistance of RSs. Without enough RSs, UE may experience unpredictable problems in synchronizing with the RBS or decoding control signals, thereby having a negative impact on services. Moreover, reducing RSs may prevent UE from entering into the terminal DTX mode and then reduce the battery life. The debate to introduce the extended cell DTX approach into the standard is still ongoing.

9.4.2 Frequency-domain approaches

Two energy-saving approaches are normally used in the frequency domain. The bandwidth reduction approach adapts the bandwidth with the traffic load. Less bandwidth is used when the downlink traffic load is low. As shown in Figure 9.3, to maintain the same power spectral density (PSD), smaller bandwidth requires less output power. The advantage of this approach is that it is suitable for low traffic load. However, this approach does not shut down the PA. As an operating PA consumes much more than an inactive PA, the energy saving from this approach may be marginal. Moreover, a PA normally operates at an optimized point with a given output power range. Reducing the output power will degrade the PA efficiency and in turn the gain for energy saving.

Carrier aggregation is the second approach used for frequency-domain energy saving. It assumes carriers are aggregated to groups and served by an individual PA. The idea is to shut down the associated PA when the corresponding aggregated carriers are not scheduled for traffic. This approach relies on the implementation of PAs in the RBS. It is only applicable to an RBS that has aggregated carriers and has a separate PA attached to each carrier.

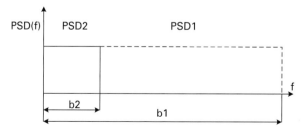

Figure 9.3 Bandwidth change with same power spectral density.

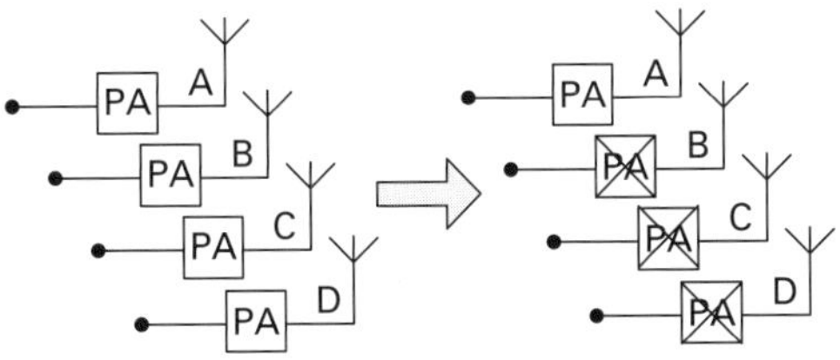

Figure 9.4 Reduce antenna number for energy saving.

In general, frequency-domain solutions have less impact on UE. However, due to the constraints mentioned above, the efficiency of frequency-domain solutions is limited.

9.4.3 Spatial-domain approaches

Reducing the number of antenna is the most commonly used energy-saving approach in the spatial domain. It is illustrated in Figure 9.4. In this case, if the branches of antenna are reduced from 4 to 1, the energy consumption of a transceiver is reduced to 1/4, as PAs associated with those branches can be shut down. This approach can be used in the low-traffic condition. The reduction of antenna branches decreases the total output power and makes the cell size smaller. An additional mechanism is then needed to maintain the strength of the control signals at the cell edge. The antenna number-reduction approach may lead to service degradation or interruption as the antenna reconfiguration causes delay. It is suggested that this approach is used for semi-static load.

The cell switch-off approach [12] is a system-level approach that works in an area covered by multiple cells. The system-level approach means there is no need to modify the lower layer components in the RBS. When the traffic load in a given area is low, some cells can be shut down and the served UEs are handed over to the remaining working cells. Those inactive cells can be turned on again during the busy time. There are two ways to switch cells on/off. One is the signaling directly between BSs. The other is the dedicated control from the operations, administration, and maintenance (OAM) layer of the system.

A special case of the cell switch-off approach is called the hierarchical cell structure (HCS) approach [13], in which always-on macro-cells are deployed for basic coverage and small cells are used for capacity boost. Cells for capacity boost only operate when the traffic load is high in macro-cells.

While the cell switch-off approach tries to make a good balance between performance and energy saving, it has limitations. First, frequently switching cells on/off affects services in UEs. Its use should be limited to a semi-static manner. Secondly, switching off cells may reduce the battery life of served UEs as they have to connect other cells far away. Thirdly, if switching off a cell creates an uncovered area, remaining active neighboring cells need to increase their power to cover this area. This may result in a marginal gain in energy saving.

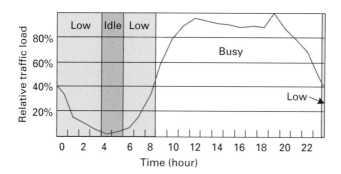

Figure 9.5   Daily downlink traffic load pattern.

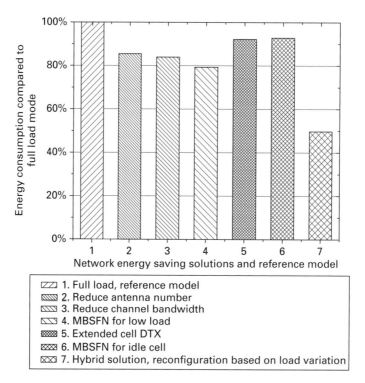

Figure 9.6   Relative power consumption of energy-saving solutions.

9.4.4 Performance comparison

We use the results from [4] to show the actual performance of different energy-saving approaches under a realistic daily traffic pattern. The daily traffic pattern is shown in Figure 9.5, which includes 7 hours of low-load time and 2 hours of idle time. It should be noted that different traffic patterns lead to different performance results.

Figure 9.6 shows the relative energy-saving performance of different approaches as compared to the full-load case. The energy-saving approaches for no-load scenarios, which are extended cell DTX and MBSFN for no load, achieve energy saving of less than 8%. This is because the idle time only occupies a very small fraction of time during a day. The energy-saving approaches working for low-load scenarios provide better energy-saving performance than those only working in idle scenarios. Among three low-load approaches, MBSFN for low-load provides the highest performance. The bandwidth reduction approach outperforms the antenna number-reduction approach. The highest gain in energy saving is achieved by a hybrid approach, which reconfigures the bandwidth, antenna numbers, and carriers according to the traffic load in a semi-static way. More than 50% of energy saving is achieved in this approach.

9.5 Layered structure for energy saving

A layered structure-based wireless access network is a composite wireless network with two to several layers of radio access technologies (RAT) to offer the same services to end users. The layer is used as a logical term to indicate the heterogeneity between layers, and is characterized by the RAT, the reach range of the RBS, the deployment strategy, and the topology in the layer. A layered structure with two layers is illustrated in Figure 9.7. Normally layers are defined by their access nature, for instance, wireless wide-area networks (WWAN) as one layer and wireless local-area networks (WLAN) as the other. In different layers, a homogenous RAT can be applied, e.g. LTE for macro-cell, micro-cell, and femto-cell. A layered structure is able to boost capacity and coverage, and improve radio resource utilization.

9.5.1 System model and assumptions

The layered structure is proposed as an energy-efficient architecture. The studied layered LTE structure, as shown in Figure 9.7, is composed of two layers: macro-cell and

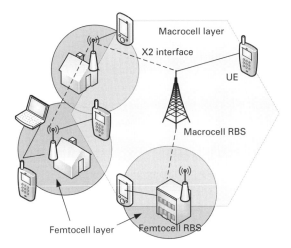

Figure 9.7 A layered structure with two layers.

femto-cell. The RBSs in the macro-cell, or in other words, eNodeBs, are deployed hexagonally as in a typical cellular system. The distance between two adjacent eNodeBs is $2R_m$ and the range of an eNodeBs is therefore R_m. The studied macro-cell is denoted as C_0. The maximum radiated power by an eNodeB is set to P_{Max}^m per sector. The RBSs in the femto-cell, namely, HeNodeBs, are normally installed without cell planning. We assume that they are uniformly distributed in the region covered by C_0. In the rest of the chapter, we use eNodeB and macro-cell RBS, as well as HeNodeB and femto-cell RBS, interchangeably. The femto-cells are denoted as F_j where $j \in [1, M]$ and M is the number of femto-cells. The range of a femto-cell, denoted by R_f, is much smaller than R_m. The maximum radiated power of a HeNodeB is P_{Max}^f. The user terminal, or UE in LTE, denoted by U_l with $l \in [1, K]$, are uniformly located in C_0. K is total number of UEs in the region of C_0.

The RBSs of a macro-cell are assumed to be able to adjust their power consumption according to the downlink traffic load. This is a reasonable assumption since the RBSs of LTE can provide a certain degree of power adaptation [14, 15]. As the transmitting PA of an RBS works only for the downlink, we only consider the downlink traffic in this chapter. Except for the signaling, we assume only the downlink traffic in both macro-cell and femto-cell. The interference model is built upon it. The interference is also related to the spectrum access model of the macro-cell and femto-cell. There are two spectrum access models for femto-cells: spectrum sharing or spectrum allocation. The former uses a flexible spectrum use (FSU) model with macro-cells, while the latter allocates dedicated spectrum to femto-cells. The spectrum-allocation model is assumed in this chapter.

We assume the open access mode in femto-cells, in which a femto-cell permits the access of every UE in its coverage range. This mode reveals the upper bound of energy-saving gain in the layer structure. However, the proposed approach allows the use of the closed access and hybrid access modes. A UE is allowed to hand over between macro-cells and femto-cells, according to the energy-saving goal. We assume a semi-stationary mobility model of UEs, in which the mobility of UEs will not affect their association with femto-cells.

The traffic model determines the energy-saving gain of the proposed structure. The assumption is that each UE has its own requirement on quality-of-service (QoS), which maps to a certain amount of traffic load at the RBS. We assume that for each UE in a macro-cell, the downlink traffic load is the same. If a UE moves from a macro-cell to a femto-cell, the associated traffic load with that UE is removed from the macro-cell. We admit that the traffic model in a real system is rather complex. However, it is common sense that fewer UEs in a macro-cell generate less load.

The energy for control signaling is assumed to be fixed and relatively small compared to that consumed by the downlink data traffic. Hence this part of the energy is not considered in the calculation of the energy-saving gain.

9.5.2 Energy-consumption model of RBS

The energy-consumption model depicted in Figure 9.1 is used for the macro-cell and femto-cell RBS. It should be noted that the analytical models for these two types of RBS are quite different due to their implementation and required power. An eNodeB consumes much more power than a HeNodeB, and its power consumption is more sensible to the load condition since the power consumption of a PA is proportional to the traffic load. The power consumption of a HeNodeB, on the other hand, is independent of the load and almost fixed.

The power-consumption model of a macro-cell can be defined in different ways. In the Celtic OPERA-NET project [14], it is defined as

$$P_{RBS} = P_P + m \times (P_{TRX} + k \times P_{RF}/c), \quad (9.8)$$

where P_{RBS} is the power consumption of an RBS, P_P is the power of the processing unit, m is the number of antenna sectors, P_{TRX} is the fixed part of radio transmission power, k is the load factor, P_{RF} is radiated power, and c is a conversion factor representing radio frequency (RF) load dependency.

In [15], the power consumption of an eNodeB is modeled as

$$P_{RBS} = P_0 + \Delta \times P_{RF}, \quad (9.9)$$

where P_0 is the power of P_{RBS} at 1% of maximum P_{RF}, Δ is the power increasing factor, and P_{RF} is the output power of an RBS. For an LTE macro-cell, we assume $P_0 = 712$ W, $\Delta = 14.5$, and the maximum of P_{out} is 3×40 W for 3 sectors.

According to [15], the power consumption of a HeNodeB is rather simple, as the power consumption of the radio part is almost fixed. We have

$$P_{HeNB} = P_{fix} + P_{radio}, \quad (9.10)$$

where P_{HeNB} is the power of a HeNodeB, P_{fix} is the power consumed by other parts except the radio, and P_{radio} is the power consumed by the radio. According to some measurement results, we assume that P_{fix} consumes 2/3 of P_{HeNB}, and P_{radio} consumes the remaining. We assume $P_{HeNB} = 10$ W and obtain $P_{fix} = 6.5$ W and $P_{radio} = 3.5$ W.

Considering a macro-cell and the femto-cells in its coverage as a whole, the EE metric is

$$\eta_{EE} = \frac{S_{C_0} + \sum_{i=1}^{M} S_{F_i}}{P_{C_0} + \sum_{i=1}^{M} P_{F_i}} = \frac{S_{C_0} + \sum_{i=1}^{M} S_{F_i}}{P_0 + \Delta \times P_{RF} + M \times P_{F_i}}, \quad (9.11)$$

where S_{C_0} and S_{F_i} are the overall throughput of macro-cell C_0 and femto-cell F_i, P_{C_0} and P_{F_i} are the average power of macro-cell C_0 and femto-cell F_i, respectively. The unit of η_{EE} is bit/joule. To achieve energy saving, we need to maximize η_{EE} without compromising the QoS of UEs.

9.5.3 Energy-aware handover mechanism

To improve the EE defined in (9.11), we need to know when a UE should be handed over in the layered structure. Let ΔS and ΔP be the reduction of overall throughput and power after the handover, respectively. It should satisfy

$$\frac{S_{C_0} + \sum_{i=1}^{M} S_{F_i} - \Delta S}{P_{C_0} + \sum_{i=1}^{M} P_{F_i} - \Delta P} > \frac{S_{C_0} + \sum_{i=1}^{M} S_{F_i}}{P_{C_0} + \sum_{i=1}^{M} P_{F_i}}. \quad (9.12)$$

We have

$$\Delta P > \frac{P_{C_0} + \sum_{i=1}^{M} P_{F_i}}{S_{C_0} + \sum_{i=1}^{M} S_{F_i}} \times \Delta S. \quad (9.13)$$

Note that ΔP is actually the reduction of power in C_0, and ΔP and ΔS can be minus. Assuming the signaling interface between femto-cell and macro-cell, if a UE is in the coverage of a femto-cell, the macro-cell C_0 can estimate ΔP and ΔS, and collect other values in (9.13) from femto-cells. The handover decision is made by the macro-cell to guarantee that the EE as a whole is improved. Note that a handover decision is always made at the macro-cell even it is from femto-cell to macro-cell, since only the RBS in the macro-cell knows its proprietary power-consumption model. A handover from femto-cell to macro-cell occurs when significant co-channel interference can be reduced. To allow such a handover, a femto-cell has to report its request to neighboring macro-cells.

As we can see, if no interference is considered between macro-cell and femto-cells, and the target femto-cell is assumed to be always able to provide the coming UE with at least the same throughput as in the macro-cell, (9.13) always suggests a handover to the femto-cell. However, if the spectrum is shared between macro-cell and femto-cell, the throughput is affected by the interference generated after the handover. Additional power is needed in the macro-cell to compensate the interference in order to maintain the throughout. In this case, it is not always good to issue a handover to a femto-cell and (9.13) should be carefully evaluated.

The energy-aware handover mechanism works as follows. First, a UE joins the network by scanning pilot signals of RBSs and selecting the one with the best channel quality. The associated RBS can be a macro-cell or a femto-cell. During the lifetime of the network, the UE continuously scans the channels and detects other RBSs. The scanning results are reported to its host RBS. If the host RBS is a femto-cell, it will relay the scanning results to the neighboring macro-cell where the UE is under its coverage. Once a macro-cell RBS collects sufficient information and finds a handover to/from other RBS may improve the EE, it contacts the target RBS through the signaling interference and negotiates the handover. If through the admission control QoS can be guaranteed in the target RBS, the handover is initialized. After the handover, the macro-cell RBS will monitor the gain in energy saving through the changes in load. If the handover creates strong co-channel interference, the spectrum sharing coordination or a new handover will be performed by the involved RBSs to continuously improve the EE.

Table 9.1. Simulation parameters

Bandwidth	10 MHz
R_m	1 km
P^m_{Max}	43 dBm
Power model of eNodeB	$P = 712 + 14.5 \times P_{RF}$ W
Path loss of eNodeB	$L = 128.1 + 37.6 \log_{10}(R)$, dB
	R in km [16]
R_f	200 m
P^f_{Max}	15 dBm
Power model of HeNodeB	$P = 10$ W
Path loss of HeNodeB	$l = 37 + 32 \log_{10}(r) + nL_w$ dB
	r in meter, $L_w = 6.9$ dB is attenuation
	per wall [17]
Spectrum share model	Separate spectrum allocation

9.5.4 Simulation study

Simulations are used to show the energy saving achieved by the layered structure. We study the case where there is only one eNodeB and multiple HeNodeBs and UE is randomly distributed in the coverage of the eNodeB. The simulation parameters are listed in Table 9.1.

We assume that there is no interference between eNodeB and HeNodeB, and leave the study of spectrum share model between eNodeB and HeNodeB for future work. The power consumption is compared in two scenarios: eNodeB only, and eNodeB with HeNodeBs. The QoS model is the following. The achievable rate of the eNodeB at the cell edge is obtained, which is $R = 18.9$ Mbits/s. Assuming that the number of UEs is K, each UE receives a rate of R/K. When a UE is handed over to a HeNodeB, it receives the same rate of R/K. The rates of all other UE remain the same.

The total power consumed by all eNodeB and HeNodeBs of the layered structure is shown in Figure 9.8. Our first observation is that if the power of HeNodeBs is considered in the calculation of gain in energy saving, only the 5 HeNodeBs case consumes less power than the eNodeB-only case. The deployment of a femto-cell brings the energy cost. Seen from Figure 9.9, as the number of HeNode Bs increases, the saving of the total RF power from the layered structure diminishes. Considering each HeNodeB consumes a fixed amount of power, as the number of HeNodeBs increases, the associated energy cost increases proportionally, and then diminishes the energy-saving gain obtained from eNodeB by handover. From Figure 9.8 we also notice that the total power almost keeps the same regardless of the number of UEs. This is because the overall rate of the layered structure is fixed to R.

It is of interest to show the gain in energy saving obtained by the eNodeBs. In some scenarios, HeNodeBs have been deployed for other purposes, therefore their energy consumption should not be counted in the energy-saving calculation. Figure 9.10 shows

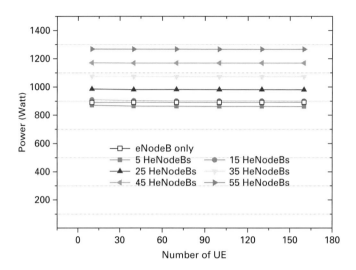

Figure 9.8   Total power-consumption of the layered structure as function of UE number.

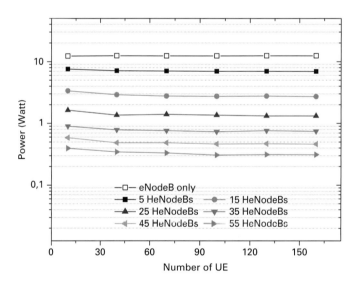

Figure 9.9   Total RF power of RBSs in layered structure as function of UE number.

the power saving achieved by eNodeB when the layer structure is used. It is easy to see the saving gain increases as the number of HeNodeBs increases. However the increase rate of the saving gain becomes smaller as the number of HeNodeBs becomes large. It also shows that the saving gain keeps almost the same under different numbers of UEs. As the overall rate of the whole layered structure remains the same, the increase of the UE number reduces the rate per UE. The overall rate and therefore the RF power of the eNodeB remains the same regardless of the UE number.

9.6 Layered structure for energy saving

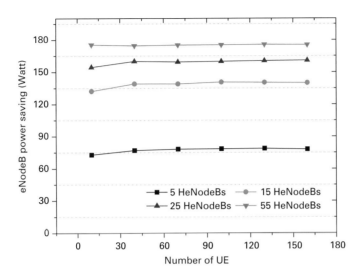

Figure 9.10 eNodeB energy-saving gain as a function of UE number.

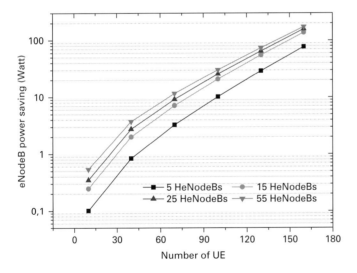

Figure 9.11 eNodeB energy-saving gain as a function of UE number in the fixed-rate case.

We also show the gain in energy saving of eNodeB when each UE receives the same rate in different UE number cases. Figure 9.10 shows the results when each UE receives a rate of 118 kbits/s. Different from Figure 9.10, the gain in saving increases as the number of UEs increases. The reason is that in this case the load of the eNodeB is proportional to to the number of UEs. It shows that when the load of eNodeB is low, the gain in energy saving from the layered structure is less significant.

9.6 Conclusion

We have provided an overview of the network energy-saving techniques proposed for the LTE system. Moreover, a layered structure has been proposed as an energy-efficient architecture for LTE, with the objective of improving the energy-saving performance of the system/network based on those network energy-saving techniques. The energy saving of the layered structure comes from the energy-aware load balancing between macro-cells and femto-cells. The study shows that under the separate spectrum allocation mode the proposed architecture can reduce the energy consumption of macro-cells. We do not study the co-channel spectrum access between macro-cells and femto-cells and the corresponding handover mechanism here and leave this study for future work.

As current energy-saving techniques used to improve RBS efficiency are mainly applied in a single system, the proposed layered structure approach is a solution to target the energy-efficiency problem from a broad view. To fundamentally improve the EE of wireless systems, we argue that a holistic view is required. Fundamental research problems remain in this area.

The first step to solve those problems is to develop a clear understanding of the energy consumption in current wireless networks in a wide range of scenarios. Most likely any energy-saving solution comes with a certain cost. The energy saving achieved in some components of a network may cause additional energy consumption in other components. Solutions need to be carefully evaluated from a holistic view, with all impacts taken into account.

Another challenge is to derive energy-efficiency metrics accurately and efficiently quantify the energy consumption. Those metrics are the key to evaluating energy-saving solutions. The basic form of energy-efficiency metrics is bit/joule or joule/bit. It is not always easy to get that figure when the measurement is not possible. Other metrics, for instance, km^2/W or subscribers/W for cellular system [5], can be used. Note that measuring those metrics may cost additional energy.

One of the key research problems in energy saving is to make energy consumption scale with the traffic load, and further with services. It is a goal that can only be achieved using a holistic solution across all layers of protocol stack and heterogenous networks. Novel architecture and methods are demanded, which most likely cross the boundary of currently isolated systems.

The most promising way to improve the EE of wireless systems may lie in cognitive approaches, in which learning is used for better adaptation [18]. Cognitive approaches can be used at the spectrum level for spectrum coexistence and interference management. It is specially important for the dynamic spectrum access (DSA) scenario. Moreover, beamforming through multi-input multi-output (MIMO) antennas, powered by cognitive techniques, has great potential to isolate interference. Interference mitigation directly results in energy saving. Cognitive approaches can also be applied at the network layer to integrate heterogenous networks and let EE become one of the primary optimization objectives to provide service across those networks.

References

[1] R. Tafazolli and V. Mattila, "eMobility mobile and wireless communications technology platform: strategic research agenda," Version 7, 2008. [Online]. Available: www.emobility.eu.org, accessed May 2010.

[2] S. Armour, *et al.*, "Green radio: sustainable wireless networks," *IET Magazine, Readers Articles*, 2009.

[3] H. Karl (ed.), "An overview of energy-efficiency techniques for mobile communication systems," TKN, Technical University Berlin, Tech. Rep. TKN-03-XXX, 2003. [Online]. Available: www.tkn.tu-berlin.de/publications/papers/TechReport_03_017.pdf, accessed May 2010.

[4] "Energy saving techniques to support low load scenarios," 3GPP R1-101084, Huawei, 2.2010.

[5] "Energy efficiency of wireless access network equipment," ETSI TS102706, 2009. [Online]. Available: www.etsi.org, accessed May 2010.

[6] "Flexi EDGE Base Station Follows ETSI Energy Efficiency Standards," Nokia Siemens Networks, 2010. [Online]. Available: www.nokiasiemensnetworks.com/news-events/press-room/press-releases/nokia-siemens-networks-first-to-publish-wireless-access-network, accessed July 2010.

[7] "Network and Telecom Equipment - Energy and Performance Assessment," ECR Initiative, 2010. [Online]. Available: www.ecrinitiative.org/pdfs/ECR_2_1_1.pdf, accessed July 2010.

[8] M. Alouini and A. Goldsmith, "Area spectral efficiency of cellular mobile radio systems," *IEEE Transactions on Vehicular Technology*, vol. 48, no. 4, pp. 1047–1066, 1999.

[9] F. Richter, A. Fehske, and G. Fettweis, "Energy efficiency aspects of base station deployment strategies in cellular networks," in *Proc. of IEEE VTC Fall 2009*, Anchorage, USA, 2009.

[10] "Network energy saving," 3GPP R1-100199, Motorola, 1.2010.

[11] "Extended cell DTX for enhanced energy-efficient network operation," 3GPP R1-095011, Ericsson, 11.2009.

[12] "Overview to LTE energy saving solutions to cell switch off/on," 3GPP R1-100162, Huawei, 1.2010.

[13] "Considerations on energy saving solutions in heterogeneous networks," 3GPP R3-092478, Ericsson, 10.2009.

[14] "WP2.1 E2E mobile network power consumption analysis," Celtic OPERA-NET project, 2009.

[15] "D2.3 Energy efficiency analysis of the reference systems, areas of improvements and target breakdown," EU FP7 EARTH project, 2010. [Online]. Available: https://www.ict-earth.eu/publications/deliverables/deliverables.html, accessed March 2011.

[16] "Physical layer aspects of UTRA high speed downlink packet access," 3GPP TR 25.848, 2001.
[17] D. Cichon and D. Kurner, "EURO-COST 231 Final Report," Celtic OPERA-NET project, 1998.
[18] H. Zhang, "Cognitive radio for green communications and green spectrum," in *COMNETS 2008* in conjunction with *CHINACOM 2008*, Hangzhou, China, August 2008.

10 Power management for base stations in a smart grid environment

Xiao Lu, Dusit Niyato, and Ping Wang

10.1 Introduction

The overall contribution of cellular network operators to the entire human CO_2 emissions is estimated at 2.5% in the US [1]. About 60% – 80% originates from wireless base stations (BSs) [2]. As current cellular network architectures are designed to cope with peak load and degraded conditions, underutilization of them occurs most of the time. A recent study [3] shows that the average power-consumption of the traditional BS amounts to nearly 850 W, with only up to 40 W power consumed to transmit from the antennas and the rest wasted even during idle operation. This result indicates that there is much room for power savings in today's cellular networks.

In this chapter, we consider the problem of power management for BSs with a renewable power source in a smart grid environment. In Section 10.2, we first provide an introduction to green wireless communications with the focus on two closely related research fields, i.e. renewable power source and smart grid. Then, we provide an overview of the power-management approaches for BS, which consists of two major directions, i.e. BS power control and smart BS operation. The former is achieved at the equipment level, while the latter can be realized at the system/network level. Afterwards, we discuss some challenges and open issues with regard to power management for BS. In Section 10.3, we present the power-consumption model for a BS. Specifically, the power-consuming components are first introduced and analyzed. Moreover, we present two power-consumption models, one for macro BSs that contain a static power-consumption part only, and the other for micro BSs that additionally consist of a dynamic power-consumption part. The power-consumption models of macro BSs and micro BSs allow us to characterize, quantify, and compare different deployment strategies when realistic input parameters are available. In Section 10.4, we propose an adaptive power-management approach for wireless BS with a renewable power source in a smart grid environment.

While the main power supply for the BS is from the electrical grid, an alternative is from a solar panel. An adaptive power-management approach is used to coordinate between the electrical grid and the solar panel, which is energy efficient and allows for greater penetration of variable renewable power sources in a green communication system. With smart grid technology, the adaptive power management can be informed of the instantaneous power price from the electrical grid, and then adjustment of the power buying can be performed accordingly. However, in such an environment, many parameters are uncertain (e.g. generated renewable power, power price from electrical grid, and power consumption of BS that depends on the traffic load). Therefore, a stochastic optimization problem is formulated and solved to achieve the optimal decision on power management. The performance evaluation results show that with the optimal policy of the proposed adaptive power management, the power cost of a BS can be minimized. We believe that the proposed solution based on the studied communication architecture is a major step towards green wireless communications.

10.2 Power management for wireless base station

This section focuses on the power management for a BS, which is a strategic target for green wireless communications. In the following, we first provide a background on green wireless communications and the related issues. Then, we provide an overview of the power-management approaches for BS. Finally, we outline some challenges and open issues.

10.2.1 Green communications in centralized wireless networks

Until now, wireless communications systems have been well developed and optimized in terms of spectrum efficiency [4], transmission reliability [5], and users' satisfaction [6] from a variety of mobile applications. However, the increasing power cost and higher volume of tele-traffic demand have posed new challenges to wireless communications systems recently [7]. Owing to economical, environmental, and marketing reasons, there is an immediate need for "green" wireless communications, which is a set of concepts, designs, and approaches to improve the power efficiency of wireless systems, while meeting the quality-of-service (QoS) of mobile applications and services. The advancement of green wireless communications will benefit the network operator not only in saving the power cost through better power-efficiency per service, but also in providing the environmental responsibility through minimizing the environmental impact (e.g. by using renewable power resources to reduce CO_2 emissions). In addition, a wireless communications system has to match its own power demand to the change of power supply side in a future-generation power grid, known as "smart grid" [1], which will become more dynamic and distributed. Given all these requirements, the issue of power management for wireless systems is becoming increasingly crucial and needs to be addressed accordingly. Over the years, valuable research efforts have been made in the wireless industry, aiming for environment-friendly power solutions which lead to green wireless communications. Green wireless communications will contribute to the reduction of our global carbon footprint and enable mutual broader impacts across related fields, among which, renewable power sources [8] and smart grid are attracting growing interest. In the following, we present an introduction to the major concerns related to renewable power sources and the smart grid.

Renewable power sources and green wireless communications
Pushed by the climate change deriving from huge power consumption as a result of rapid industrial development, renewable power resources are emerging as an attractive

alternative energy because of their low pollution and sustainable accessibility. In green wireless communications, renewable power resources can be used to replenish the energy of wireless BSs and/or network nodes as replacements of traditional fossil energy.

Manufacturers and network operators have started developing and deploying wireless BSs with renewable power sources [9]. For example, Ericsson and Telecom Italia developed and tested the Eco-Smart solution that uses a solar panel to fully power the cell site. Vodafone, China Mobile, and Huawei jointly performed various experiments on the renewable power sources including solar panel, wind power generator, and the hybrid system for wireless BSs [10]. The experiments focus on the implementation verification, power reliability, and cost reduction. However, renewable power sources (e.g. solar, wind, hydro, geothermal, tidal energy, and biomass) are typically featured as weather-driven, unevenly distributed in a geographic region, non-scheduled, and relatively unpredictable. Since the amount of renewable power generation is known to fluctuate and be intermittent, efficient power management to match the power-replenishment rate with actual power demand has become one of the primary concerns for the applications of renewable power sources.

In wireless networks, power saving often requires a degradation in network performance (i.e. higher latency and lower throughput) [11]. Compromise between power saving and network performance therefore should be taken care of in the design of efficient power-management schemes. A number of interesting works have been carried out to address this issue in applying renewable power sources in wireless networks. Compromise between throughput and power constraints for a wireless network employing renewable power sources was studied in [12]. Optimal scheduling algorithms were presented to maximize the throughput and total transmission data. More recently, [13] investigated the compromise between power saving and prevention of node outage. An energy-aware resource provisioning algorithm was proposed for power provisioning in solar-powered wireless mesh network.

Smart grid and green wireless communications
High-voltage, long-distance transmission, and large-scale centralized electricity generation are the two basic causes of power inefficiency in a traditional electrical grid system. To deal with this problem, the concept of smart grids has been proposed to improve the power efficiency and reliability of the electrical grid by using information and communications technology (ICT). The key features of a smart grid related to the green wireless communications are demand-response (DR), demand-side management (DSM), decentralized power generation, and price signaling. With DR and DSM, the power generators and consumers can interact to improve the efficiency of the power supply and consumption. For example, the operation and power consumption of the deferrable load (e.g. heating and pumping) can be adjusted according to the generator capability. These two concepts will be further introduced in Section 10.4.2.

With decentralized power generation, the power generation can be performed by consumers and small power plants (e.g. solar panel and wind turbine). As a result, consumers will be less dependent on the main electrical grid, resulting in reduced power cost and reduced impact from power failure. With price signaling, the consumers will be aware of the current power price and the generators can use a lower power price to encourage consumers to use the electrical power during the off-peak period (e.g. night time or weekend). Consequently, the peak load will be reduced, which results in lower investment for the infrastructure (e.g. transmission line and substation).

The enabling technologies for integrating smart grids and wireless communications have attracted a lot of research attention [14]. On one hand, wireless communications technologies will be key components in smart grids to communicate a variety of data and measurement among power generators, transmission lines, distribution substations, and consumer loads. On the other hand, smart grids can be used to support green wireless communications for the better use of power to provide a wireless service to mobile units. A similar concept was explored in the context of "green computing" [15]. In a green computing environment, the data center can schedule the service request (i.e. data processing) according to the power supply from the electrical grid. Also, efforts have been made on theoretical analysis. In wireless networks, each wireless BS/node powered by a smart grid might be selfish in improving its own performance in terms of capacity or QoS. In this context, how to improve power saving without adversely affecting the capacity and QoS is one of the main concerns. Recent progress in wired distributed computing theory [16] has provided fundamental models for the coordinated management and load balancing of wireless BSs underneath a smart grid, which has potential for addressing the concern.

10.2.2 Approaches for power management in a base station

In past years, cellular network operators have dedicated valuable efforts to streamlining all the network components across BSs [17], mobile units [18, 19], and backhaul networks [20] for environmental as well as economic reasons. The focus of this section is on power management in the BSs, since BSs consume the most significant portion of power used to run cellular wireless networks. Power management for BSs has been extensively studied over the past few decades. The existing approaches can be achieved in two ways: BS power control and smart BS operation. The former is done at the equipment level, while the latter can be realized at the system/network level. In the following, we present an overview of power-management approaches that are applicable to BSs, which is of great significance towards green communications.

BS power control
Power control is a necessary feature in cellular wireless networks and the key for the management of interference, energy, and connectivity [21]. With power control, a BS transmits and receives at an appropriate power level to minimize the interference as well as to satisfy the required QoS. The QoS requirement is mainly to satisfy some received power and signal quality requirements at the receiver side (i.e. mobile unit).

BS power control is often formulated as an optimization problem. According to the optimization variables, the existing approaches of BS power control can be grouped into three types: opportunistic BS power control, which considers only transmit power as an optimization variable; joint BS power control and beamforming, which additionally concerns beamforming vectors; and joint BS power control and BS assignment, which considers both transmit power and connectivity in an optimization formulation.

The objective of the opportunistic BS power control approach is to adjust the transmit power and increase transmission rate when the channel is of good quality and avoid using the channel when the channel quality is below a threshold. To exploit variations in channel quality, the instantaneous and history information of channel quality are leveraged opportunistically. Opportunistic BS power control enables priority in channel access for mobile units. The concept of opportunistic BS power control can be explained in the context of a cognitive radio network [22], where primary users (users with higher priority) are always able to maintain their signal-to-interference ratio (SIR) requirement. However, because of the variation in channel quality and user mobility, the secondary users (users with low priority) have to transmit their data opportunistically without violating the SIR requirement of primary users. Opportunistic BS power control that maximizes the total downlink utilities of a cell was studied in [23, 24]. In opportunistic BS power control, a BS schedules the channel and transmit power for selected mobile units to transmit to. The most commonly adopted scheduling approaches for BS mainly include proportional fair scheduling [24, 25] and general utility-based scheduling [26].

Joint BS power control and beamforming is an approach used for power management in multiple antenna systems. With the introduction of multiple-input multiple-output (MIMO) technology, deploying multiple antennas at both BS and mobile unit provides better diversity in power control. To compromise among the hardware limitation of a mobile unit, network performance, and complexity, beamforming technology with a multi-antenna BS and single-antenna mobile unit was advocated in [27, 28]. In the downlink, the beamforming vectors spread signals over the antenna array prior to transmission [27]. The beamforming vectors are coupled and must be optimized jointly, which makes joint BS power control and beamforming a complicated problem. Approaches to address the problem have been proposed in [29]–[32] for fixed SIR requirement, and in [33] for max-min SIR fairness. The key to these solutions is the use of the uplink-downlink duality concept [34]. The idea of uplink-downlink duality is that, under the same power constraint, the optimal beamforming vectors in the downlink are also the optimal beamforming vectors in the uplink [34]–[36]. This duality theorem leverages the better-understood knowledge in uplink beamforming for the downlink problem. Based on the duality theorem, downlink BS power control problems can be transformed into their dual uplink domains and efficiently solved.

BS assignment [37] involves reassignment of BSs to mobile units to deal with the dynamic connectivity patterns between mobile units and BSs. Joint BS power control and BS assignment aims to determine the BS assignment that minimizes the required transmit power with fixed SIR targets on each link. By jointly considering power and connectivity as the network resources to be optimized, improvement in user experience could be expected. However, the joint optimization problem is difficult to solve because of the coupling in SIR on each link. For example, the power adjustment caused by the BS assignment for a mobile unit from one BS to another may create greater interference for those mobile units near the latter BS. To overcome the difficulty, solutions were proposed in [38]–[40] by applying the uplink-downlink duality mentioned above.

Smart BS operation

Smart BS operation refers to the BS power management at the system/network level to alleviate the power inefficiency that results from the fact that BSs are typically deployed and operated continuously based on peak-hour traffic estimates. Existing efforts addressing the energy-efficient operation of BSs mainly include BS mode switching and cooperative relaying.

In the BS mode-switching approach, the BS can operate in two modes, i.e. active and sleep. This approach is applied in the cellular wireless networks where BSs are densely deployed, e.g. urban areas. The idea of BS mode switching is to dynamically minimize the number of active BSs to meet the traffic variation in the network. The overlapped coverage of the BSs results in a small number of the mobile units associated with each BS and high traffic dynamics among BSs [41], which makes it feasible for BS mode switching to save power. To achieve this, some BSs are switched off during the periods when they are under-used because of low traffic load, leaving their radio coverage and service quality taken care of by the BSs that remain active. A case study in an urban UMTS network shows that a little amplification in the transmit power of active BSs is negligible compared to the total power-consumption saved by switching-off the BSs [42]. Thus, the wide implementation of BS mode switching has the potential to substantially reduce the power consumption of wireless cellular networks. BS mode switching approaches have been proposed for BSs in cooperative 2G/3G networks [43, 44], 2G, and HSPA networks [45], and the Long Term Evolution (LTE) system [46]. Except for the studies of BSs in wide area networks, BS mode switching approaches for small cell BSs, e.g. femto-cell BS and pico-cell BS, have also been investigated and developed [47, 48]. Moreover, joint consideration of BS mode switching problem and mobile unit association for energy-delay tradeoffs with timescale separation has been studied in [49].

The objective of cooperative relaying is to minimize transmission power-consumption of wireless cellular networks, subject to the average transmission rate and signal-to-interference-plus-noise ratio (SINR) requirement of a mobile unit's traffic. Cooperative relaying can be used in cellular wireless networks with intra-cell orthogonal medium access control (e.g. TDMA, FDMA, or orthogonal CDMA), and decode-and-forward relaying adopted by either mobile units or relay stations (RSs). RSs can be considered as mini BSs that can transmit and receive signals from mobile units and have wireline connections to BSs. In the literature, two different types of relaying network architecture have been investigated. One uses a mobile RS as a signal relay between BSs and mobile units [50, 51]. The mobile RSs can also be other mobile equipment with additional features to function as relays. In the other case, RSs are fixed at certain locations in the cell [52]–[54]. When a BS needs to transmit a signal to mobile units, RSs placed in between are chosen for signal relaying. It is proven that cooperative relaying solves the problem of asymmetrical coverage [55] and "dead spots" [56], and can increase system coverage [57] and capacity gains [58], as well as reducing path loss and power consumption [59]. With regard to the effect of power-consumption saving, case studies have been made in [60] for cooperative broadcasting in dense wireless cellular networks, in [61] by joint power minimization with routing in an ad hoc network, and in [62] for a pair of source and destination nodes in a network with a number of RSs.

10.2.3 Open research issues

In the previous section, we have provided an overview of the existing approaches to power management for wireless BSs. In this section, we outline some open research issues related to advancing the power management for wireless BSs. The open research issues are presented accordingly with regard to our previous classification of the BS power-management approaches.

The open research issues in BS power control are summarized as follows:

- In opportunistic BS power control, feedback of channel quality measurement is the key to selecting mobile units to transmit. However, the feedback is sometimes inaccurate and delayed. How to characterize the impact of the limited information from feedback of channel quality remains an open issue for opportunistic BS power control.
- Joint BS power control and beamforming can be formulated in two ways: minimization of total transmit power for fixed SIR and maximization of total utility for variable SIR. Existing solutions have solved the former problem by adding a beamforming vector update to power-control algorithms [29]–[32]. As for the latter problem, solutions have been attempted only in special cases [33, 63]. Joint power control and beamforming for maximizing a general concave utility function are yet to be fully studied.
- The main challenge in joint BS power control and BS assignment lies in the coupling in SIR on each link, combined with the integer constraints introduced by connectivity variables, which is hard to solve. Also, joint power control, beamforming, and BS assignment has been proposed for further optimizing network performance [64]. The joint optimization problems require more research efforts.

The open research issues in smart BS operation are summarized as follows:

- In BS mode switching, frequent mode switching should be prevented, considering factors in implementation, e.g. signal overhead processing, working stability, and equipment lifetime. How to achieve the best tradeoff between those factors in implementation and power saving requires further study. Besides, more research efforts on the effects of BS switch-off on coverage are needed, taking the consideration of technological aspects such as cell breathing [65] and antenna tilting [41] into account.
- As for cooperative relaying, channel knowledge is crucial for RS selection. Most of the studies on cooperative relaying assume perfect channel knowledge at a mobile unit and BS side. However, it is hard to acquire accurate knowledge about the channel, e.g. fast-fading channels only remain stationary for a small amount of time. Thus, the existing cooperative relaying approaches would not work well in a real implementation. How to cope with the positioning of relaying when the channel knowledge is inaccurate is a challenge. Another important issue is the incentive mechanism for cooperative relaying, which is also associated with implementation. Although from the collective point of view, the total power consumed from relay transmission by the RSs and mobile units is lower than that from direct transmission by BS, a mobile unit will not relay traffic for another one unless it also benefits. It is necessary to study the mechanism to optimize the incentive for a mobile unit to participate in relaying before implementing cooperative relaying approaches.

10.3 Power-consumption model for a base station

Today's cellular networks mainly consist of a large amount of conventional macro BSs and a small number of recently adopted micro BSs. Macro BSs are strong in power and usually cover an area with a radius of about 500 meters up to 2500 meters and a degree of coverage of at least 90% in urban areas [66]. The average power-consumption of a macro BS is determined by its covered area and the degree of coverage. Comparatively, a micro BS is a small-coverage, low-power, lower-cost cellular BS with the target to expand coverage and improve the capacity of a wireless cellular network [67]. Compared to the macro BS, the area covered by a micro BS generally enjoys much higher average SINR due to advantageous path loss and shorter propagation distance. Due to spectrum scarcity and the high bandwidth requirements of the mobile units, a much more dense wireless BS deployment is expected in future wireless cellular networks, particularly with regard to traffic coverage. With the increasing demand of high QoS in wireless cellular network, it is expected that a significant number of micro BSs will be deployed in the future [68]. Therefore, a significantly increased demand in power consumption for wireless cellular networks can be envisioned. This also indicates that the environmental problem associated with wireless BS will become an important issue in the coming years.

This section deals with power consumption in macro and micro BSs. In particular, power-consumption models for these BSs are introduced with a focus on energy consumption at the component level. Power consumption is one of the most important figures of metrics to measure the energy efficiency of deployment strategies applied in wireless cellular networks. The power-consumption models allow us to characterize, quantify, and compare different deployment strategies when realistic input parameters are available.

10.3.1 Components of a base station

A BS is equipment communicating with the mobile units and the backhaul network. In a wireless BS, there are typically several power-consuming components. Figure 10.1 gives an overview of these components. These components include

- Power amplifier (PA) is responsible for amplifying input power. To maximize PA efficiency, a PA is expected to work in a state in which the peak value of the signal corresponds with the possible peak power of the PA.
- Digital signal processing is responsible for system processing and coding.
- A/D converter converts an input analog voltage or current to a digital number proportional to the magnitude of the voltage or current.
- Transceiver is in charge of receiving and sending the microwave signal to mobile units.
- Signal generator produces a microwave signal.
- Antenna is used to transmit or receive microwave signals.
- Feeder is the component to feed the microwave signal to the rest of the antenna structure.

10.3 Power-consumption model for a base station

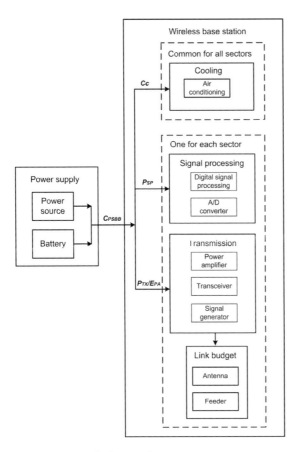

Figure 10.1 Block diagram of a base station.

Since the power consumption of an A/D converter is less than 5% of a macro BS's input power [69], it is not considered separately and is assumed to be included in the signal processing part. The power consumption of the signal processing part is denoted as P_{SP}. The transmission part consists of PA, transceiver, and signal generator, which totally consume a power of P_{TX}. The power consumption of the antenna and feeder is included in the link budget.

Also, a BS contains equipment that is commonly used by all sectors such as cooling, which is responsible for dissipating the heat generated by the functioning of the components. The power consumption of cooling mainly depends on environmental conditions and is defined as C_c. The values are typically between zero (i.e. free cooling) and 40% [69]. Besides, a wireless BS is powered by a power supply and battery backup component. Power loss occurs within this component during power transmission. The power consumption of this component, denoted as P_{PSBB}, is typically between 10% and 15% [69] depending on the technology employed.

The main contributors to a BS's power consumption typically include the use of the PA with the corresponding link budget; different methods of cooling, e.g. air conditioning, air circulation, and free cooling; and site sharing, especially regarding infrastructure; and number of carrier frequencies.

For simplicity in modeling, each component of a wireless BS can be assumed to consume a constant value of power, except for the PA, the power consumption of which depends on the input power of the antenna. A parameter, PA efficiency E_{PA}, is defined as the ratio of transmit power to direct current input power to characterize the power consumption of a PA.

10.3.2 Assumptions and power-consumption model for a macro base station

The power consumption of a BS varies over time, depending on the mode it is working in. Two modes can be distinguished; specifically, a BS can work in an operational or non-operational mode. In an operational mode, there are low traffic and peak traffic modes. Furthermore, a transition time is needed for a BS to switch between different modes, e.g. turning on the power of a BS consumes a dedicated amount of time and power. Since the time spent by a BS in each mode is usually greater than all the transition time, it is assumed that the transition time is disregarded in the model for both the macro BS and micro BS.

The power consumption of a BS is composed of two parts: static and dynamic power-consumption. The former describes the consumption that already exists in an empty BS, while the latter depends on the dynamic load situation. A recent measurement [3] shows that the power consumption of a macro BS is barely relevant to the traffic load. For example, over a period of several days, about a 3% and 2% variation in power consumption for a UMTS and a GSM BS is observed, respectively, while the data traffic of both BSs varies between no load and peak load level. Thus, a macro BS is characterized to be independent of the load level. Also, since the amount of dynamic power is negligible, the dynamic power-consumption is disregarded and the power consumption of macro BSs is considered to be static.

Considering all the different components of a macro BS, the power consumption can be calculated as follows:

$$P_{BS,Macro} = N_{Sector} \cdot N_{PApSec} \cdot \left(\frac{P_{TX}}{E_{PA}} + P_{SP} \right) \cdot (1 + C_C) \cdot (1 + C_{PSBB}), \quad (10.1)$$

where the parameters are denoted in Table 10.1.

10.3.3 Assumptions and power-consumption model for a micro base station

It is assumed that a micro BS consists of only one sector which contains one PA. Neither battery backup nor air conditioning is typically needed for power supply and cooling, respectively, in a micro BS. Compared to a macro BS, the design size of a micro BS is much more compact, which results in a limited transmit power and coverage area. Thus, the power consumption is considerably smaller.

Since the number of mobile units in a micro-cell is statistically varying, the traffic load variation, and thus the power consumption are dynamic. The power adaptivity of

Table 10.1. Table of notations

Symbol	Definition
N_{Sector}	Number of sectors
N_{PApSec}	Number of PAs per sector
P_{TX}	Transmit power
E_{PA}	PA efficiency
C_c	Cooling loss
P_{SP}	Power consumption for signal processing overhead
C_{PSBB}	Power supply and battery backup loss of macro BS
P_{MTX}	Maximum transmit power per PA
$C_{TX,static}$	Static transmit power
$P_{SP,static}$	Power for static signal precessing
C_{PS}	Power supply loss of micro BS
C_{TX,N_L}	Dynamic transmit power per link
P_{SP,N_L}	Dynamic signal processing per link
N_L	Number of active connections

the PA is the key for energy efficiency in a micro-cell. For example, for a given mobile unit density, there is a high probability that no one is located in a micro-cell; lowering the transmit power in this case will lead to abundant power saving. Therefore, micro BSs are usually equipped with more efficient PAs to cope with the traffic load variation. It is assumed that the PAs in micro BSs will be able to adapt their power consumption to traffic load conditions.

The power consumption of a micro BS can be modeled as

$$P_{BS,Micro} = P_{static,Micro} + P_{dynamic,Micro}, \qquad (10.2)$$

where $P_{static,Micro}$ and $P_{dynamic,Micro}$ represent the static and dynamic power consumption, respectively, in a micro BS.

Similar to that of a macro BS, the static power consumption of a micro BS can be calculated as

$$P_{static,Micro} = \left(\frac{P_{MTX}}{E_{PA}} \cdot C_{TX,static} + P_{SP,static}\right) \cdot (1 + C_{PS}), \qquad (10.3)$$

where the parameters are explained in Table 10.1.

The digital part (i.e. digital signal processing) is the main cause of the dynamic part of the power consumption of a micro BS. It is assumed that the digital part scales according to the number of active connections. Thus, the dynamic power consumption can be calculated as

$$P_{dynamic,Micro} = \left(\frac{P_{MTX}}{E_{PA}}(1 - C_{TX,static}) \cdot C_{TX,static} + C_{TX,N_L} + P_{SP,N_L}\right)$$
$$\cdot N_L \cdot (1 + C_{PS}), \qquad (10.4)$$

where the parameters are explained in Table 10.1.

10.4 Optimization of power management in a smart grid environment

The fact that the current power network (i.e. the electrical grid) needs to provide enough generation, transmission, and distribution capacities for peak rather than for average demand leads to a surplus of power supply and a large amount of waste for most of the time, as load demand fluctuates periodically. Smart grid technology has been proposed to address the inefficiency of the electrical grid, which has become a social issue due to resource depletion and increase in power cost. A smart grid is an auto-balancing, self-monitoring electrical grid that integrates various generation concepts and technologies [70]. In smart grids, consumers adjust their consumption based on their demand and market information to optimize the use of power sources and minimize the negative impact on the environment (e.g. CO_2 emission).

One of the strategical approaches to improve efficiency and reducing power waste is to match the supply [71]. However, with the steady rise in the proportion of renewable power sources, power supply nowadays is also becoming highly time-variant, which makes matching the supply increasingly challenging. To address the challenge, we study a green communication system model where a BS is provisioned with a combination of the electrical grid and a renewable power source. Taking into account the real-time availability of a partly unpredictable supply, we design a reliable and efficient DR program, called adaptive power management, with the purpose of facilitating the adoption of renewable power resource generation by matching the load demand to an ever-changing power supply.

10.4.1 System model

The system model of adaptive power management for wireless BS is shown in Figure 10.2. The components in this system model are as follows:

- *BS*: Wireless BS or access point is a centralized device used to provide wireless services to mobile units. As introduced before, wireless BS is the main power consumer, the power consumption of which depends on the type of BS (i.e. macro BS or micro BS) and traffic load (i.e. the number of ongoing connections from active mobile units).
- *Electrical grid*: An electrical grid is an interconnected network composed of power lines and distribution substations for electricity delivery from generators to consumers. The electrical grid acts as the main source of power supply to the wireless BS. The power supplied from the electrical grid is charged at a price per kWh (kilowatt-hours).
- *Renewable power source*: Renewable power is provided from natural resources such as sunlight and wind that are replenishable. As a result, the variable cost of renewable power is usually cheaper than that from the electrical grid. However, owing to the unpredictable availability of natural resources, the amount of power generated is typically highly time-variant. Therefore, the renewable power source is considered to be an alternative power supply to the electrical grid. The maximum amount of power (i.e. capacity) generated from the renewable power source is denoted by R kW (kilowatt).
- *Power storage*: A battery is the power storage device for a wireless BS. A battery can be charged by the power from the renewable power source when it is available or from the electrical grid when the power price is low. A battery has a limited maximum

10.4 Optimization of power management in a smart grid environment

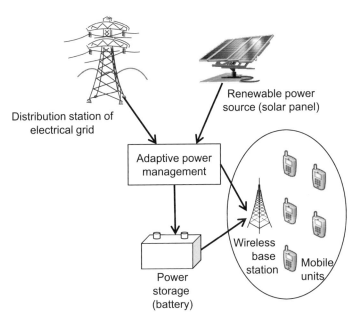

Figure 10.2 System model of adaptive power-management for a base station powered by smart grid.

capacity for power storage, denoted by B kW. Note that the power stored in a battery can decrease even without consumption. This is referred to as the self-discharge phenomenon [72]. The self-discharge rate per time unit is denoted by L.

- *Adaptive power-management controller*: An adaptive power-management controller is implemented with a DR program to make the decision on the power supply from the renewable power source and the electrical grid to the battery and wireless BS. The design goal of the adaptive power-management controller is to make an optimal decision to minimize the power cost, subject to the constraint that the power demand of wireless BS should be met. The details of how to realize optimal decision making will be presented later in this section.

While the electrical grid is owned by the utility company, the wireless BS, power storage, renewable power source, and adaptive power-management controller belong to the network operator, with the objective of minimizing the power cost. The information to be exchanged and maintained among the above components to support adaptive power-management for wireless BS includes: the price of power from the electrical grid, the amount of power generated from the renewable power source, the battery storage, and the power consumption of the wireless BS. This information needs to be measured and reported periodically to the adaptive power-management controller. We assume that the communications infrastructure to transfer this information is available, as this communication capability will become the norm in future smart grids [73]. To enable information exchange in this case, broadband access (e.g. ADSL) and a local area network (e.g. Ethernet) can connect the adaptive power-management controller with the electrical grid and renewable power source as well as with the wireless BS.

Adaptive power-management can be considered as a DR program, the main aim of which is to provide an incentive to mobile units to defer their power consumption from peak periods to off-peak periods. In the environment of the system model under consideration, adaptive power-management can control the power buying (i.e. consumption) from the electrical grid and from renewable power sources given the varied price and the amount of renewable power generation. In addition, the adaptive power-management controller will defer buying power when the price is high (e.g. peak hour) and shift the consumption (e.g. charging battery) to the off-peak duration such as night time. Micro BSs are considered in which the power consumption depends on the traffic load (i.e. dynamic power-consumption is significant compared to static consumption). The model described in (10.2) is used.

10.4.2 Demand-response for base station in smart grid

An important feature of a smart grid is that consumers work as an integral part of the power system, participating in system operation and management. With the aim of balancing the power consumption from peak to off-peak period, reducing the cost of infrastructure to accommodate peak demand, it is critical for a smart grid to enable the power consumers (e.g. wireless BS) to adjust their own power consumption. In this context, two concepts, DR and DSM, are basically perceived as the solutions to achieve the aim.

DR refers to the short-term changes targeted for the critical hours during a period (e.g. a day/year), when the demand is high or when the reserve margin is low; while DSM is the long-term changes in power consumption or customer behavior achieved by demand-side efforts. By enabling mobile units to take actions in response to the supply conditions, both DR and DSM turn the electrical grid from a vertically controlled structure into a collaborative environment where the grid system is not only governed by the generation side, but also affected by the participating mobile units.

This section focuses on DR scheduling and implementation in a smart grid system. Under a smart grid, the objective of DR is to average out the load shape, shifting as much of the flexible demand as possible away from the peak period into less stressful periods (i.e. through power storage). This requires careful balancing of the relationship between the electrical grid operators and mobile units. The advantage of DR is twofold. First, it alleviates the maximum power-generation capacity required by a utility in order to avoid instability of the electrical grid, occurrences of equipment failures, blackouts, or brownouts. Second, by shaping the power usage to remain relatively constant over time, it avoids repeatedly starting and stopping power-generating units.

The management of DR in a smart grid involves four main participants: a power-balancing authority that orders (i.e. offers or requests DR), a DR aggregator that integrates individual DR resources, a distribution utility/system operator that operates the distribution grid where DR events are happening, and power consumers that are subscribed to the DR programs. In the system model introduced in Section 10.4.1, an adaptive power-management controller is a combination of the power-balancing authority and the DR aggregator, while the electrical grid and wireless BS work as the distribution utility/system operator and power consumer, respectively.

10.4 Optimization of power management in a smart grid environment

The scheduling process of DR management is initiated from the balancing authority, the order from whom is signalled to DR aggregators covering the whole or part of the grid territory. DR aggregators compare the order with the DR availability and select the participant units by considering each one's compliance factor, which is statistically calculated. Then expected DR actions can be determined by DR aggregators and are returned to the balancing authority as feedback.

According to the party that initiates the demand reduction, DR programs can broadly be of three types: incentive-based DR, rate-based DR, and demand reduction bid.

- Incentive-based DR is initiated by the utility or DR aggregator to provide an incentive (i.e. reward), for a wireless BS to decrease its power consumption during peak periods. Through shifting load from peak to off-peak periods, a wireless BS can release the demand stress it places on the electrical grid. In turn, the participating wireless BS could receive a discounted charge or separate incentive reward.
- Rate-based DR charges the price of electricity on a real-time basis and lets the wireless BS voluntarily respond to the varying electricity price. Under this type of DR program, the wireless BS would have to pay a high price for peak periods and a low price for off-peak periods.
- Demand reduction bid is to encourage wireless BSs to reduce load demand at prices for which they are willing to be curtailed. The bid is initiated and sent from the wireless BS to the utility or DR aggregator, normally with the information of the amount of demand reduction and the price asked for.

In the following section, we introduce an adaptive power-management approach for a wireless BS in a smart grid environment, which can be regarded as an Incentive-based DR program.

10.4.3 Optimization formulation for power management

The challenge of adaptive power-management for wireless BS rests with the uncertainty in the smart grid environment and system. To address this issue, a stochastic optimization problem can be formulated and solved to realize the best decision making of power buying for the adaptive power-management controller. The general objective is to minimize the cost, while meeting the demand.

Uncertainty

A variety of uncertainties exist for the power management for wireless BSs, which are listed as follows.

- *Renewable power source*: Due to the weather-driven nature of renewable power resources, the amount of power generated from renewable power sources such as solar and wind generators is highly time-variant [74]. For example, solar energy depends on the amount of sunlight. Unpredictable factors in weather, like cloud and rain, can reduce the amount of generated power.
- *Power price from electrical grid*: The power price from the electrical grid fluctuates within a certain range depending on the instantaneous system conditions (e.g. demand)

of the electrical grid [75], which is unpredictable. For example, the power price can be high in a certain time interval (i.e. peak hour), and the consumers need to be informed instantly (i.e. by the price-signaling feature in a smart grid).

- *Traffic load of wireless BS*: The traffic load is random for two reasons. First, the connection arrival (i.e. newly initiated and handoff units) can be varied (e.g. due to the mobility), which results in a random number of ongoing connections N in a wireless BS [76]. Second, the connection demand of a wireless BS varies depending on the usage condition (e.g. a special event that results in peak load). As a result, the power consumption, which can be calculated from the power-consumption model introduced in Section 10.3.3, is also random.

The uncertainty can be represented by the "scenario," which is the realization of the random variable. The scenario takes its value from the corresponding space, which is commonly assumed to be a finite discrete set. For example, during a certain time interval, the power price can be taken from a set of 20 and 30 cents per kWh (i.e. off-peak and peak hour prices, respectively). The scenario can be also defined over multiple intervals. For example, with four intervals in one day, the first scenario can be defined as {20, 20, 30, 20} cents per kWh for the power prices in the morning (6:00–12:00), afternoon (12:00–18:00), evening (18:00–24:00), and at night (0:00–6:00), respectively. Alternatively, the second scenario can be defined as {20, 30, 30, 20} cents per kWh, which means in this case the peak-hour price is offered in the afternoon and evening. With multiple random parameters, the scenario is defined as a composite value of the generated renewable power, the power price from the electrical grid, and the power consumption of the wireless BS. For example, one scenario denoted by ω is defined as follows: For morning, afternoon, evening, and night, the generated renewable powers are {200, 300, 0, 0} W, the power prices are {20, 20, 30, 20} cents per kWh, and the power consumptions are {200, 250, 300, 200} W, respectively. A possible way to obtain the scenarios is to extrapolate from historical data, e.g. the traffic load history and the corresponding power price from the electrical grid. Meanwhile, the weather forecast can be used to determine the scenario for the generated renewable power.

The probability distribution associated with the scenarios of the generated renewable power, the power price from the electrical grid, and the power consumption of the wireless BS can be estimated. Given the observation period (e.g. 90 days), the number of days for the observed scenario can be counted. The corresponding probability can be then calculated by dividing this number of days of a certain period by the duration of the observation period (i.e. 90 days). For example, if the number of days for the power price scenario {20, 20, 30, 20} is 27 days, while the number of days for scenario {20, 30, 30, 20} is 63 days, then the probabilities for the first and second power price scenarios are $27/90 = 0.3$ and $63/90 = 0.7$, respectively. The same method can be applied for the scenarios for the generated renewable power and power consumption.

Stochastic programming formulation
An optimization problem based on multi-period linear stochastic programming can be formulated and solved to obtain the decision of an adaptive power-management controller

under uncertainty. Stochastic programming is an extension of deterministic mathematical programming, and can be used to model the optimization problem where there is uncertainty in the parameters [77]. An advantage of stochastic programming is that you do not need complete knowledge of the parameters. Instead, stochastic programming incorporates the probability distribution of random parameters, which can be statistically estimated, into the optimization formulation. A feasible policy for the possible cases (i.e. scenarios) can be obtained from the optimal solution of stochastic programming. This optimal solution or policy, which is a mapping from the scenario to the decision, can minimize the expectation of the objective (i.e. cost). To get the optimal solution for stochastic programming, the equivalent deterministic mathematical program can be formulated and efficiently solved by the standard methods (e.g. the interior-point method). For example, a linear stochastic programming problem can be transformed into the deterministic linear programming problem and solved to obtain the optimal solution.

Although other approaches (e.g. Markov decision process, robust optimization, and chance-constrained programming) can also cope with the optimization problem with uncertainty, they are not necessarily suitable for the cost optimization of adaptive power management for wireless BSs. For the Markov decision process, the stochastic process of the random parameters must have the Markov property, i.e. the state (i.e. scenario) of the random parameter depends on the current state rather than the past one. This Markov property may not be held in many situations in a smart grid environment (e.g. power price from the electrical grid). For robust optimization, the solution is obtained only for the worst-case scenario with which performance can be unrealistically poor due to the consideration of the extreme case. For chance-constrained programming, with optimal solution, the constraint violation will be bounded by the threshold. However, only complex analysis exists for the basic probability distribution (e.g. normal distribution).

Therefore, stochastic programming is the best approach for adaptive power-management, since this approach can be used to obtain an optimal solution which ensures that all constraints will be met. An efficient method can be applied to obtain the optimal solution for possible scenarios in which the expectation of cost given the uncertainty is minimized.

For the optimization model, we consider a T-time-interval decision horizon, with the length of each time interval to be one hour. Within a time interval, the power price from the electrical grid is constant. Then, a multi-time-interval stochastic programming model for adaptive power-management can be formulated as follows:

$$\min_{x_{t,w}} \sum_{t=1}^{T} \mathbb{E}(x_t P_t + s_t L) = \sum_{t=1}^{T} \sum_{\omega \in \Omega} Pr(\omega)(x_{t,\omega} P_{t,w} + s_{t,w} L) \qquad (10.5)$$

$$\text{subject to} \quad s_{t,\omega} + x_{t,w} + R_{t,w} = s_{t+1,w} + C_{t,w} + y_{t,w}, \quad t = 1, \cdots, T-1, \omega \in \Omega \qquad (10.6)$$

$$s_{t,\omega} \leq B, \quad t = 1, \ldots, T, \omega \in \Omega \qquad (10.7)$$

$$s_{1,\omega} = B_1, \quad s_{T,\omega} = B_T \qquad (10.8)$$

$$x_{t,\omega} \geq 0, \quad s_{t,\omega} \geq 0, \quad y_{t,\omega} \geq 0, \quad t = 1, \ldots, T, \quad \omega \in \Omega. \qquad (10.9)$$

The objective and constraints of the optimization formulation defined above are as follows:

- (10.5) is an objective to minimize the expected cost due to power buying from the electrical grid and battery loss due to self-discharging over the entire decision horizon (i.e. $t = 1,\ldots,T$). $s_{t,\omega}$ represents the amount of power stored in the battery at the beginning of time interval t in scenario ω. $x_{t,\omega}$ and $P_{t,\omega}$ are denoted as the amount of power buying from the electrical grid and the power price during time interval t in scenario ω, respectively. $\mathbb{E}(\cdot)$ is the expectation which is over all scenarios in space Ω given the corresponding probability $Pr(\omega)$ of scenario $\omega \in \Omega$.
- (10.6) is a constraint to balance the power input and output of a time interval t. The power input of a time interval t includes the power stored $s_{t,\omega}$ at the beginning of time interval t, power buying $x_{t,\omega}$, and generated renewable power $R_{t,\omega}$ during time interval t. The power output of a decision time interval t includes the power remaining in the battery $s_{t+1,\omega}$ at the end of time interval t (i.e. at the beginning of time interval $t+1$), the power consumption of a wireless BS in the current time interval $C_{t,\omega}$, and excess power $y_{t,\omega}$. Note that the excess power is used to represent the amount of power input exceeding the power consumption and battery capacity.
- (10.7) is a constraint of power storage that the power in the battery must be lower than or equal to the capacity B.
- (10.8) denotes the initial and termination condition constraints, where B_1 and B_T are the power storage in the battery at the first and last time intervals, respectively.
- (10.9) states the constraint of non-negative value of power.

The above multi-time-interval stochastic programming model can be transformed to a linear programming problem, which can be solved by applying a standard method to obtain a linear programming solution [77]. The solution value, denoted by $x^*_{t,\omega}$, is the amount of power buying from the electrical grid at time interval t given scenario ω. When the realization of the scenario of generated renewable power, power price, and power consumption is observed, this solution is applied.

10.4.4 Performance evaluation

Parameter setting

Adaptive power-management for a micro BS is considered with the parameter setting of the micro BS similar to that in [69]. The static power-consumption is 194.25 W, while the dynamic power-consumption coefficient is 24 W per connection. The transmission range of the micro BS is 100 meters, in which the transmit power calculated as in [69] is applied to ensure the reliable connectivity of the mobile units. The maximum number of connections of the micro BS is 25.

We consider the solar panel as a renewable power source. The capacity of the solar panel is 450 Wh. The battery capacity is 2 kW. The initial and termination power levels of the battery are assumed to be 500 W. The self-discharge rate of the battery is 0.1% per hour. We consider the randomness of the power price, generated renewable power, and the traffic load of the micro BS. For power price, two scenarios are considered, i.e. peak and off-peak hour prices, whose average power prices are 20 and 12 cents per kWh, and the corresponding probabilities are 0.6 and 0.4, respectively. For a renewable power source, two scenarios are considered, i.e. clear sky and cloudy, whose average generated power from 6:00–18:00 are 292 Wh and 150 Wh, and the corresponding probabilities are 0.6 and 0.4, respectively. For the traffic load of the micro BS, five scenarios are considered, i.e. heavy uniform, medium uniform, light uniform, heavy morning, and heavy evening, and the corresponding probabilities are 0.1, 0.1, 0.2, 0.2, and 0.4, respectively. For heavy uniform, medium uniform, and light uniform scenarios, the traffic load is uniform and the mean connection arrival rates are 0.56, 0.22, and 0.15 connections per minute, respectively. For heavy morning and heavy evening scenarios, the connection arrival rate is at the peak during 8:00–11:00 and 17:00–21:00, whose mean connection arrival rate is 0.8 connections per minute. The adaptive power-management scheme is optimized for a 24-hour time interval.

Numerical results

Figure 10.3 shows the different average power over the optimization period. In this case, the renewable power source (i.e. the solar panel) can generate power only when sunlight is available. Therefore, the adaptive power-management controller has to optimize the power storage in the battery and the power buying from the electrical grid to meet the requirements of the micro BS. We observe that the battery is charged with renewable power. The power is bought from the electrical grid occasionally for the micro BS (e.g. when the renewable power and storage power are unavailable between 23:00 and 5:00) or to charge the battery (e.g. at 8:00). Given the average power shown in Figure 10.3, the power cost of this micro BS with adaptive power-management is 9.19 dollars per month.

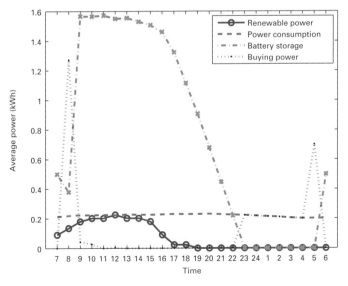

Figure 10.3 Average generated renewable power, power consumed by wireless base station, battery storage, and buying power from smart grid.

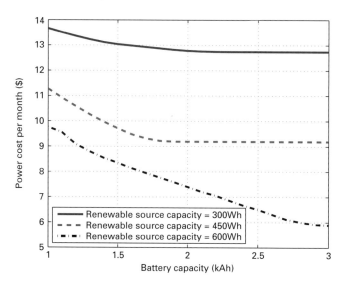

Figure 10.4 Power cost per month under different battery capacities.

For comparison purposes, we consider a simple power-management scheme in which the power is bought from the electrical grid when renewable power is unavailable. In this case, the power storage in the battery is maintained at a constant level (i.e. 1 kW). The power cost per month of this simple power-management scheme is 14.7 dollars per month. Clearly, the proposed adaptive power-management scheme achieves lower cost and lower power consumption from the electrical grid. Even though the cost saving for one micro BS may be marginal (i.e. 5.51 dollars per month or about 37.48%), this cost saving can be significant when micro BSs experience wide deployment. Furthermore, the reduction of power demand from the electrical grid, which is mostly generated from traditional fossil fuels (e.g. coal and oil), will help to decrease the CO_2 emissions, which is the main aim of green wireless communications.

We then study the impact of the battery capacity on the power cost. Figure 10.4 shows the power cost per month under different battery capacities and different renewable power source capacity. With the increase in battery capacity, the adaptive power-management controller can store more power when the power price is low, i.e. when renewable power is generated or when the power price from the electrical grid is cheap. As a result, the power cost per month decreases. However, when a certain value of capacity is reached, the power cost becomes constant because all the generated renewable power or the power with cheap price from the electrical grid can be stored in the battery and there is sufficient for future demand. Consequently, further increasing the battery power capacity over this threshold will not help to reduce the power cost, while the cost of battery will be higher.

In addition, as expected, the power cost of buying from the electrical grid decreases as the capacity of the renewable power source increases (Figure 10.4). However, it is also important to note that the cost of increasing the capacity of the renewable power

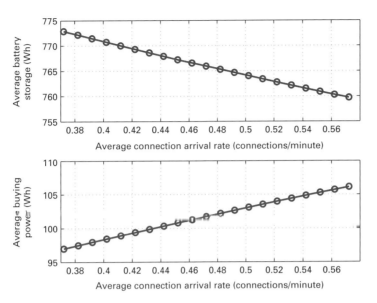

Figure 10.5 Average buying power and battery storage under different connection arrival rates.

source cannot be ignored. For example, the average price of a solar panel with a capacity of 150 W is 230 dollars. Therefore, the optimal deployment of battery capacity and renewable power source capacity is an important issue and requires further studies in future work.

Next, we investigate the effect of the traffic load of the micro BS on the power consumption. Figure 10.5 shows the average power stored in a battery and power buying from the electrical grid under different connection arrival rates. As expected, when the connection arrival rate increases, the micro BS consumes more power. However, with the renewable power source, the consumed power can be supplied partly from the electrical grid. However, the power of battery storage increases as the traffic load decreases due to higher power consumption.

Then, the impact of the threshold in call admission control (CAC) to the QoS performance and power cost is investigated. With the guard channel CAC [78], the threshold is used to reserve the channels for handoff connections, since the mobile units are more sensitive to the dropping of the handoff connection than the blocking of the new connection. With guard channel CAC, the new connection is accepted if the current number of ongoing connections is less than the threshold, and will be rejected otherwise. As expected, as the threshold becomes larger, more new connections are accepted and can perform data transmission. Consequently, the new connection blocking probability decreases (Figure 10.6). However, the handoff connection dropping probability increases, since fewer channels are reserved. We observe that as the threshold increases, there are more ongoing connections with the micro BS due to more accepted new connections. Therefore, the power consumption increases, and the amount of power-cost saving (compared to that without CAC) decreases (Figure 10.6). This result can be used to optimize the

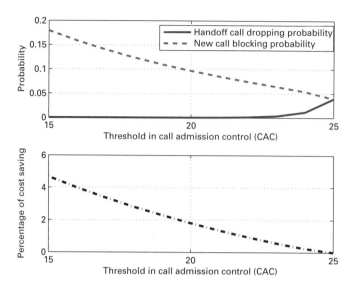

Figure 10.6 Handoff call dropping probability, new call blocking probability, and percentage of cost saving under different threshold in call admission control (CAC).

parameter (i.e. threshold) of CAC. For example, if the objective is to minimize the handoff connection dropping probability and maximize the power-cost saving subject to the new connection dropping probability to be less than 0.1, then, the threshold should be set to 20.

From the above results, it is clear that, given various uncertainties including renewable power generation, power price, and traffic load of the wireless BS, the proposed adaptive power-management can make optimal decisions to minimize the cost of power consumption. The optimization formulation will be useful for the design of the resource management of the wireless system in green wireless communications.

10.5 Conclusion

The growing concern of global environmental change has caused a revolution in the way power is used. In the wireless industry, green wireless communications have recently gained increasing attention and are expected to play a major role in reducing electrical power consumption. In particular, actions to promote power saving in wireless communications with regard to environmental protection are becoming imperative. The renewable power source and the smart grid have emerged as two major research fields that are closely associated with, and push the promotion of, green wireless communications. With the integration of these state-of-the-art green technologies and computing technologies, future wireless communications systems will become more smart, autonomous and open. To counter this trend, we contribute a study on power management for wireless BSs in a smart grid environment.

First, an introduction to green wireless communications has been given. Then, for understanding and optimizing the operation of wireless BS, the power-consumption models of macro and micro BSs have been presented. Finally, an adaptive power-management approach for a wireless BS with a renewable power source in a smart grid environment has been introduced. With a stochastic optimization method, the power cost of a wireless BS can be minimized while meeting the demand of mobile units.

References

[1] Global e-sustainability initiative, GeSI "Smart 2020: enabling the low carbon economy in the information age," June 2008.

[2] M. A. Marsan et al., "Optimal energy savings in cellular access networks," in *Proc. of First International Workshop on Green Communications*, Dresden, Germany, June 2009.

[3] A. Corliano and M. Hufschmid, "Energieverbrauch der mobilen Kommunikation," Bundesamt fur Energie, Ittigen, Switzerland, Technical Report, in German, February 2008.

[4] T. A. Weiss and F. K. Jondral, "Spectrum pooling: an innovative strategy for the enhancement of spectrum efficiency," *IEEE Communications Magazine*, vol. 42, no. 3, pp. 8–14, August 2004.

[5] A. Willig, "Redundancy concepts to increase transmission reliability in wireless industrial LANs," *IEEE Transactions on Industrial Informatics*, vol. 1, no. 3, pp. 173–182, August 2005.

[6] U. Toseef et al., "User satisfaction based resource allocation in future heterogeneous wireless networks," in *Proc. of Annual Communication Networks and Services Research Conference*, pp. 217–223, Ottawa, ON, Canada, May 2011.

[7] Ericsson, "Green power to bring mobile telephony to billions of people," 2008. [Online]. Available: www.ericsson.com/ericsson/press/videos/2008/081215-green-power.shtml

[8] A. Kumar et al., "Sustainability in wireless mobile communication networks through alternative energy resources," *International Journal of Computer Science and Technology*, vol. 1, no. 2, pp. 196–201, December 2010.

[9] J. Gozalvez, "Green radio technologies [Mobile Radio]," *IEEE Vehicular Technology Magazine*, vol. 5, no. 1, pp. 9–14, March 2010.

[10] M. Belfqih et al., "Joint study on renewable energy application in base transceiver stations," in *Proc. of International Telecommunications Energy Conference*, Seoul, Korea, October 2009.

[11] Y. He and R. Yuan, "A novel scheduled power saving mechanism for 802.11 Wireless LANs," *IEEE Transactions on Mobile Computing*, vol. 8, no. 10, pp. 1368–1383, February 2009.

[12] F. Zhang and S. T. Chanson, "Improving communication energy efficiency in wireless networks powered by renewable energy sources," *IEEE Transactions on Vehicular Technology*, vol. 54, no. 6, pp. 2125–2136, November 2005.

[13] G. H. Badawy, A. A. Sayegh, and T. D. Todd, "Energy provisioning in solar-powered wireless mesh networks," *IEEE Transactions on Vehicular Technology*, vol. 59, no. 8, pp. 3859–3871, October 2010.

[14] B. Heile, "Smart grids for green communications [industry perspectives]," *IEEE Wireless Communications*, vol. 17, no. 3, pp. 3–6, June 2010.

[15] A. H. Mohsenian-Rad and A. Leon-Garcia, "Coordination of cloud computing and smart power grids," in *Proc. of IEEE Smart Grid Communications Conference*, Gaithersburg, USA, October 2010.

[16] W. Jai and W. Zhou, *Distributed Network Systems*. New York: Springer, 2005.
[17] F. Richter, A. J. Fehske, and G. P. Fettweis, "Energy efficiency aspects of base station deployment strategies for cellular networks," in *Proc. of IEEE Vehicular Technology Conference*, Anchorage, Alaska, USA, September 2009.
[18] H. Kim and G. de Veciana, "Leveraging dynamic spare capacity in wireless systems to conserve mobile terminals energy," *IEEE/ACM Trans. on Networking*, vol. 18, no. 3, pp. 802–815, June 2010.
[19] G. W. Miao et al., "Cross-layer optimization for energy-efficient wireless communications: a survey," *Wiley Journal Wireless Communications and Mobile Computing*, vol. 9, no. 4, pp. 529–542, April 2009.
[20] L. Chiaraviglio, M. Mellia, and F. Neri, "Energy-aware backbone networks: a case study," in *Proc. of First International Workshop on Green Communications*, Dresden, Germany, June 2009.
[21] M. Chiang et al., "Power control in wireless cellular networks," *Foundations and Trends in Networking*, vol. 2, no. 4, pp. 381–533, 2008.
[22] H. B. Salameh, M. Krunz, and O. Younis, "Cooperative adaptive spectrum sharing in cognitive radio networks," *IEEE/ACM Transactions on Networking*, vol. 18, no. 4, pp. 1181–1194, August 2010.
[23] J. W. Lee, R. R. Mazumdar, and N. B. Shroff, "Opportunistic power scheduling for dynamic multi-server wireless systems," *IEEE Transactions on Wireless Communications*, vol. 5, no. 6, pp. 1506–1515, June 2006.
[24] D. Tse and P. Viswanath, *Fundamentals of Wireless Communications*. Cambridge University Press, 2005.
[25] Y. Liu and E. Knightly, "Opportunistic fair scheduling over multiple wireless channels," in *Proc. of IEEE International Conference on Computer Communications*, San Francisco, CA, USA, April 2003.
[26] J. Lee, R. Mazumdar, and N. Shroff, "Joint resource allocation and base-station assignment for the downlink in CDMA networks," *IEEE/ACM Transactions on Networking*, vol. 14, no. 1, pp. 1–14, February 2006.
[27] B. Hochwald and S. Vishwanath, "Space-time multiple access: linear growth in the sum rate," in *Proc. of Allerton Conference on Communication, Control and Computing*, Monticello, Illinois, USA, October 2002.
[28] M. Sharif and B. Hassibi, "A comparison of time-sharing, DPC, and beamforming for MIMO broadcast channels with many users," *IEEE Transactions on Communications*, vol. 55, no. 1, pp. 11–15, January 2007.
[29] F. Rashid-Farrokhi, K. J. R. Liu, and L. Tassiulas, "Transmit beamforming and power control for cellular wireless systems," *IEEE Journal of Selected Areas in Communications*, vol. 16, no. 8, pp. 1437–1450, October 1998.
[30] E. Visotsky and U. Madhow, "Optimum beamforming using transmit antenna arrays," in *Proc. of IEEE Vehicular Technology Conference*, Houston, USA, May 1999.
[31] M. Bengtsson and B. Ottersten, "Optimal downlink beamforming using semidefinite optimization," in *Proc. of Allerton Conference on Communication, Control and Computing*, Allerton House, Illinois, USA, September 1999.
[32] D. Samuelsson, M. Bengtsson, and B. Ottersten, "An efficient algorithm for solving the downlink beamforming problem with indefinite constraints," in *Proc. of IEEE International Conference on Acoustics, Speech, and Signal Processing*, Philadelphia, USA, March 2005.

[33] M. Schubert and H. Boche, "Solution of the multi-user downlink beamforming problem with individual SIR constraints," *IEEE Transactions on Vehicular Technology*, vol. 53, no. 1, pp. 18–28, January 2004.

[34] H. Boche and M. Schubert, "A general duality theory for uplink and downlink beamforming," in *Proc. of IEEE Vehicular Technology Conference*, Boston, MA, USA, September 2000.

[35] W. Yu, "Uplink-downlink duality via minimax duality," *IEEE Transactions on Information Theory*, vol. 52, no. 2, pp. 361–374, February 2006.

[36] P. Viswanath and D. Tse, "Sum capacity of the multiple antenna gaussian broadcast channel and uplink-downlink duality," *IEEE Transactions on Information Theory*, vol. 49, no. 8, pp. 1912–1921, August 2003.

[37] R. D. Yates and C.-Y. Huang, "Integrated power control and base station assignment," *IEEE Transactions on Vehicular Technology*, vol. 44, no. 3, pp. 638–644, August 1995.

[38] F. Rashid-Farrokhi, K. Liu, and L. Tassiulas, "Downlink power control and base station assignment," *IEEE Communications Letters*, vol. 1, no. 4, pp. 102–104, July 1997.

[39] F. Rashid-Farrokhi, K. Liu, and L. Tassiulas, "Downlink and uplink capacity enhancement in power controlled cellular systems," in *Proc. of IEEE Vehicular Technology Conference*, Phoenix, AZ, USA, May 1997.

[40] M. Bengtsson, "Jointly optimal downlink beamforming and base station assignment," in *Proc. of IEEE International Conference on Acoustics, Speech, and Signal Processing*, Salt Lake City, UT, USA, May 2001.

[41] D. Willkomm *et al.*, "Primary user behavior in cellular networks and implications for dynamic spectrum access," *IEEE Communications Magazine*, vol. 47, no. 3, pp. 88–95, March 2009.

[42] L. Chiaraviglio *et al.*, "Energy-aware UMTS access networks," *International Workshop on Green Wireless*, Apland, Finland, September 2008.

[43] L. Saker, S.-E. Elayoubi, and H. O. Scheck, "System selection and sleep mode for energy saving in cooperative 2G/3G networks," in *Proc. of IEEE Vehicular Technology Conference*, Anchorage, Alaska, USA, September 2009.

[44] L. Saker and S-E. Elayoubi, "Sleep mode implementation issues in green base stations," in *Proc. of International Symposium on Personal, Indoor and Mobile Radio Communications*, Istanbul, Turkey, September 2010.

[45] L. Saker, S-E. Elayoubi, and T. Chahed, "Minimizing energy consumption via sleep mode in green base station," in *Proc. of IEEE Wireless Communication & Networking Conference*, Sydney, Australia, April 2010.

[46] W. Rui, J. S. Thompson, and H. Haas, "A novel time-domain sleep mode design for energy-efficient LTE," in *Proc. of International Symposium on Communication, Control and Signal Processing*, Limassol, Cyprus, March 2010.

[47] I. Haratcherev, C. Balageas, and M. Fiorito, "Low consumption home femto base stations," in *Proc. of IEEE International Symposium on Personal, Indoor and Mobile Radio Communications*, Tokyo, Japan, September 2009.

[48] I. Ashraf, F. Boccardi, and L. Ho, "Power savings in small cell deployments via sleep mode techniques," in *Proc. of IEEE International Symposium on Personal, Indoor and Mobile Radio Communications Workshop*, Istanbul, Turkey, September 2010.

[49] K. Son *et al.*, "Base station operation and user association mechanisms for energy-delay tradeoffs in green cellular networks," USC CENG Technical Report (CENG-2010-11), August 2010.

[50] V. Sreng, "Capacity enhancement through two-hop relaying in cellular radio systems," Master's thesis, Carleton University, January 2002.

[51] J. Vidal et al., "Multihop networks for capacity and coverage enhancement in TDD/UTRAN," *Mediterranean Ad Hoc Networking*, Sardegna, Italy, September 2002.

[52] H. Hu and H. Yanikomeroglu, "Performance analysis of cellular networks with digital fixed relays," *Wireless World Research Forum*, pp. 27–28, NY, USA, October 2003.

[53] A. Molina, E. Tameh, and A. Nix, "The optimization of fixed relay location to enhance the performance of a 3G microcellular network," *IST Mobile Communications Summit*, pp. 27–30, Lyon, France, June 2004.

[54] O. Simeone et al., "Throughput of low-power cellular systems with collaborative base stations and relaying," *IEEE Transactions on Information Theory*, vol. 54, no. 1, January 2008.

[55] J-Y. Song, H-J. Lee, and D-H. Cho, "Power consumption reduction by multi-hop transmission in cellular networks," in *Proc. of IEEE Vehicular Technology Conference*, Los Angeles, CA, USA, September 2004.

[56] G. N. Aggelou and R. Tafazolli, "On the relaying capability of next generation GSM cellular network," *IEEE Personal Communications*, vol. 8, no. 1, pp. 40–47, February 2001.

[57] T. J. Harrold and A. R. Nix, "Intelligent relaying for future personal communication systems," *IEE Colloquium on Capacity and Range Enhancement Techniques for Third Generation Mobile Communications and Beyond*, February 2000.

[58] N. Badruddin and R. Negi, "Capacity improvement in a CDMA system using relaying," in *Proc. of Wireless Communication & Networking Conference*, pp. 243–248, Atlanta, GA, USA, March 2004.

[59] B. Wang and D. Zhao, "Downlink power distribution in a wireless CDMA network with cooperative relaying," in *Proc. of IEEE International Conference on Communications*, Dresden, Germany, June 2009.

[60] B. Sirkeci-Mergen and A. Scaglione, "On the power efficiency of cooperative broadcast in dense wireless networks," *IEEE Journal on Selected Areas in Communications*, vol. 25, no. 2, pp. 497–507, February 2007.

[61] Z. Yang, J. Liu, and A. Host-Madsen, "Cooperative routing and power allocation in ad-hoc networks," in *Proc. of IEEE Global Telecommunications Conference*, St. Louis, MO, USA, November 2005.

[62] J. Luo et al., "New approaches for cooperative use of multiple antennas in ad hoc wireless networks," in *Proc. of IEEE Vehicular Technology Conference*, Los Angeles, California, USA, September 2004.

[63] M. Schubert and H. Boche, "Iterative multiuser uplink and downlink beamforming under individual SINR constraints," *IEEE Transaction on Signal Processing*, vol. 53, no. 7, pp. 2324–2334, July 2005.

[64] F. Rashid-Farrokhi, L. Tassiulas, and K. J. R. Liu, "Joint optimal power control and beamforming in wireless networks using antenna arrays," *IEEE Journal of Selected Areas in Communications*, vol. 46, no. 10, pp. 1313–1324, October 1998.

[65] P. Bahl et al., "Cell breathing in wireless LANs: algorithms and evaluation," *IEEE Transactions on Mobile Computing*, vol. 6, no. 2, pp. 164–178, December 2007.

[66] F. Richter and G. P. Fettweis, "Cellular mobile network densification utilizing micro base stations," in *Proc. of IEEE International Conference on Communications*, Cape Town, South Africa, May 2010.

[67] Keynote Speech, "Green radio and cognitive radio," in *Proc. of International Conference on Cognitive Radio Oriented Wireless Networks and Communications*, Hannover, Germany, June 2009.

[68] V. Chandrasekhar, J. Andrews, and A. Gatherer, "Femtocell networks: a survey," *IEEE Communications Magazine*, vol. 46, no. 9, pp. 59–67, September 2008.

[69] O. Arnold et al., "Power consumption modeling of different base station types in heterogeneous cellular networks," in *Proc. of 19th Future Network & Mobile Summit*, Florence, Italy, June 2010.

[70] C. Wei, "A conceptual framework of smart grid," *Power and Energy Engineering Conference*, Wuhan, China, March 2010.

[71] B. Kirby and E. Hirst, "Load as a resource in providing ancillary services," Technical report, Oak Ridge National Laboratory, 1999.

[72] T. Liu, W. G. Pell, and B. E. Conway, "Self-discharge and potential recovery phenomena at thermally and electrochemically prepared RuO_2 super capacitor electrodes," *Electrochimica Acta*, vol. 42, no. 23, pp. 3541–3552, Ottawa, ON, Canada, April 1998.

[73] *The smart grid: an introduction.* The US Department of Energy, 2008.

[74] K. Ponnambalam et al., "Comparison of methods for battery capacity design in renewable energy systems for constant demand and uncertain supply," in *Proc. of IEEE International Conference on the European Energy Market*, Madrid, Spain, June 2010.

[75] F. J. Heredia, M. J. Rider, and C. Corchero, "Optimal bidding strategies for thermal and generic programming units in the day-ahead electricity market," *IEEE Transactions on Power Systems*, vol. 25, no. 3, pp. 1504–1518, August 2010.

[76] B. M. Epstein and M. Schwartz, "Predictive QoS-based admission control for multiclass traffic in cellular wireless networks," *IEEE Journal on Selected Areas in Communications*, vol. 18, no. 3, pp. 523–534, March 2000.

[77] J. R. Birge and F. Louveaux, *Introduction to Stochastic Programming.* Springer, July 1997.

[78] Y. Fang and Y. Zhang, "Call admission control schemes and performance analysis in wireless mobile networks," *IEEE Transactions on Vehicular Technology*, vol. 51, no. 2, pp. 371–382, March 2002.

11 Cooperative multicell processing techniques for energy-efficient cellular wireless communications

Mohammad Reza Nakhai, Tuan Anh Le, Auon Muhammad Akhtar, and Oliver Holland

11.1 Introduction

The emerging applications for mobile internet in different areas such as education, health care, smart grids, and security are growing very fast. Due to the integration of these applications with people's lives it is inevitable that over then the next ten years there will be huge increases in the required user data rate per area and the spectral efficiency. On the other hand, delivering a higher data rate per area requires more transmission power, which is constrained not only by safety limits but also by the importance of the global warming issues and need for green communications. Therefore, high-speed transmission would mean a diminishing coverage range, as otherwise, an enormous increase of transmission power is required by both mobile terminals and base stations to maintain the current cell size and achieve the ambitious targets of the beyond-IMT advanced technologies.

Cell splitting, i.e. dividing large cells into a number of smaller cells, is a promising method that can significantly increase both the capacity and the coverage in future cellular networks. Since the divided cells have to operate with full spectrum reuse across all base stations, i.e. due to the scarcity of bandwidth resources, co-channel interference becomes the major issue in cell splitting. On the other hand, the spatial dimension can be more effectively exploited as a complementary communication resource to the traditional ones. In a cellular network, spatial dimensions are available either locally, i.e. through multiple antennas at BSs, terminals, relays, vehicles, or in a distributed configuration such as an array of multiple antenna BSs within a cluster of cells. Spatial dimensions provide for spatial multiplexing of multiple simultaneous data streams to enhance the network capacity and throughput if they can be scheduled to be mutually interference-free. The science behind the exploitation of distributed spatial dimensions is rooted in the theory of multiple-input-multiple-output (MIMO) communications that has witnessed incredible advances in the last decade or so. The key advantage of MIMO in offering significant enhancements both in capacity and coverage without requiring additional power can be paired with coordination across the antenna arrays of a number of base stations to find power-efficient solutions to major challenges in cell splitting.

Recently, the idea of multicell processing (MCP) in cellular networks has been identified as an effective technique to overcome intercell interference and substantially improve the capacity. In MCP, a coordinated virtual architecture is mapped over a cellular infrastructure such that each and every mobile user is collaboratively served by its surrounding base stations rather than only by its designated base station. In this architecture, base stations are equipped with multiple antennas but user terminals can have either single or multiple antennas. Using coordinated scheduling and/or beamforming among a number of local base stations enables the network to constructively overlay the desired signals at an intended user and eliminate or sufficiently mitigate them at the other unintended users. Ideally, in this way, each user within a cell feels free of intercell interference and, hence, can potentially achieve the highest capacity with the lowest energy consumption under the reuse-one regime, i.e. while all the available spectrum is fully reused within the adjacent cells.

Although the MCP overcomes intercell interference and, hence, in theory, provides the ground for achieving a high throughput at a low energy cost, there are still many problems in transferring from theory to practical implementation. The most important practical issue is the increased computational complexity due to the extra signal processing load required by the MCP. On the other hand, however, the MCP architecture facilitates the integration of some cognitive functionalities leading to energy efficiency in cellular networks. These functionalities can be taken from the well-researched cognitive radio area and integrated with the MCP to address these practical issues. Flexible system design is one of the cognitive functionalities that allows the use of signal processing algorithms with different levels of complexity according to the needs and circumstances. For instance, users within a cluster can be served by full or partial coordination among BSs or even with no coordination, based on their location. Sharing information via backhaul links among the BSs could be limited to the channel state information (CSI) only or could include users' data. In the former case, as a result of coordinating beamforming and scheduling, each BS sends data to its local user, while in the latter case, data transmission jointly takes place by the coordinating BSs towards a single user.

Learning from the environment and the user behavior, which is a core idea of cognitive radio, can be exploited in MCP to turn off some antennas and their RF hardware in certain low-traffic times of the day to improve the energy efficiency of the network. Establishing and maintaining a robust and fast wireless backhaul, inter-connecting the coordinating BSs in MCP, is a challenging issue. However, in an infrastructure that arises from cell splitting, unoccupied UHF frequency bands with very good propagation characteristics can be used to establish robust wireless links between the neighboring BSs.

This chapter formulates four downlink multicell processing algorithms and develops a framework to evaluate and compare their power efficiency in the presence of imperfect backhaul. As global interconnectivity among all base stations in the network is impractical and unrealistic, this chapter focuses on cluster-based multicell processing and develops a channel model for a cluster of three base stations. It is assumed that intercluster interference is mitigated through dynamic coordination and control information sharing between the clusters at an extra scheduling cost. For instance, users at the cluster edge can be orthogonalized using coordinated resource allocation between the adjacent clusters. The details of inter-cluster interference mitigation techniques are beyond, the scope of this chapter. Furthermore, this chapter assumes that the channel state information between each user and all of the coordinating base stations is available to all of the coordinating base stations. In practice, a downlink channel state between a base station and a user can be estimated at the user terminal by measuring the strength of a pilot signal sent by the base station. Then, the user feeds back the estimated channel state to the base station through a reverse channel. The details of providing users' channel state information at the coordinating base stations is not discussed in this chapter.

It is assumed that each base station is directly connected to the backbone network via a finite-capacity backhaul link. As full data sharing among the base stations assumes high-capacity backhaul links, which may not be available in practice, this chapter assumes additional wireless backhaul links between the base stations to avoid high data traffic between the base stations and the core network. This chapter introduces a fast backhaul protocol to exchange information among three base stations of a cluster. Finally, this chapter presents a power-efficient cooperative routing algorithm between two distant source and destination base stations and compares its performance against a non-cooperative power-efficient routing algorithm.

11.2 Cell splitting

The transmission power level from both base stations and mobile terminals is constrained by two important issues, namely, human safety limits and the requirement of green communications. However, huge increases in network capacity and coverage are envisaged due to the fast-growing demand for mobile internet over the next decade. Hence, maintaining a higher data rate per user terminal would mean diminishing coverage, as high-speed transmission requires more power. Therefore, one needs to drop the coverage range, as the transmitted power by the base stations and the mobile terminals are both currently standing at the human safety limit and well above the green communications targets. Cell splitting, i.e. dividing the existing macro-cells into a number of smaller cells, can maintain the capacity and coverage while accounting for power safety limits at mobile terminals and avoiding high-power macro-cell base stations.

Figure 11.1 illustrates a macro-cell divided into four tiers of smaller cells. Consider a definition for the power-saving gain in cell splitting as $G = \frac{P}{mP_0}$, where m is the number of small cells each having a central base station with power P_0, and P is the power required by the macro-cell base station to maintain the same received power at any given point on the macro-cell edge as the power P_0 of a local small cell base station. It was shown in [1] that G varies with path-loss exponents, with various numbers of tiers N as the parameter, according to Figure 11.2. It should be noted that Figure 11.2 is plotted in the worst case where the base stations of all small cells are active with transmission power P_0. However, cell splitting offers the flexibility of shutting down a base station when no active user is present in the corresponding small cell area.

The scarcity of the available spectrum enforces an efficient reuse of bandwidth resources across small cells of a divided macro-cell and, inevitably, the problem of intercell interference emerges as the major challenge in cell splitting. Intercell interference in cellular networks is influenced by two major factors, namely, path loss and cell size. According to the path-loss power model [2], the attenuation of the transmitted power over a distance d is proportional to $d^{-\alpha}$, where α is the path-loss exponent.

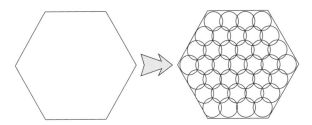

Figure 11.1 Dividing a cell into 4 tiers.

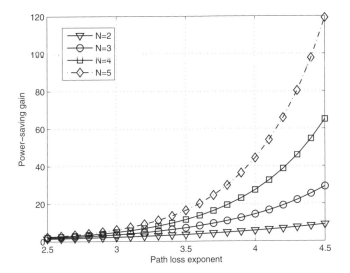

Figure 11.2 Power-saving gain against path-loss exponent.

For a fixed cell size, intercell interference decreases exponentially with the increasing path-loss exponent, as a result of a decreasing interference from an interfering transmitter. However, shrinking the cell size decreases the distance among the neighboring BSs and, hence, increases intercell interference.

In an MCP network, since the received signals at the BSs are jointly decoded in the uplink and every user terminal receives a useful signal from the cooperating BSs in the downlink, intercell interference is not only eliminated but it can also be beneficially used to provide diversity and multiplexing gains. However, achieving an energy-efficient MCP network depends on whether the energy consumed on the backbone network and the joint processing can be balanced by the energy saved as a result of eliminating the intercell interference.

11.3 A multicell processing model

11.3.1 Transmission and channel model

We assume conventional cells with three sectors covered by multiple antenna base stations. As shown in Figure 11.3, we define a group of three sectors having mutually

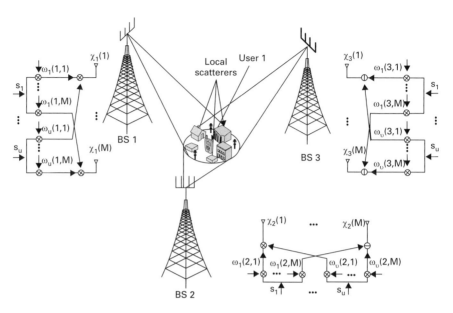

Figure 11.3 System model.

common boundaries and belonging to three adjacent conventional cells as a virtual cell. A virtual cell is a hexagonal cell with a multiple antenna base station located on alternate vertices. Thus, up to three geographically distributed antenna arrays (DAA) can coordinate to serve user terminals anywhere within a virtual cell. The available spectrum is globally reused within a virtual cell and co-channel interference at the boundaries of adjacent virtual cells can be managed by time-sharing with power coordination or beam coordination between the co-located multiple antennas of the corresponding adjacent sectors. Within a virtual cell, it is assumed that each user belonging to a sector receives dominant interference only from the base stations of the other two sectors of the same virtual cell. Hence, the model described in this section and used throughout this chapter considers interference from the outer tiers of base stations surrounding a virtual cell on a user within the virtual cell as noise. Although this assumption is made to reduce the complexity of coordinated signal processing, as well as to simplify the backhaul analysis and design, the channel modeling procedure and design strategies described hereafter in this chapter can be extended for larger clusters of coordinating base stations.

In a frequency-division duplex (FDD) system, mobile terminals broadcast their estimated global channel state information (CSI) through feedback channels and the coordinating base stations either decode them directly or receive them in whole or partially via backhaul links. In distributed multicell processing, the global CSI at transmitter, i.e. CSIT, of all users within a virtual cell should be available to all base stations. Depending on the coordination level, users' data may also be required to be exchanged among the base stations via backhaul links.

Figure 11.3 illustrates a scenario where three coordinating base stations, each equipped with a linear antenna array of M elements, jointly transmit to the ith user of U

simultaneous single antenna users within a virtual cell. It is assumed that each user i is surrounded by Q randomly positioned local-scatterers, that are at the far-field distances from the BSs, and there is no line-of-sight (LoS) transmission from the BSs to the user. Thus, wavefronts originated from each one of the serving BSs hit all of the Q local scatterers of each user. The spacing between the antenna elements of a BS is negligible with respect to the distance of the BS from the scatterers. Hence, rays departing from M antenna elements of a BS towards a scatterer can be assumed to have the same fading coefficients.

Let $s_i, i = 1, \cdots, U$, be the intended symbol for the ith user and $\mathbf{x}_p = \begin{bmatrix} x_p(1) & x_p(2) & \cdots & x_p(M) \end{bmatrix}^T$, where $x_p(k)$ is the transmitted signal by the kth antenna element of the pth BS, $p = 1, 2, 3$. The transmitted signal by three BSs can be written as

$$\begin{bmatrix} \mathbf{x}_1 \\ \mathbf{x}_2 \\ \mathbf{x}_3 \end{bmatrix} = \begin{bmatrix} \mathbf{w}_1(1) & \mathbf{w}_2(1) & \cdots & \mathbf{w}_U(1) \\ \mathbf{w}_1(2) & \mathbf{w}_2(2) & \cdots & \mathbf{w}_U(2) \\ \mathbf{w}_1(3) & \mathbf{w}_2(3) & \cdots & \mathbf{w}_U(3) \end{bmatrix} \begin{bmatrix} s_1 \\ \vdots \\ s_U \end{bmatrix}, \tag{11.1}$$

where $\mathbf{w}_i(p) = \begin{bmatrix} w_i(p,1) & w_i(p,2) & \cdots & w_i(p,M) \end{bmatrix}^T$ is the beamforming vector of the ith user at the pth BS and $w_i(p,k)$ is the corresponding beamforming coefficient of the kth antenna element. The received signal at user i is given by

$$y_i = \mathbf{h}_i \mathbf{x} + z_i, \tag{11.2}$$

where $\mathbf{x} = \begin{bmatrix} \mathbf{x}_1^T & \mathbf{x}_2^T & \mathbf{x}_3^T \end{bmatrix}^T$,

$$\mathbf{h}_i = \begin{bmatrix} \varrho_{i,1} & \varrho_{i,2} & \cdots & \varrho_{i,Q} \end{bmatrix} \begin{bmatrix} \xi_{i,1}(1)\mathbf{h}_{i,1}(1) & \xi_{i,1}(2)\mathbf{h}_{i,1}(2) & \xi_{i,1}(3)\mathbf{h}_{i,1}(3) \\ \xi_{i,2}(1)\mathbf{h}_{i,2}(1) & \xi_{i,2}(2)\mathbf{h}_{i,2}(2) & \xi_{i,2}(3)\mathbf{h}_{i,2}(3) \\ \vdots & \vdots & \vdots \\ \xi_{i,Q}(1)\mathbf{h}_{i,Q}(1) & \xi_{i,Q}(2)\mathbf{h}_{i,Q}(2) & \xi_{i,Q}(3)\mathbf{h}_{i,Q}(3) \end{bmatrix}, \tag{11.3}$$

$\varrho_{i,t} = e^{-j\frac{2\pi c}{\lambda}\tau_{i,t}}$ is the phase shift due to the time delay $\tau_{i,t}$ between the tth scatterer and the ith user, $\xi_{i,t}(p) = a_i(p)\sqrt{S_{i,t}(p)L_{i,t}(p)}$ models the path loss and the large-scale fading coefficients with $S_{i,t}(p) = 10^{-\frac{x}{10}}$ and $L_{i,t}(p)$ being the log-normal shadow fading coefficient, i.e. $x \sim N(0, \sigma_S^2)$, and the path-loss coefficient, respectively, between the pth BS and the tth scatterer of user i. The coefficients $\xi_{i,t}(p)$ include the effect of user distribution in cellular network in the MCP channel model. The controlling coefficient $a_i(p)$ is either 1, if the ith user is allocated to be served by the pth BS, or zero, otherwise. Furthermore in (11.3), the row vector $\mathbf{h}_{i,t}(p) = \begin{bmatrix} h_{i,t}(p,1) & h_{i,t}(p,2) \cdots h_{i,t}(p,M) \end{bmatrix}$, where $h_{i,t}(p,k)$ is the channel between the tth scatterer of the ith user and the kth antenna element of the pth BS, and finally z_i is the zero mean circularly symmetric complex Gaussian (ZMCSCG) random variable, i.e. $z_i \sim N(0, \sigma_N^2)$, modeling the additive white Gaussian noise at the ith user's receiving point.

Let $l_i(p)$ be the distance from BS p to user i, and $l_{i,\min} = \min_p l_i(p)$. We can write

$$h_{i,t}(p,k) = F_{i,t}(p)e^{j\frac{2\pi}{\lambda}[l_i(p)-l_{i,\min}]}e^{j\frac{2\pi\Delta}{\lambda}(k-1)\sin[\theta_i(p)+\phi_{it}(p)]}, \qquad (11.4)$$

where $F_{i,t}(p)$ is the complex Gaussian fading coefficient between BS p and scatterer t of user i with variance σ_F^2, λ is the carrier wavelength, Δ is the spacing between the BS antenna elements within a sector, $\theta_i(p)$ is the angle of departure with respect to the broadside of BS p for user i, and $\phi_{it}(p)$ is the angular offset of the scatterer t with respect to $\theta_i(p)$. We assume that the local scatterers are distributed randomly around each user i and the resulting angle spread has a normal distribution with standard deviation of σ, i.e. $\phi_{it}(p) \sim N(0, \sigma^2)$. The channel coefficient of the kth ray of the DAA in (11.4) differs from the channel coefficient of the kth ray of a conventional linear antenna array in factor $e^{j\frac{2\pi}{\lambda}[l_i(p)-l_{i,\min}]}$, which represents the phase difference between the geographically separated BSs. Note that (11.4) reduces to the kth ray channel of the conventional antenna array by substituting $l_{i,\min} = l_i(p)$.

Let $\mathbf{R}_i = \mathbf{E}\left(\mathbf{h}_i^H \mathbf{h}_i\right)$ denote the spatial channel covariance matrix between the DAA, i.e. the distributed antenna array formed by the coordinating BSs, and the ith user. Assume that each scatterer sees the antenna arrays of different BSs under independent fading coefficients and vice versa, i.e. the channels between any two different scatterers and any same multi-antenna base station fade independently. Then, it can be verified by direct substitution from (11.3) that $\mathbf{R}_i$ is a block diagonal matrix, i.e., $\mathbf{R}_i = \text{diag}\left[\mathbf{R}_i(1), \mathbf{R}_i(2), \mathbf{R}_i(3)\right]$, where

$$\mathbf{R}_i^{[m,n]}(p) = \sum_{t=1}^{Q} L_{i,t}(p)\sigma_F^2 e^{\frac{0.5(\ln 10\sigma_s)^2}{100}} e^{j\frac{2\pi\Delta}{\lambda}(n-m)\sin\theta_i(p)} e^{-2\left[\frac{\pi\Delta\sigma}{\lambda}(n-m)\cos\theta_i(p)\right]^2} \qquad (11.5)$$

is the (m,n)th element of $\mathbf{R}_i(p)$ and $m, n \in [1, M]$. The $M \times M$ matrix $\mathbf{R}_i(p)$ indicates the spatial covariance matrix between the base station p and the user i and is substituted by a zero matrix when the pth BS is not allocated to transmit to user i, i.e. $a_i(p) = 0$ in (11.3).

11.3.2 User-position-aware multicell processing

In this part, we outline a user-position-aware (UPA) algorithm that assigns users to one or more BSs based on their locations within a virtual cell. This method facilitates the use of signal processing algorithms with flexible levels of complexity according to the users' needs. The UPA algorithm determines the controlling coefficients $a_i(p)$ in (11.3) based on zonal classification, i.e. Figure 11.4, which is briefly described below.

The radius Ω of the QoS guarantee circle for a cell is determined by the path-loss exponent, the user's targeted SINR, and the transmit power limit at the BS. The QoS circles are shown as the outer-cell circles in Figure 11.4. An intersection of two QoS circles and the nearest BS defines the radius Ω_o of the inner-cell circle in Figure 11.4. We distinguish three zones, as follows. Zone 1 is the area within the inner-cell circle, i.e. Z_1, Z_2, and Z_3. Users in zone 1 are supported by the nearest array of the same BS only. Zone 2 is the area bound by virtual-cell borders, two cells' inner-cell circles, and the outer-cell circle of the third cell, i.e. Z_{12}, Z_{13}, and Z_{23}. Users in zone 2 are served by the

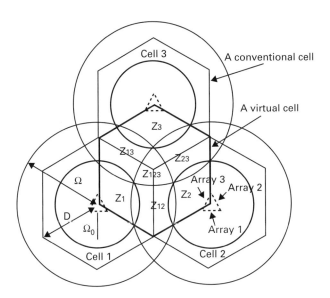

Figure 11.4 A virtual cell with zonal classification for the user-position-aware algorithm.

two nearest arrays of two different BSs. Zone 3 is the common area of three outer-cell circles, i.e. Z_{123}. Users in zone 3 are supported by three arrays of three different BSs. In this way, flexible levels of coordination, i.e. from no coordination to full coordination, can be scheduled across a virtual cell to guarantee a consistent quality of service. As users move towards the cell edges, the UPA algorithm switches the virtual cell into a cooperative mode that may involve two or three base stations on the basis of the user's zonal position.

11.4 Multicell beamforming strategies

In multicell beamforming (MBF), base stations coordinate at the signal level, i.e. share users' data via backhaul links, and possess full global CSIT. The objective is to find a set of beamforming vectors or precoders for a number of simultaneously active users such that the overall transmit power in a virtual cell is minimized while a prescribed SINR target is maintained at each user. In this section, we develop solutions to three multicell beamforming strategies.

11.4.1 MBF using instantaneous CSIT

The optimization problem of finding beamforming vectors can be formulated as

$$\min_{\mathbf{w}_i} \sum_{i=1}^{U} \mathbf{w}_i^H \mathbf{w}_i \quad (11.6)$$

$$\text{subject to} \quad \text{SINR}_i \geq \gamma_i, \ 1 \leq i \leq U,$$

where

$$\text{SINR}_i = \frac{|\mathbf{h}_i \mathbf{w}_i|^2}{\sum_{u=1, u \neq i}^{K} |\mathbf{h}_i \mathbf{w}_u|^2 + \sigma_N^2}, \tag{11.7}$$

$\mathbf{w}_i = [\mathbf{w}_i(1)^T \quad \mathbf{w}_i(2)^T \quad \mathbf{w}_i(3)^T]^T$ and γ_i is the target SINR at user i. Introducing a slack variable P_0 and denoting $\mathbf{W} = [\mathbf{w}_1 \quad \mathbf{w}_2 \quad \cdots \quad \mathbf{w}_U]$, $\mathbf{H} = [\mathbf{h}_1^T \quad \mathbf{h}_2^T \quad \cdots \quad \mathbf{h}_U^T]^T$, we can rewrite the problem (11.6) as

$$\min_{\mathbf{W}, P_0} \quad P_0$$

$$\text{subject to} \quad \frac{|\mathbf{e}_i^T \mathbf{HW} \mathbf{e}_i|^2}{\sum_{u=1, u \neq i}^{U} |\mathbf{e}_i^T \mathbf{HW} \mathbf{e}_u|^2 + \sigma_N^2} \geq \gamma_i, \ 1 \leq i \leq U \tag{11.8}$$

$$\text{Tr}\left[\mathbf{W}^H \mathbf{W}\right] \leq P_0,$$

where $\mathbf{e}_i$ defines a column vector with all zeros except a one at the ith element.

It can be easily checked that for an optimal $\mathbf{W}$ satisfying problem (11.8), $\mathbf{W}\text{diag}\left[e^{j\psi_1}, \cdots, e^{j\psi_U}\right]$, where $\psi_i, i \in \{1, \cdots U\}$ is an arbitrary phase, is also an optimal solution. Therefore, one can design the beamformer matrix $\mathbf{W}$ up to an arbitrary phase scaling so that the scalar $\mathbf{e}_i^T \mathbf{HW} \mathbf{e}_i$ is always non-negative and real. Rearranging the SINR constraint in (11.8), one can write $\left(1 + \frac{1}{\gamma_i}\right) |\mathbf{e}_i^T \mathbf{HW} \mathbf{e}_i|^2 \geq \sum_{u=1}^{U} |\mathbf{e}_i^T \mathbf{HW} \mathbf{e}_u|^2 + \sigma_N^2, 1 \leq i \leq U$. Therefore

$$\sqrt{1 + \frac{1}{\gamma_i}} \mathbf{e}_i^T \mathbf{HW} \mathbf{e}_i \geq \left\| \begin{bmatrix} (\mathbf{HW})^H \mathbf{e}_i \\ \sigma_N \end{bmatrix} \right\|, \ 1 \leq i \leq U, \tag{11.9}$$

which can be written in a second-order-cone form as

$$\begin{bmatrix} \sqrt{1 + \frac{1}{\gamma_i}} \mathbf{e}_i^T \mathbf{HW} \mathbf{e}_i \\ (\mathbf{HW})^H \mathbf{e}_i \\ \sigma_N \end{bmatrix} \succeq_K 0, \ 1 \leq i \leq U, \tag{11.10}$$

where $a \geq \|\mathbf{a}\|$ is denoted by $[a \ \mathbf{a}^T]^T \succeq_K 0$ and $\|\cdot\|$ indicates the standard Euclidean norm. According to [3], the second-order-cone constraint in (11.10) can be cast in a semi-definite (also known as linear matrix inequalities) form as

$$\mathbf{A}_i = \begin{bmatrix} \sqrt{1 + \frac{1}{\gamma_i}} \mathbf{e}_i^H \mathbf{HW} \mathbf{e}_i & [\mathbf{e}_i^H \mathbf{HW} \quad \sigma_N] \\ \begin{bmatrix} (\mathbf{HW})^H \mathbf{e}_i \\ \sigma_N \end{bmatrix} & \sqrt{1 + \frac{1}{\gamma_i}} \mathbf{e}_i^H \mathbf{HW} \mathbf{e}_i \mathbf{I} \end{bmatrix} \succeq 0, \text{ for } 1 \leq i \leq U, \tag{11.11}$$

where $\mathbf{A} \succeq 0$ is used to indicate that $\mathbf{A}$ is a positive semi-definite matrix and $\mathbf{I}$ is an identity matrix with a suitable size.

The power constraint in problem (11.8) can also be rewritten as $\|\text{vec}(\mathbf{W})\| \leq \sqrt{P_0}$ which can be recast in a second-order-cone form and then in a semi-definite form as

$$\mathbf{B} = \begin{bmatrix} p_0 & \text{vec}^H(\mathbf{W}) \\ \text{vec}(\mathbf{W}) & p_0 \mathbf{I} \end{bmatrix} \succeq 0, \tag{11.12}$$

where the operator $\text{vec}(\mathbf{W})$ stacks all columns of $\mathbf{W}$ into a long-column vector and $p_0 = \sqrt{P_0}$. From (11.11) and (11.12), one can rewrite the optimization problem (11.8) in semi-definite programming (SDP) form [4] as

$$\begin{aligned} &\min_{\mathbf{W}, p_0} && p_0 \\ &\text{subject to} && \mathbf{A}_i \succeq 0, \ \mathbf{B} \succeq 0, \ 1 \leq i \leq U. \end{aligned} \tag{11.13}$$

11.4.2 MBF using second-order statistical CSIT

The optimization problem (11.6) can also be solved using the ith user DAA spatial channel covariance matrix $\mathbf{R}_i$, as follows. In this case, one can write

$$\text{SINR}_i = \frac{\mathbf{w}_i^H \mathbf{R}_i \mathbf{w}_i}{\sum_{u=1, u \neq i}^{U} \mathbf{w}_u^H \mathbf{R}_i \mathbf{w}_u + \sigma_N^2}. \tag{11.14}$$

Let us define the Hermitian positive semi-definite matrix $\mathbf{F}_i$ as $\mathbf{F}_i = \mathbf{w}_i \mathbf{w}_i^H$. Introducing a slack variable P_o, using the rotational property of trace operator, i.e. $\text{Tr}(\mathbf{AB}) = \text{Tr}(\mathbf{BA})$, and rearranging the modified constraint (11.14), one can rewrite the original optimization problem (11.6) as

$$\begin{aligned} &\min_{\mathbf{F}_i, P_o} && P_o \\ &\text{subject to} && \left(1 + \frac{1}{\gamma_i}\right) \text{Tr}[\mathbf{R}_i \mathbf{F}_i] - \sum_{j=1}^{U} \text{Tr}[\mathbf{R}_i \mathbf{F}_j] - \sigma_N^2 \geq 0, \\ &&& P_o - \sum_{i=1}^{U} \text{Tr}[\mathbf{F}_i] \geq 0, \\ &&& \mathbf{F}_i = \mathbf{F}_i^H \succeq 0, \ 1 \leq i \leq U. \end{aligned} \tag{11.15}$$

The third constraint in (11.15) is to satisfy the condition that $\mathbf{F}_i$ is a Hermitian positive semi-definite matrix. The problem stated in (11.15) is in standard SDP form and equivalent to the original optimization problem if $\text{rank}[\mathbf{F}_i] = 1$. However, it is shown in [5] that if (11.15) is feasible, there is at least one solution satisfying the condition of $\text{rank}[\mathbf{F}_i] = 1$ for all i. Let ϵ_i and $\mathbf{x}_i$ be the eigenvalue and the corresponding eigenvector of the rank 1 matrix $\mathbf{F}_i$. Then, one can easily show that the ith beamforming vector is given as $\mathbf{w}_i = \sqrt{\epsilon_i} \mathbf{x}_i$.

11.4.3 An iterative MBF using second-order statistical CSIT

Using theorem 1 in [6], one can easily verify that the solution to the problem stated in (11.6) with SINR_i as given by (11.14), is identical with the solution to the following dual-uplink problem

$$\min_{p_i} \sum_{i=1}^{U} p_i$$
$$\text{subject to} \quad \arg\max_{\hat{\mathbf{w}}_i} \frac{p_i \hat{\mathbf{w}}_i^H \mathbf{R}_i \hat{\mathbf{w}}_i}{\sum_{t=1, t \neq i}^{U} p_t \hat{\mathbf{w}}_i^H \mathbf{R}_t \hat{\mathbf{w}}_i + \sigma_N^2 \hat{\mathbf{w}}_i^H \hat{\mathbf{w}}_i} \geq \gamma_i, \quad (11.16)$$
$$1 \leq i \leq U,$$

where $p_i = \lambda_i \sigma_N^2$ and $\hat{\mathbf{w}}_i$, with $\hat{\mathbf{w}}_i^H \hat{\mathbf{w}}_i = 1$, are the dual-uplink power and a dual-uplink beamforming vector for user i, respectively, and λ_i is the ith Lagrange multiplier associated with the ith constraint in the optimization problem (11.6) using (11.14) as SINR_i.

Defining $\mathbf{p} = \begin{bmatrix} p_1 & p_2 & \cdots & p_U \end{bmatrix}^T$ and $\mathbf{Q}_i(\mathbf{p}) = \left(\sum_{t=1, t \neq i}^{U} p_t \mathbf{R}_t + \sigma^2 \mathbf{I} \right)$, one can rewrite the optimization problem (11.16) in the following compact form

$$\min_{p_i} \sum_{i=1}^{U} p_i \quad (11.17)$$
$$\text{subject to} \quad \mathbf{p} \succeq \Gamma \mathbf{t}(\mathbf{p}),$$

where $\succeq$ defines element-wise inequality, $\Gamma = \text{diag}[\gamma_1, \gamma_2, \cdots, \gamma_U]$ and $\mathbf{t}(\mathbf{p}) = \begin{bmatrix} t_1(\mathbf{p}) & t_2(\mathbf{p}) & \cdots & t_U(\mathbf{p}) \end{bmatrix}^T$ with

$$t_i(\mathbf{p}) = \arg\min_{\hat{\mathbf{w}}_i} \frac{\hat{\mathbf{w}}_i^H \mathbf{Q}_i(\mathbf{p}) \hat{\mathbf{w}}_i}{\hat{\mathbf{w}}_i^H \mathbf{R}_i \hat{\mathbf{w}}_i}, \quad (11.18)$$

as the effective interference function of user i. One can verify the fact that the uplink vector $\hat{\mathbf{w}}_i^\star$ maximizing the argument of the ith constraint in (11.16) is identical to the vector $\hat{\mathbf{w}}_i^\star$ minimizing the argument in (11.18). Vector $\hat{\mathbf{w}}_i^\star$ is determined as the eigenvector associated with the maximum eigenvalue, i.e. the dominant eigenvector, of the matrix $\mathbf{G}_i = \mathbf{Q}_i^{-1}(\mathbf{p}) \mathbf{R}_i$.

It is shown in [7] that the function $\mathbf{t}(\mathbf{p})$ is standard as it satisfies the positivity, monotonicity, and scalability criteria for $\mathbf{p} \succeq 0$. According to [8], if $\mathbf{t}(\mathbf{p})$ is standard and Γ is a diagonal matrix of positive elements, the solution to the dual-uplink problem in (11.17) can be found via the following iterative function:

$$\mathbf{p}(n+1) = \Gamma \mathbf{t}(\mathbf{p}(n)). \quad (11.19)$$

11.4 Multicell beamforming strategies

Algorithm 11.1 Iterative downlink algorithm for MBF

1: Define a stopping point ϵ.
2: $n = 1$.
3: Initialize $\mathbf{p}(n) \succeq 0$.
4: For $1 \leq i \leq U$, find $\hat{\mathbf{w}}_i(n)$ as the dominant eigenvector of the matrix $\mathbf{G}_i(n) = \mathbf{Q}_i^{-1}(\mathbf{p}(n))\mathbf{R}_i$ and calculate $t_i(\mathbf{p}(n)) = \frac{\hat{\mathbf{w}}_i^H(n)\mathbf{Q}_i(\mathbf{p}(n))\hat{\mathbf{w}}_i(n)}{\hat{\mathbf{w}}_i^H(n)\mathbf{R}_i\hat{\mathbf{w}}_i(n)}$.
5: Update $\mathbf{p}(n+1) = \Gamma\mathbf{t}(\mathbf{p}(n))$.
6: $n = n + 1$.
7: Repeat steps 4 to 6 until $\|\mathbf{p}(n+1) - \mathbf{p}(n)\|_2 \leq \epsilon$.
8: $p_i^\star = \mathbf{p}(n+1)$ and $\hat{\mathbf{w}}_i^\star = \hat{\mathbf{w}}_i(n+1)$.
9: The optimal downlink beamforming vector for user i is given as $\mathbf{w}_i^\star = \sqrt{\alpha_i}\hat{\mathbf{w}}_i^\star$.

As shown in [7], the optimal downlink beamforming vector for user i can be expressed as $\mathbf{w}_i^\star = \sqrt{\alpha_i}\hat{\mathbf{w}}_i^\star$, where α_i and $\hat{\mathbf{w}}_i^\star$ are, respectively, the ith scaling factor and the optimal solution to the dual uplink problem (11.16). The iterations to find the downlink parameters for the MBF are summarized in algorithm 11.1.

Figure 11.5 characterizes the speed of convergence of algorithm 11.1 in terms of the variation of the Euclidean norm of power residue, i.e. $\|\mathbf{p}(n) - \mathbf{p}^\star\|_2$, where $\mathbf{p}^\star$ is the optimal dual-uplink power vector, versus the number of iterations n. As shown in Figure 11.5, the iterations converge faster to the optimal point with more antenna elements at base stations and lower target SINR values at user terminals for a given number of active users in the system.

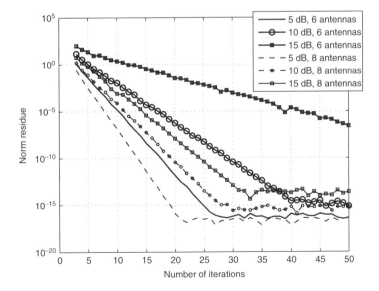

Figure 11.5 Convergence for 6 users per virtual cell with different required SINRs and number of antenna elements.

11.5 Coordinated beamforming

In coordinated beamforming (CBF), while the user terminals are served by their local base stations only, a number of base stations coordinate at beamforming level to minimize their mutual intercell interferences. In CBF, the backhaul overhead is lighter than MBF because only global CSIT is required by each one of the coordinating base stations. In order to be consistent with MBF and for simplicity, we limit our attention within a virtual cell by accounting for only dominant interference and considering interference from the outer tiers of base stations surrounding the virtual cell as noise. However, the analysis given in this section can be straightforwardly extended to larger clusters of coordinating base stations. Iterative algorithms have been proposed for solving the CBF optimization problem, e.g. [6]. In this section, we formulate the CBF optimization in the standard semi-definite programming (SDP) form.

Let the index $i(p)$ denote user i served by base station p and $\mathbf{w}_{i(p)} = [w_i(p, 1) \, w_i(p, 2) \cdots w_i(p, M)]^T$ indicate the beamforming vector of the ith user at the pth BS, $w_i(p, k)$ be the corresponding beamforming coefficient of the kth antenna element and $\mathbf{w}_i = \left[\mathbf{w}_{i(1)}^T \, \mathbf{w}_{i(2)}^T \, \mathbf{w}_{i(3)}^T \right]^T$. In general, the number of active users in different sectors of a virtual cell may be different. A zero vector for $\mathbf{w}_{i(p)}$ in $\mathbf{w}_i$ indicates an absent user i from the corresponding sector/cell p. Hence, without loss of generality, we assume a total number of U users per sector of a virtual cell.

Let $s_{i(p)}$ denote the ith user's symbol at the pth base station and $\mathbf{s}_p = \left[s_{1(p)} \cdots s_{i(p)} \cdots s_{U(p)} \right]^T$. Let $\mathbf{x}_p = \mathbf{w}_{i(p)} \mathbf{s}_p$ be the transmitted signal by the pth base station and $\mathbf{x} = \left[\mathbf{x}_1^T \, \mathbf{x}_2^T \, \mathbf{x}_3^T \right]^T$. Define

$$\overline{\mathbf{H}}_{i(p)} = \begin{bmatrix} \mathbf{h}_{i(p)}(1) & \mathbf{0} & \mathbf{0} \\ \mathbf{0} & \mathbf{h}_{i(p)}(2) & \mathbf{0} \\ \mathbf{0} & \mathbf{0} & \mathbf{h}_{i(p)}(3) \end{bmatrix}, \quad (11.20)$$

where $\mathbf{h}_{i(p)}(p')$ is the $1 \times M$ vector channel of user i in cell p as seen by cell $p' \in \{1, 2, 3\}$, and $\mathbf{0}$ represents $1 \times M$ vector with all zeros elements. Moreover,

$$\mathbf{h}_{i(p)}(p') = \left[\varrho_{i(p),1} \, \varrho_{i(p),2} \cdots \varrho_{i(p),Q} \right] \begin{bmatrix} \xi_{i(p),1}(p') \mathbf{h}_{i(p),1}(p') \\ \xi_{i(p),2}(p') \mathbf{h}_{i(p),2}(p') \\ \vdots \\ \xi_{i(p),Q}(p') \mathbf{h}_{i(p),Q}(p') \end{bmatrix}, \quad (11.21)$$

where $\mathbf{h}_{i(p),t}(p') = \left[h_{i(p),t}(p', 1) \, h_{i(p),t}(p', 2) \cdots h_{i(p),t}(p', M) \right]$, $h_{i(p),t}(p', k)$ is the channel between the tth scatterer of the ith user of cell p and the kth antenna element of the p'th BS. Using (11.4) with $l_{i,\min} = l_i(p)$, one can get the expression for $h_{i(p),t}(p', k)$. The received signal by user $i(p)$ is given by

$$y_{i(p)} = \sum_{q=1}^{3} \mathbf{e}_q^T \overline{\mathbf{H}}_{i(p)} \mathbf{x} + z_{i(p)}, \quad (11.22)$$

11.5 Coordinated beamforming

where $\mathbf{e}_q$ is a 3×1 unit vector with 1 in row q and zero elsewhere and $z_{i(p)} \sim N(0, \sigma_N^2)$ is white ZMCSCG noise at user $i(p)$. Finally, the optimization problem for CBF is formulated as

$$\min_{\mathbf{w}_i} \sum_{i=1}^{U} \mathbf{w}_i^H \mathbf{w}_i$$

$$\text{subject to } \frac{\left|\mathbf{e}_p^T \overline{\mathbf{H}}_{i(p)} \mathbf{w}_i\right|^2}{\sum_{u=1, u \neq i}^{U} \left|\mathbf{e}_p^T \overline{\mathbf{H}}_{i(p)} \mathbf{w}_u\right|^2 + \sum_{q=1, q \neq p}^{3} \sum_{m=1}^{U} \left|\mathbf{e}_q^T \overline{\mathbf{H}}_{i(p)} \mathbf{w}_m\right|^2 + \sigma_N^2} \geq \gamma_{i(p)},$$

$$\text{for all } 1 \leq p \leq 3, \, 1 \leq i \leq U$$

(11.23)

where $\gamma_{i(p)}$ is the target SINR at user $i(p)$. Using a slack variable P_o, we can rewrite (11.23) as follows:

$$\min_{\mathbf{W}, P_o} P_o$$

$$\text{subject to } B_{i(p)} \geq \gamma_{i(p)},$$

$$\text{Tr}\left[\mathbf{W}^H \mathbf{W}\right] \leq P_0$$

$$\text{for all } 1 \leq p \leq 3, \, 1 \leq i \leq U$$

(11.24)

where $\mathbf{W} = [\mathbf{w}_1 \, \mathbf{w}_2 \cdots \mathbf{w}_U]$ and

$$B_{i(p)} = \frac{\left|\mathbf{e}_p^T \overline{\mathbf{H}}_{i(p)} \mathbf{W} \mathbf{e}_i\right|^2}{\sum_{u=1, u \neq i}^{U} \left|\mathbf{e}_p^T \overline{\mathbf{H}}_{i(p)} \mathbf{W} \mathbf{e}_u\right|^2 + \sum_{q=1, q \neq p}^{3} \sum_{m=1}^{U} \left|\mathbf{e}_q^T \overline{\mathbf{H}}_{i(p)} \mathbf{W} \mathbf{e}_m\right|^2 + \sigma_N^2}.$$

Notice that $\mathbf{e}_p$ and $\mathbf{e}_q$ are 3×1 unit vectors while $\mathbf{e}_i$, $\mathbf{e}_u$, and $\mathbf{e}_m$ are $U \times 1$ unit vectors. Rearranging the SINR constraint in (11.24), we have
$\left(1 + \frac{1}{\gamma_{i(p)}}\right) \left|\mathbf{e}_p^T \overline{\mathbf{H}}_{i(p)} \mathbf{W} \mathbf{e}_i\right|^2 \geq \sum_{q=1}^{3} \sum_{m=1}^{U} \left|\mathbf{e}_q^T \overline{\mathbf{H}}_{i(p)} \mathbf{W} \mathbf{e}_m\right|^2 + \sigma_N^2$ for all $1 \leq p \leq 3$, $1 \leq i \leq U$. Then, we can write

$$\sqrt{1 + \frac{1}{\gamma_{i(p)}}} \mathbf{e}_p^T \mathbf{P}_i \mathbf{H}_p \mathbf{W} \mathbf{e}_i \geq \left\| \begin{bmatrix} \text{vec}(\mathbf{P}_i \mathbf{H}_p \mathbf{W}) \\ \sigma_N \end{bmatrix} \right\|$$

(11.25)

for all $1 \leq p \leq 3, 1 \leq i \leq U$, where $\mathbf{H}_p = \left[\overline{\mathbf{H}}_{1(p)}^T \, \overline{\mathbf{H}}_{2(p)}^T \cdots \overline{\mathbf{H}}_{U(p)}^T\right]^T$, $\mathbf{P}_i = \left[\cdots \vdots \mathbf{0} \vdots \mathbf{I} \vdots \mathbf{0} \vdots \cdots\right]$ is a $3 \times 3U$ permutation block matrix with 3×3 identity matrix $\mathbf{I}$ as the ith block and blocks of 3×3 all zero matrices $\mathbf{0}$ elsewhere. Similar to the MBF scheme, one can recast (11.25) in semi-definite form as follows.

$$\mathbf{L}_{i(p)} = \begin{bmatrix} \sqrt{1 + \frac{1}{\gamma_{i(p)}}} \mathbf{e}_p^T \mathbf{P}_i \mathbf{H}_p \mathbf{W} \mathbf{e}_i & \left[\text{vec}^H (\mathbf{P}_i \mathbf{H}_p \mathbf{W}) \quad \sigma_N\right] \\ \begin{bmatrix} \text{vec}(\mathbf{P}_i \mathbf{H}_p \mathbf{W}) \\ \sigma_N \end{bmatrix} & \sqrt{1 + \frac{1}{\gamma_{i(p)}}} \mathbf{e}_p^T \mathbf{P}_i \mathbf{H}_p \mathbf{W} \mathbf{e}_i \mathbf{I} \end{bmatrix} \succeq 0,$$

(11.26)

for all $1 \leq p \leq 3$, $1 \leq i \leq U$. Finally, the power constraint in (11.24) is written in semi-definite form as

$$\mathbf{N} = \begin{bmatrix} p_0 & \text{vec}^H(\mathbf{W}) \\ \text{vec}(\mathbf{W}) & p_0 \mathbf{I} \end{bmatrix} \succeq 0, \quad (11.27)$$

where $p_0 = \sqrt{P_o}$. We now recast the CBF problem in SDP form as

$$\begin{aligned} \min_{\mathbf{W}, p_0} \quad & p_0 \\ \text{subject to} \quad & \mathbf{L}_{i(p)} \succeq 0, \mathbf{N} \succeq 0, \text{ for all } 1 \leq p \leq 3, \ 1 \leq i \leq U. \end{aligned} \quad (11.28)$$

11.6 Backhaul protocol

This section presents a wireless backhaul protocol to exchange information among the base stations of a virtual cell to establish intercell coordination. We develop [9] expressions for power consumption in the backhaul and an overall sum-rate analysis evaluating the effective sum capacity in the presence of an imperfect wireless backhaul in the downlink of a virtual cell.

11.6.1 A protocol for information circulation in the backhaul

The implementation of the beamforming techniques for the virtual cell demands backhaul to circulate information amongst three BSs. Taking into account the broadcast essence of wireless communications, we assert a fast protocol, i.e. the ring protocol, for information exchange among three BSs using network coding. Interested readers are referred to [9] for details of the protocol and throughput analysis. The protocol is briefly explained as follows.

In order to exchange information among the three BSs the ring protocol uses three steps. In step 1, BSs 1 and 2 send their messages to BS 3. BS 3 decodes and then combines the received messages using bitwise XOR. The combined message is broadcast by BS 3 in step 2. Finally, in step 3, BS3 sends its message to the other two BSs. The broadcast rates in steps 2 and 3 are chosen to be at the rate of the weaker link so that the messages can be decoded at both BSs.

Figure 11.6 shows the achievable maximum backhaul spectral-efficiency for the ring protocol against γ_1 with different values of γ_2 in linear scale. In this figure, γ_1 and γ_2 indicate, respectively, the signal-to-noise ratios (SNRs) between the BS2 and BS3 links and the BS1 and BS 3 links. It is clear from Figure 11.6 that the ring model has the best performance when the two links are comparable in terms of SNRs, i.e. γ_1 and γ_2. In the case of imbalanced link quality, i.e. links with incomparable SNRs, the maximum backhaul spectral-efficiency of the ring protocol strongly depends on the SNR of the weaker link. In order to improve the maximum backhaul spectral-efficiency, the SNR of the weaker link should be increased. In the case that the SNR of the weaker link can not be improved, the overall power-consumption can be reduced instead by reducing the

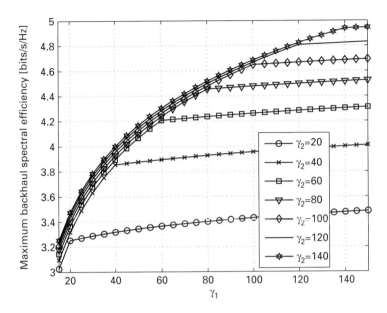

Figure 11.6 Achievable maximum backhaul spectral-efficiency for the ring model against γ_1 with different values of γ_2 in linear scale.

SNR of the stronger link in step 1 and making it comparable with the SNR of the weaker one. However, Figure 11.6 shows that this procedure results in a minor sacrifice in the backhaul spectral-efficiency.

11.6.2 Power calculation for the ring protocol

Let us assume that the base stations exchange information in the backhaul using isotropic antennas in free space. The transmit power required by a base station to attain an SNR level of γ at a receiving base station, i.e. $P(\gamma)$, is given by

$$P(\gamma) = \gamma \sigma_{bs}^2 \left(\frac{4\pi d_{bs}}{\lambda_c} \right)^2, \quad (11.29)$$

where λ_c is the backhaul carrier wavelength, d_{bs} is the distance between the BSs, and σ_{bs}^2 is the ZMCSCG noise variance at the BS. Using (11.29) and the expression for the overall time duration T required by the ring protocol to circulate information in the backhaul from [9], one can calculate the backhaul average power as

$$\overline{P} = P(\gamma_1) + \frac{2C(\gamma_1)}{4C(\gamma_1) + C(\gamma_1 + \gamma_2)} P(\gamma_2), \quad \text{if } \gamma_1 \leq \gamma_2 < \gamma_1 + \gamma_1^2, \quad (11.30)$$

$$\overline{P} = P(\gamma_1) + \frac{C(\gamma_1)}{C(\gamma_1) + C(\gamma_1 + \gamma_2)} P(\gamma_2), \quad \text{if } \gamma_2 > \gamma_1 + \gamma_1^2, \quad (11.31)$$

where $C(\gamma) = B\log_2(1+\gamma)$ and B is the bandwidth used for the backhaul.

11.6.3 An effective sum-rate

Referring to the zonal classification of users in Figure 11.4, we allow coordination among two and three BSs in zones 2 and 3 and define the corresponding backhaul rates of R_{bh2} and R_{bh3}, respectively. Let V_{csi}, V_2, and V_3 denote the total number of CSIT bits corresponding to a number of simultaneous multiple users in the network, the total number of data bits to be circulated in zones 2 and 3, respectively. The time duration to circulate information in the backhaul is

$$T_1 = \psi \left(\frac{V_2}{R_{\text{bh2}}} + \frac{V_3}{R_{\text{bh3}}} \right) + \frac{V_{\text{csi}}}{R_{\text{bh3}}}, \qquad (11.32)$$

where $\psi = 1$ for MBF, i.e. signal level coordination, and $\psi = 0$ for CBF, i.e. beamforming level coordination.

Let R_i be the downlink rate excluding the backhaul latency effect for user i, and v_i be the corresponding number of data bits delivered in a time span of v_i/R_i seconds. Assume, user m requires the longest time span among the other $U-1$ users, i.e. $T_2 = \max_i (\frac{v_i}{R_i}) = \frac{v_m}{R_m}$. Then, an overall duration of $T = T_1 + T_2$ seconds is required to exchange information, i.e. users' data and CSIT, in the backhaul and, then, deliver a total number of $V = V_2 + V_3$ bits to U simultaneous users. Note that, here, we ignore the processing time and power from our calculations. We can write

$$T = V \frac{\chi R_{\text{bh3}} + R_m [\varphi + \psi \{1 + \rho (\beta - 1)\}]}{R_{\text{bh3}} R_m}, \qquad (11.33)$$

where $\varphi = V_{\text{csi}}/V$, $\chi = v_m/V$, $\rho = V_2/V$, and $\beta = R_{\text{bh3}}/R_{\text{bh2}}$.
Let $R_{\text{eff}} = \frac{V}{T}$ define the effective sum rate, i.e. including the downlink and the backhaul latency, of U simultaneous users. Then

$$R_{\text{eff}} = \frac{R_{\text{bh3}} R_m}{\chi R_{\text{bh3}} + R_m [\varphi + \psi \{1 + \rho (\beta - 1)\}]}. \qquad (11.34)$$

11.7 Performance evaluation

In this section, we evaluate and compare the aforementioned beamforming strategies under ideal backhaul, i.e. delay-less, error-free, and high speed with no power-consumption channels interconnecting the base stations, and imperfect backhaul with latency, power consumption, and limited capacity.

11.7.1 Performance evaluation under ideal backhaul

The simulation setup uses six antenna elements per sector with an antenna spacing of a half of a wavelength and a downlink carrier frequency of 2 GHz. A standard deviation

11.7 Performance evaluation

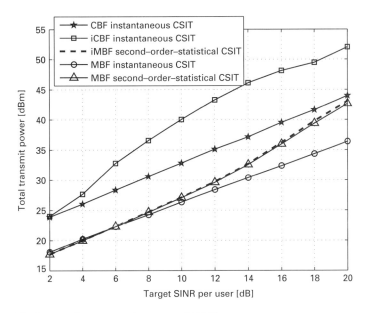

Figure 11.7 Total transmit power against targeted SINR per user.

of 2° for the angular spread due to five random scatterers around each user terminal is assumed. We assume $N = 512$ Gaussian parallel MISO subchannels between a base station and a user terminal, where the fading coefficients for each subchannel are 1×6 randomly generated ZMCSCG variables with unit variance. The noise power spectral density for all users is -174 dBm/Hz, the noise figure is 5 dB and a subchannel bandwidth is 15 kHz wide. The array antenna gain at the BSs is set at 15 dBi. We use $128.1 + 37.6\log_{10}(l)$, where l is in km, as the path-loss model. A standard deviation of 8 dB is assumed for log-normal shadowing. Also, any two neighboring BSs are located 3 km apart from one another. We examine these schemes with three single-antenna users per virtual cell, i.e. one user per sector. Users are randomly dropped in the critical zones of 2 and 3 of the virtual cell, i.e. Z_{12}, Z_{13}, Z_{23}, and Z_{123} in Figure 11.4. Each simulation cycle to find the average transmitted power per subchannel is averaged over 100 random user locations, with 10 000 channel realizations per location. The SeDuMi solver [10] is used to attain the optimal solutions.

Figure 11.7 shows the variation of the sum-transmit power of BSs in a virtual cell against targeted SINR levels at user terminals for different beamforming strategies. It is shown that the MBF solution, i.e. solution to the optimization problem (11.15), can be closely approached by the iterative MBF, i.e. iMBF, algorithm with 50 iterations, when the second-order statistical CSIT is used. While such a perfect match does not exist between the CBF solution, i.e. solution to the optimization problem (11.23), and the iterative CBF algorithm, i.e. iCBF, in [6], despite the same CSIT being used by both schemes. This performance gap can be explained as follows. The feasibility region of the iCBF optimization problem in [6] is more restricted than the CBF optimization problem in (11.23). The downlink beamforming vectors of the iCBF are found as the multiplication of some scalars by the corresponding virtual uplink beamforming vectors which are in turn found by the MMSE solution, i.e. [6]. Limitation of the dual-uplink solution to MMSE can be interpreted as an additional constraint to the original optimization problem (11.28), while this additional constraint is not imposed when solving (11.28) for the CBF.

Figure 11.7 also shows that the MBF is more power-efficient than the CBF and, furthermore, MBF using the instantaneous CSIT outperforms MBF using the second-order statistical CSIT in terms of power efficiency. The advantage of MBF in being more power-efficient is due to more effective co-channel interference management by the MBF strategies than by the CBF, and this effectiveness improves as the accuracy of available CSIT increases.

11.7.2 Performance evaluation under limited backhaul

In this section, we draw the ratio $\frac{P_{\text{CBF}}}{P_{\text{MBF}}}$, where P_{CBF} and P_{MBF} are the total power-consumptions due to the CBF and the MBF schemes, respectively, versus the effective sum-rate, i.e. (11.34), at various MBF backhaul rate constraints in Figure 11.8 using the following steps.

1. Assuming equal target SINRs for all users, find the corresponding transmit power and maximum rate for each scheme using Figure 11.7 and $R_m = B\log_2(1+\text{SINR}_m)$, respectively, where B indicates the transmission bandwidth.
2. Given a limited backhaul rate for MBF, i.e. $R_{\text{bh3}}(\text{MBF})$, and R_m from step 1, set $\psi = 1, \varphi = 0.2, \chi = 1/3, \rho = 0, \beta = 2/3$, and use (11.34) to find R_{eff}.

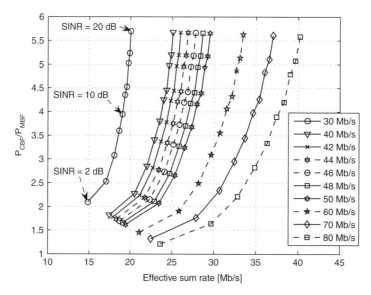

Figure 11.8 Illustration of total power-consumption ratios of the CBF over the MBF schemes versus the effective sum-rate with various MBF backhaul rate constraints.

3. Given R_m and R_{eff} from steps 1 and 2, use (11.34) with $\psi = 0$ to find the backhaul rate for CBF, i.e. $R_{\text{bh}3}(\text{CBF})$.
4. Using $R_{\text{bh}3}(\text{MBF})$ and $R_{\text{bh}3}(\text{CBF})$ and B, find the SNR values, i.e., γ_1 and γ_2, for the backhaul channels of each scheme from Figure 11.6.
5. For each scheme, use the corresponding values of γ_1 and γ_2 in (11.30) or (11.31) to find the backhaul power required for the MBF and the CBF schemes. Then, the total power of each scheme is the sum of their corresponding transmit power, i.e. calculated in step 1, and the backhaul power.

The downlink bandwidth for data transmission is 7.68 MHz wide. Backhaul bandwidths of 7.5, 10, 10.5, 11, 11.5, 12, 12.5, 15, 17.5, and 20 MHz with a backhaul spectral efficiency of 4 bits/s/Hz are used. An equal noise power spectral density of -174 dBm/Hz is assumed at all BSs and a carrier frequency of 2.4 GHz is used for the backhaul links.

The results shown in Figure 11.8 confirm that the MBF is more power-efficient than the CBF even when the backhaul effects are taken into consideration.

11.8 Cooperative routing

Consider a scenario where a user in the source cell needs to transfer a message to a user in the destination cell, as shown in Figure 11.9. An energy-efficient way of communicating between the two users is through multi-hop routing. In the uplink, when a user transmits a message to a BS, neighboring BSs can either receive it directly by overhearing or via a broadcast channel initiated by the source BS. Then, the neighboring BSs join the source BS to transmit the message cooperatively to the next hop destination. In this section, we introduce a power-aware cooperative routing (PACR) algorithm and compare its

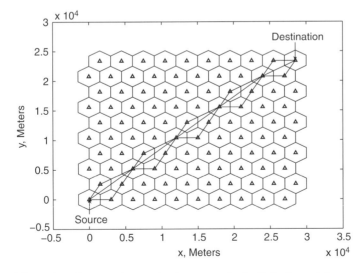

Figure 11.9 Illustration of cooperative routing in a network of cellular base stations.

energy-saving potential against the power-saving routing (PSR) algorithm in [11] that does not use cooperative transmission.

11.8.1 Power-aware cooperative routing algorithm

Authors in [11] introduced the PSR algorithm to find the location of an intermediate virtual node, i.e. between the source and the final destination, as the closest point to the next hop destination, as follows:

$$d_{opt} = \sqrt[\alpha]{\frac{C\mu_{pa}}{B(\alpha-1)}}, \qquad (11.35)$$

where C is the increase in circuit power consumed by the transmitter and the receiver as a result of transmission between two BSs, μ_{pa} is the power amplifier efficiency of the transmitting BS, and α is the path-loss exponent. Furthermore, $B = \gamma_r \sigma^2 (4\pi/\lambda)^2$, where γ_r is the minimum SNR requirement at the final destination for successful decoding of the transmitted message, and σ^2 is the noise variance at the destination. Using (11.35) and the process of finding the virtual node [11], we introduce the PACR algorithm, as described in the following.

A source BS broadcasts its data to its $l-1$ adjacent BSs. Then, l BSs join together to transmit cooperatively to a destination base station r. Assuming equal power-amplifier efficiency of μ for each transmitting BS and a target SNR of γ_r at the destination, one can show, using the results in [12], that the overall transmit power by l base stations is given by

$$P_t = \frac{B}{\sum_{k=1}^{l} \frac{1}{d_{kr}^\alpha}}, \qquad (11.36)$$

where d_{kr}, $k = 1, \cdots, l$ is the distance from the kth transmitting BS to the destination r. Hence, the overall power-consumption of the power amplifiers of l transmitting BSs can be written as

$$P_c = \frac{B}{\mu \sum_{k=1}^{l} \frac{1}{d_{kr}^\alpha}}. \qquad (11.37)$$

Let us define a supernode located at the position of the closest transmitting BS to the destination. Let this supernode be equivalent to the l cooperating BSs in terms of maintaining the same target SNR, i.e. γ_r, at the destination while consuming the same power, i.e. P_c, with a power-amplifier efficiency of μ_s. We assume without loss of generality that d_{1r} is the closest distance to the destination, i.e. $d_{1r}^\alpha \leq d_{2r}^\alpha \leq \cdots \leq d_{lr}^\alpha$. Equating the total power consumption of l cooperative BSs from (11.37) with the equivalent power consumption of the supernode, one can write

$$\frac{B}{\mu \sum_{k=1}^{l} \frac{1}{d_{kr}^\alpha}} = \frac{B}{\mu_s} d_{1r}^\alpha. \qquad (11.38)$$

11.8 Cooperative routing

Solving (11.38) for μ_s, we get

$$\mu_s = \mu \left(1 + \sum_{k=2}^{l} \frac{d_{1r}^{\alpha}}{d_{kr}^{\alpha}}\right). \tag{11.39}$$

It can be seen from (11.39) that the highest power-amplifier efficiency of the supernode is obtained when all of the l transmitting BSs are distant enough from the destination BS so that they can be seen at the same distance as the closest node from the destination, i.e. at d_{1r}. Hence, the power-amplifier efficiency of the supernode is upper bound as

$$\mu_s \leq l\mu. \tag{11.40}$$

Furthermore, as the distance between the destination BS and the closest transmitting BS diminishes, i.e. $d_{1r} \to 0$, then $\mu_s \to \mu$, which means that the performance of the cooperative transmission reduces to that of the non-cooperative one. Using $\mu_{pa} = \mu_s = l\mu$, i.e. the condition where cooperation is most efficient, and $C = (l+1)E$, where E is the circuit power consumption of each operating BS, in (11.35), one can calculate the optimum distance from the supernode to a virtual node, i.e. an imaginary node that is closest to the next hop destination, as $d_{opt} = \sqrt[\alpha]{\frac{(l+1)El\mu}{B(\alpha-1)}}$. This virtual node is located d_{opt} apart from the supernode and on the line connecting the supernode to the destination BS. The next hop (BS) destination is chosen as the nearest hop to this virtual node. Let us denote this next hop destination as BS(n), i.e. the nth hop destination found during the routing process, and its relative distance from the corresponding supernode as d_n. Then, the cooperative transmission to BS(n) is energy-efficient if

$$\frac{B}{\mu_s} d_n^{\alpha} + (l+1)E \leq \frac{B}{\mu} d_n^{\alpha} + 2E, \tag{11.41}$$

or equivalently

$$d_n \geq \sqrt[\alpha]{\frac{(l-1)E}{B} \left(\frac{\mu_s \mu}{\mu_s - \mu}\right)}. \tag{11.42}$$

Otherwise, when (11.42) does not hold, then non-cooperative transmission is more energy-efficient. The aforementioned procedure is summarized in Algorithm 11.2.

In a simulation setup, we implement PACR and PSR algorithms in a network of 400 base stations where the source and the destination BSs are located in the opposite corners of the network, as shown in Figure 11.9. Figure 11.10 shows the ratio $\frac{P_N}{P_C}$, where P_C and P_N are the powers consumed by PACR and PSR, respectively, against the BS to BS distance at a path-loss exponent of 4. The result confirms a maximum of 30% power-saving gain of the PACR algorithm over the PSR algorithm.

Algorithm 11. 2 Power-aware cooperative routing algorithm

Require: Wireless network, with source S, destination D and location of BSs.
1: Initialize $n = 0$.
2: $BS(n) \leftarrow S$.
3: **while** $BS(n) \neq D$ **do**
4:   **if** D is a first tier neighbor of $BS(n)$ **then**
5:     $BS(n+1) \leftarrow D$
6:   **else**
7:     Broadcast to $(l-1)$ closest neighbors.
8:     Find the virtual node position and select $BS(n+1)$.
9:     Select between cooperative and non-cooperative transmission modes.
10:   **end if**
11:   Update route table.
12:   $n = n + 1$.
13: **end while**

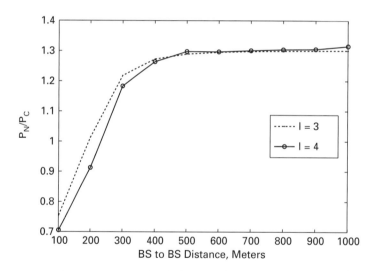

Figure 11.10 Power saving gain as a function of BS to BS distance with $\alpha = 4$, $l = 3$, and $l = 4$.

11.9 Conclusion

Cell splitting and base station cooperation are two effective means to achieve high spectral-efficiency and coverage in cellular wireless networks. A higher data rate requires more transmit power, which is constrained by human safety limits as well as green communications targets. Dividing macro-cells into a number of smaller cells provides coverage at allowed transmission energy limits and multicell processing transforms the obstructive interference dimensions into capacity achieving signals by coordinating among the base stations. As full interconnection among all base stations in the network is practically unrealistic, we have defined the concept of virtual cells comprising of three base stations to cover critical areas where high traffic demand is expected. Users within a virtual cell are served by a coordination of up to three base stations or individually by their local base stations based on their locations.

We have developed solutions to multicell processing, where user signals are globally shared by the coordinating base stations, using both instantaneous and second-order statistical CSIT, as well as an iterative solution using statistical CSIT. Then, we have presented a standard semi-definite programming formulation to the coordinated beamforming, where user signals are not shared among the coordinating base stations. Taking into account the latency and the limited capacity of the imperfect wireless backhaul links among the base stations, we have derived a relationship to calculate the effective sum-rate in the downlink, including the backhaul effects. Our performance evaluation results for the cooperative downlink in a virtual cell have shown that the multicell beamforming is more power-efficient than the coordinated beamforming, even if the backhaul constraints and burden are considered. Finally, we have studied cooperative routing of user signals between two distant base stations within the cellular network. We have developed a power-aware cooperative routing algorithm and showed that it is more power-efficient than a non-cooperative counterpart when the distance between the source and the destination base stations is adequately large.

Acknowledgment

This research has been funded by the Industrial Companies who are Members of Mobile VCE, with additional financial support from the UK Government's Engineering & Physical Sciences Research Council, whose funding support is gratefully acknowledged.

References

[1] T. A. Le and M. R. Nakhai, "Possible power-saving gains by dividing a cell into tiers of smaller cells," *IET Electronics Letters*, vol. 46, no. 16, pp. 1163–1165, Aug. 2010.

[2] A. Goldsmith, *Wireless Communications*. Cambridge University Press, 2005.

[3] A. Wiesel, Y. C. Eldar, and S. Shamai, "Linear precoding via conic optimization for fixed MIMO receivers," *IEEE Trans. Sign. Proc.*, vol. 54, no. 1, pp. 161–176, Jan. 2006.

[4] L. Vandenberghe and S. Boyd, "Semidefinite programming," *SIAM Review*, vol. 38, no. 1, pp. 49–95, Mar. 1996.

[5] M. Bengtsson and B. Ottersten, "Optimal downlink beamforming using semidefinite optimization," in *Proc. of 37th Annual Allerton Conf. Commun., Control, and Computing*, pp. 987–996, Sep. 1999.

[6] H. Dahrouj and W. Yu, "Coordinated beamforming for the multicell multi-antenna wireless system," *IEEE Trans. Wireless Commun.*, vol. 9, no. 5, pp. 1748–1759, May 2010.

[7] T. A. Le and M. R. Nakhai, "An iterative algorithm for downlink multi-cell beam-forming," in *Proc. of IEEE Global Telecommunications Conference (GLOBECOM 2011)*, pp. 1–6, Dec. 2011.

[8] R. D. Yates, "A framework for uplink power control in cellular radio systems," *IEEE JSAC*, vol. 13, no. 7, pp. 1341–1348, Sep. 1995.

[9] T. A. Le and M. R. Nakhai, "Throughput analysis of network coding enabled wireless backhauls," *IET Commun.*, vol. 5, no. 10, pp. 1318–1327, Jun. 2011.

[10] J. F. Sturm, "Using SeDuMi 1.02, a MATLAB toolbox for optimization over symmetric cones," *Optimization Methods and Software*, 1999, [Online] Available: http://sedumi.mcmaster.ca/

[11] I. Stojmenovic and X. Lin, "Power aware localized routing in wireless networks," *IEEE Trans. Parall. and Distrib. Systems*, vol. 12, no. 11, pp. 1122–1133, Nov. 2001.

[12] A. E. Khandani, J. Abounadi, E. Modiano, and L. Zheng, "Cooperative routing in static wireless networks," *IEEE Trans. Commun.*, vol. 55, no. 11, pp. 2185–2192, Nov. 2007.

Part IV

Wireless access techniques for green radio networks

12 Cross-layer design of adaptive packet scheduling for green radio networks

Ashok Karmokar, Alagan Anpalagan, and Ekram Hossain

12.1 Introduction

In a cellular wireless network, most of the energy is consumed in the radio access network [1]. Over the last decades, a significant amount of research work has focused on spectrally efficient and reliable wireless communications techniques at the physical (PHY) layer. However, the interactions among the layers (e.g. PHY-layer, radio link layer, and network layer) in the transmission protocol stack have to be taken into account to minimize the overall energy-consumption [2] in a green wireless network. The success of such a green wireless technology can be measured by energy-efficient metrics at different levels from the physical to application layer [3]. Energy efficiency across the entire system or network exploiting the layer interactions is not well understood and needs more attention. The joint optimization of the transmission scheduling and resource allocation (or management) at various layers is referred to as cross-layer optimization. Again, energy efficiency in wireless communications systems so far has primarily focused on uplink communication due to the miniaturized mobile terminals and their limited energy storage capabilities. However, with a significant portion of the wireless internet traffic being from power-hungry base stations to end user mobile devices, energy optimization in the downlink is most important for green radio networks. In this article, we present a cross-layer optimized downlink packet transmission scheduling technique for the realization of green radio networks.

After reviewing some related work on adaptive resource allocation in wireless networks, we discuss why cross-layer interaction, information exchange, and optimization are important for wireless networks, and more specifically, for green radio networks. Then we describe the models associated with cross-layer packet transmission adaptation techniques. A comparison between the policy structure and the nature of the single-layer and cross-layer optimized policies is then presented. The techniques for computing optimal policy that intelligently consider cross-layer design variables are discussed. With numerical analysis, the trade-off between the utilities (power, delay, overflow, and bit error rate) are illustrated. To this end, we discuss some outstanding research challenges in design, implementation, and evaluation of cross-layer optimized packet transmission protocols for green radio networks.

12.2 Related work

Adaptive channel-aware resource allocation schemes have proven to be powerful techniques both in increasing high throughput and in achieving high reliability over fading channels. Adaptive radio resource management has already been incorporated in many wireless standards, such as EDGE, 3GPP LTE, and IEEE 802.16. Some of these schemes require accurate channel estimation at the receiver and a reliable feedback path between that estimator and the transmitter in order to adapt resources at the transmitter. Link adaptation techniques in traditional cellular networks have been optimized in terms of spectral-efficiency, capacity or throughput, but not in terms of energy efficiency. Moreover, the traditional channel-dependent adaptive transmission techniques cannot guarantee application-specific quality-of-service (QoS) requirements such as packet delay and loss in wireless networks. The system-level throughput may suffer due to the loss of a large number of packets due to buffer overflow. Also, the transmission power is not controlled based on the buffer information. In wireless networks, intelligent decisions on transmission rate and power are crucial due to their limited availability, and also to conserve energy. Randomly varying incoming traffic also plays an important role in the scheduling of wireless resources, since practical systems are equipped with a finite storage buffer. Due to the randomness of incoming packets and channel gains, the exact scheduling policies cannot be determined using static channel-dependent optimization techniques.

Transmit power and rate adaptation can yield significant power savings and hence help reduce the energy footprint and greenhouse gas emissions when adaptation and optimization are based on cross-layer interaction and information exchange. There are some results in the literature that deal with energy optimization for delay-tolerant data applications [4]. Many cross-layer studies in the literature investigated the fundamental relationship between average power consumption and other QoS requirements. For example, authors in [5] investigated the problem of optimal trade-off between average power and average delay for a single user communicating over a memoryless block-fading channel using information-theoretic concepts. In [6], convexity properties and characterization of the delay-power region of different schedulers were discussed for independent and identically distributed (i.i.d.) traffic in fading channels. A Markov decision process (MDP)-based selection technique for user, power, and rate was proposed in [7]. In [8], the trade-off between energy and packet loss rate subject to a hard delay constraint was studied over correlated fading channels using a finite horizon MDP framework. An offline packet-scheduling scheme to minimize energy subject to a deadline constraint was considered in [9]. Energy-efficiency issues for ad hoc wireless networks are studied in some early work in the literature, e.g. in [4], [10].

Most of the work in the literature deals with the energy optimization problem considering an infinite transmission buffer and/or a fixed traffic arrival rate when data needs to be transmitted within a specified time limit. However, practical wireless systems use a finite transmission buffer with randomly varying bursty traffic arrival. In this article, we discuss a cross-layer energy optimization scheme for downlink transmission considering a finite transmission buffer and correlated bursty traffic arrival. We illustrate how with the proposed cross-layer optimized scheme for delay-tolerant traffic, the base station (BS) can save a huge amount of energy when compared with a traditional single-layer optimized policy.

12.3 Importance of cross-layer optimized design

Unlike wired/optical media, the channel gain of the wireless medium varies with time, space, and frequency randomly. Due to this random variation, the channel is in deep fade some of the time, causing erroneous reception, and it is at peak at other times, permitting maximum usage and errorless communications. Therefore, the prudent use of wireless channels is crucial. In order to achieve reliable and errorless transmission over a wireless channel, a plethora of adaptation schemes has been devised and used in different layers of the open system interconnection (OSI) protocol stack. For example, in the physical layer, the coding and modulation rate can be adapted with the channel condition. Thus, transmitting nothing or the minimum number of packets using the lowest modulation order (e.g. BPSK) and lowest rate coding (using more redundant bits) in bad channel states and, transmitting with higher-order modulation (e.g. 64-QAM) and higher rate coding (using a fewer number of redundant bits) as the channel improves. In the medium access control (MAC) layer, depending upon channel gains, the users can be chosen adaptively for transmission in a given time-slot. For example, in opportunistic scheduling, the user with the best channel gain is chosen for transmission in a particular time-slot. Various fairness strategies have been included in opportunistic scheduling in order to correct the inherent unfairness issue, where the users that are not among the best set are also scheduled for transmission in order to maintain fairness.

In a traditional wireless network, each layer of the OSI stack operates its own protocol in order to optimize a local objective based on the locally available information, as shown in Figure 12.1(a). Recently, adaptive cross-layer resource allocation

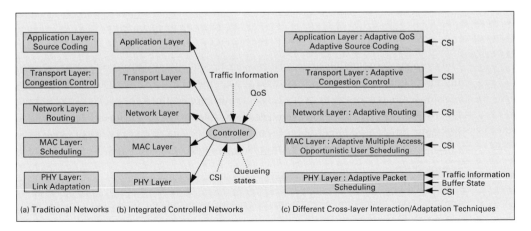

Figure 12.1 An illustration of OSI layers in traditional network, and interactions between layers in cross-layer optimized networks.

and optimization techniques that exploit inter-dependency and interaction among PHY, MAC, and higher layers attracted significant attention due to the resulting system performance improvement. An idealized cross-layer design, as shown in Figure 12.1(b), would have a centralized controller that is provided with all the channel gains, buffer occupancies, and QoS requirements for all users and applications. The scheduler will then find the optimal strategies for different layers jointly. Such a cross-layer optimized scheduler is not realizable due to their computational and other inherent complexities (e.g. timescale of operation at the different layers, amount of information exchange). A more practical approach has been studied in the literature where a layer interacts with other layers to set up its own strategy, but considering the interactions among layers, as shown in Figure 12.1(c). Different types of cross-layer interaction include the following.

- *Application layer and physical layer interaction*: The application layer adapts the QoS and source coding with the channel state information (CSI) from the physical layer.
- *Transport layer and physical layer interaction*: Unlike wired medium, packets may not only be lost due to congestion, but also because of channel fading. Adaptive congestion control in the transport layer adapts its strategy with the channel state information from the physical layer.
- *Network layer and physical layer interaction*: The network layer takes routing decisions for data traffic based on the information from the PHY channel condition.
- *MAC layer and physical layer interaction*: The MAC layer selects the user and adapts multiple access techniques based on the channel conditions of all the users in order to maximize system throughput in a fair manner.
- *Application layer, link layer, and physical layer interaction*: The PHY-layer rate and power are adapted using information from the link and application layers in addition to the channel state information.

12.4 Why cross-layer adaptation is important for green radio networks

Although separate adaptation schemes are deployed in different OSI layers, the lack of coordination among them makes the overall performance of the system non-optimal. Only proper coordination across layers can benefit the system to achieve QoS with optimized goals across layers. For some applications, the packet arrival rate at the transmission buffer is continuous, while for some other applications, the packet arrival is quite bursty in nature. Therefore, if the packet scheduler at the lower layer does not utilize the traffic information of the application it is dealing with, it may cause excessive delay (and buffer overflow when the buffering capacity is limited) and/or excessive power consumption.

An intelligent packet scheduler should be able to adjust the transmission rate at the physical layer depending not only on the channel gain, but also on the buffer while keeping the QoS requirement (such as, delay, overflow, and packet error rate). For example, when packet delay is relatively less important than transmission power, the scheduler should not hurry up transmission by using a higher power level in bad channel conditions when the buffer has relatively fewer packets. It can wait for a better channel to come. This technique achieves two goals: it satisfies the packet error rate, delay, and overflow requirement, and it does so with the lowest possible transmission power. In future green radio networks, the scheduler will have to apply similar techniques to save energy. In this article, we show how joint optimization can be used in an intelligent scheduler to reduce energy consumption.

Among all the cross-layer adaptation techniques, the rate and power adaptation techniques at the physical layer are the most important ones for green radio network design, since they minimize transmission power based on upper layer information. Therefore, without loss of generality, in this article, we concentrate on the power minimization issue that is of particular importance for green radio networks. We show how the transmitter power can be saved using cross-layer optimal policies, where the rate and power at the physical layer are adjusted to minimize power with particular delay, packet error rate, and overflow requirements striking a balance between "green need" and service requirements. We discuss dynamic programming-based decision-making techniques for packet transmission over fading channels considering both the physical layer and the data-link layer optimization goals. At the physical layer, our goal is to optimize the transmission power while satisfying a particular bit error rate (BER) requirement. On the other hand, at the data-link layer, our goal is to optimize the delay and packet loss due to overflow. Overall, the cross-layer approach is shown to be effective in conserving the energy of the system while satisfying the QoS requirements.

12.5 Cross-layer interactions, models, and actions

Let us consider a downlink cellular system as shown in Figure 12.2(a), where a BS is communicating with a user over a fading channel. Let us assume that a packet coming to the transmitter buffer from the upper layer application consists of N_p bits. At the physical layer, a block (or radio frame) consists of N_f symbols. Assume that the duration of a block is equal to N_f number of discrete time-slots, where each time-slot is T_B sec. Unless specified otherwise, we use superscript n to denote a particular variable at time-slot n. For example, s^n denotes the state of the system at time-slot n. A numbered subscript on the other hand denotes a particular value of a variable, for example, s_1 is the first system state in the set of states $S = \{s_1, s_2, \cdots, s_S\}$. When the channel is slowly varying, it can be assumed that the channel gain does not change over a time-slot.

The status of the system is described by one of the S possible system states. The job of a PHY-layer scheduler is to find the control action $u^n \in U$ for all the time-slots $n = 1, 2, \cdots, H$, where U is the set of available actions and H (in number of slots) is the duration of communications. The control actions for all the time-slots constitute the policy or contingency plan, $\pi = \{\mu^1, \mu^2, \cdots, \mu^H\}$ of the system, where μ^n is called a decision rule that maps the system state $s^n \in S$ to the action u^n. For a stationary system, the decision rule does not change over time and therefore $\mu^n = \mu$, $\forall n$. For brevity, we denote the policy by μ since the decision rules are the same for all time-slots. Our goal is to find an optimal stationary policy μ^* so that it translates the state into a corresponding

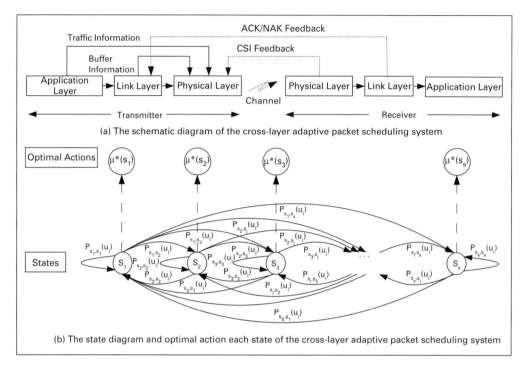

Figure 12.2 An illustration of the cross-layer adaptation system, its state diagram, and optimal policy mapping.

optimal action, $u^* = \mu^*(s)$, as shown in Figure 12.2(b). We will discuss later how the action, which corresponds to transmission rate and/or power of the problem, can be selected when adaptation is made with only the physical layer, and also with cross-layer variables. Note that P_{s_i,s_j}, $s_i, s_j \in S$ denotes the transition probabilities among the states of the system and it can be determined from the transition probabilities among traffic, channel, and buffer states.

The wireless channel gains are generally correlated, meaning that the gain in the present time-slot depends on the past time-slots. Usually a first-order Markov chain model is used to capture the memory of the slowly varying fading channel (as shown in Figure 12.3). This block fading finite-state Markov channel (FSMC), where the transitions are confined to an adjacent channel only, has a very good balance between complexity and accuracy. In this channel model, the gain of the wireless channel is divided into a finite number of states, C. The partitioning can be done in different ways, but the equal probability method, where all the channel states have the same stationary probability, is the most popular in literature, because it offers a good trade-off between the simplicity and the accuracy for modeling a wireless fading channel. We denote the channel states by $C = \{c_1, c_2, \cdots, c_C\}$, where the state is said to be in c_k when the gain lies between γ_{k-1} and γ_k.

The transition probabilities P_{c_i,c_j}, $\forall c_i, c_j \in C$ as shown in Figure 12.3 can be determined based on the number of transmitter/receiver antennas, fading severity parameters

12.5 Cross-layer interactions, models, and actions

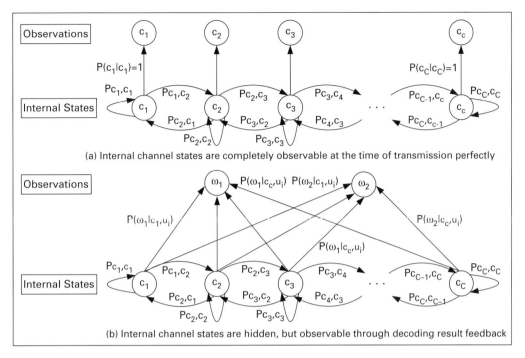

Figure 12.3 An illustration of the channel state evolution and observation for a particular action when the states are observable and hidden.

m, number of blocks (or time-slots) per second, $R_B = 1/T_B$ and the speed of the wireless user terminals relative to the BS (usually expressed in terms of Doppler frequency, f_m). The state of the channel can be estimated at the receiver and the information can be fed back to the transmitter. When the perfect channel state information is available at the transmitter before the transmission decision is taken, we usually refer to the channel as fully observable. This is shown in Figure 12.3(a), where the probability that the current estimated channel is c_k, given the current channel state is in c_k, equals 1. Sometimes, the perfect state of the channel may not be available at the time of transmission. The channel state is said to be hidden in such a scenario and it can be partially observed through positive acknowledgement (ACK) or negative acknowledgement (NAK) feedback, as shown in Figure 12.3(b). Note that the observation ω_i, $i = 1, 2$ depends on the internal channel state and the action taken. Therefore, for a particular action $u_i \in U$, the probability of ACK, $P(\omega_1|c_k, u_i)$, for example, is different for different channel states.

Usually, wireless network traffic is bursty, correlated, and randomly varying. The Markov modulated Poisson process (MMPP) model, where in any state the incoming traffic is Poisson distributed and the transitions between the states are governed by an underlying Markov chain, can be used to model this traffic. Let $F = \{f_1, f_2, \cdots, f_F\}$ denote the set of the states of the incoming traffic and P_{f_i, f_j} denote the transition probability between states. In a particular state $f_i \in F$, $0, 1, \cdots, A$ packets may arrive according to the Poisson distribution with average arrival rate λ_i packets per time-slot. In each time-slot, the transmitter selects w^n packets for transmission over the wireless channel. The number of packets selected in a block depends on the modulation scheme selected for that block and can be expressed by the following relations:

$$w^n = \left(\frac{N_f}{N_p}\right) R^n, \qquad (12.1)$$

where R^n is the number of bits in a symbol (e.g. for QPSK, 16-QAM, and 64-QAM, $R_i = 2$, 4, and 6, respectively). Since the number of arrived packets a^n and the number of packets chosen for transmission w^n are randomly varying, the buffer occupancy fluctuates between 0 to B, where B is the storage capacity of the buffer. The buffer state at time-slot n can be written as

$$b^n = \min\left\{\max\left(b_0, (b^{n-1} - \omega^{n-1} w^{n-1} + a^{n-1})\right), b_B\right\}. \qquad (12.2)$$

That is, the current buffer state depends on the previous buffer state as well as the number of packets taken, w^{n-1}, number of packet arrivals, a^{n-1}, and the acknowledgement from the receiver, ω^{n-1} for the previous time-slot $n-1$. The buffer state transition probabilities P_{b_i,b_j} are shown in Figure 12.4(a). For example, when the buffer occupancy is 5 packets (corresponding state is b_5) and 4 packets are chosen for transmission in a particular time-slot n, the new buffer state is $b^{n+1} = b_4$ with probability $P(a^n = 3 | f^n = f_i)$, $\forall f_i \in F$. A typical variation of buffer occupancy with time is also shown in Figure 12.4(b).

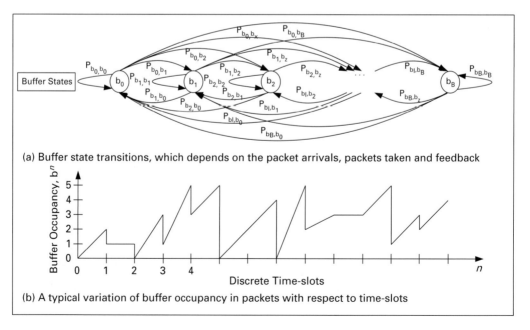

(a) Buffer state transitions, which depends on the packet arrivals, packets taken and feedback

(b) A typical variation of buffer occupancy in packets with respect to time-slots

Figure 12.4 An illustration of the buffer state evolution and variation of buffer occupancy with time in a typical wireless system.

12.6 Cross-layer vs. single-layer adaptation techniques

Let us first consider the case when all the channel states, the traffic states, and the buffer states are fully observable at the scheduler. In cross-layer packet scheduling systems, the system can be in one of $S = F \times C \times (B+1)$ states in set $S = \{s_1, s_2, \cdots, s_S\}$, where the states consist of traffic, channel, and buffer states. In the cross-layer packet scheduling technique, the goal is to find a stationary optimal transmission strategy μ^* so that the cross-layer utilities are optimized jointly. In this article, we express the utility in terms of total long-term average power, average delay, average overflow, and average BER. Without loss of generality, suppose that we are provided with a set of actions $U = \{u_1, u_2, \cdots, u_U\}$, where each action corresponds to a unique modulation rate. Note that for a particular channel state, a higher order modulation requires higher transmission power. We consider following two transmission decision scenarios, where the chosen actions for different cases are plotted as a function of buffer states and channel states assuming single traffic state for simplicity. The policy for the corresponding scenario is found by optimizing utilities, and it is shown in Figure 12.5.

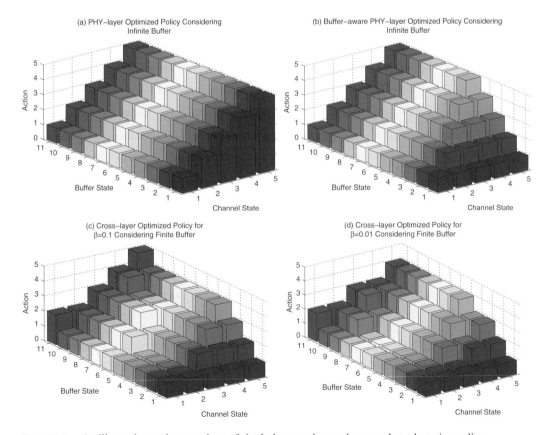

Figure 12.5 An illustration and comparison of single-layer and cross-layer packet adaptation policy structures.

- *PHY-layer optimized transmission policy*: In traditional wireless networks, each layer has its own strategy to cope with the unsuccessful transmission to maximize throughput and reliability. Generally, the transmission rate and/or power are adapted with the channel condition assuming that the buffer is always backlogged and can hold any number of incoming packets. In practice, these assumptions are not true, and consequently, PHY-layer optimized adaptive modulation and coding techniques cannot guarantee the delay or packet loss requirement. Also, it uses a large amount of energy to send a given amount of data over the horizon H, because it does not consider traffic information and buffer dynamics to schedule packet transmission over a period of time.

    For example, suppose we have five transmission rates $U = \{u_1, u_2, u_3, u_4, u_5\}$, where u_i corresponds to $i - 1$ packet transmission. Now, the channel is divided into five states and action u_i is chosen in state c_i as shown in Figure 12.5(a) [11]. That is, irrespective of the channel state, the scheduler chooses the maximum permitted rate for a particular channel state. Note that in this technique, the scheduler chooses action u_5 when the buffer does not even have four packets (the corresponding buffer state is b_5). To mitigate this problem, some works in the literature propose cross-layer design by taking into account the number of packets in the buffer. The structure of this buffer-aware PHY-layer optimized policy is shown in Figure 12.5(b), where in channel state c_i the chosen action is $u_i = \min(b_i, c_i)$, $\forall b_i < c_i$ and $u_i = c_i$, $\forall b_i \geq c_i$. However, this modified policy also cannot guarantee delay and packet overflow since it does not make the transmission decision based on buffer and traffic states in addition to the channel state; rather, it just modifies the policy when enough packets are not available for transmission.

- *Cross-layer optimized transmission policy*: In the cross-layer packet scheduling technique, the optimal transmission rate and/or power actions are determined based on both the channel gain information from the PHY-layer, and buffer occupancy information from the link-layer as well as traffic information from the application layer [12]. This policy satisfies the delay and packet overflow requirements. We will see that this cross-layer policy needs much less average transmission power over a period of time. We show two policy structures in Figure 12.5(c) and Figure 12.5(d) for two different power vs. delay trade-off settings, where the weighting parameters β_1 and β_2 determine the trade-off among utilities to be optimized jointly. More discussion on β_1 and β_2 is given in (12.4) in the following Section 12.7).

    When the value of β_1 is less than 1, it gives more importance to saving power than delay. Note that the cross-layer optimized policies are not static like the PHY-layer optimized policy shown in Figures 12.5(a) and 12.5(b). Depending on the delay and packet-loss requirements, the scheduler is able to adjust the policies dynamically. For example, in channel state c_2, the PHY-layer optimized policy never uses actions other than u_2, but for the cross-layer optimized policy, action u_1 is used when the buffer state is between b_0 and b_4 and b_0 and b_7 for β_1 equal to 0.1 and 0.01, respectively. To satisfy the delay requirement, action u_2 that needs higher power is used afterwards. The scheduler thus judiciously avoids high transmission power when possible. A similar observation can be made for channel state c_5, where the scheduler only uses action u_5

at buffer state b_{11} when $\beta_1 = 0.1$. For $\beta_1 = 0.01$, action u_5 is never used because this action is not necessary for joint optimization of the utilities.

The policy structures shown above are obtained assuming Rayleigh fading channel with unit average gain, $N_p/N_f = 1$, $\beta_2 = 0$, and $\lambda_1 = 1$. In the following, we show how to compute the optimal policy for the cross-layer scheduling problem and how much energy can be saved using cross-layer scheduling policy as compared to single-layer scheduling policy.

12.7 How to solve the cross-layer design problem

For the cross-layer scheduling problem, suppose that at the beginning of a time-slot n, the PHY-layer packet scheduler knows the state of the system perfectly with the help of the CSI feedback from the receiver, buffer occupancy from the link layer, and traffic state from the application layer. Now, it is the job of the scheduler to find an optimal action for that state. The optimal policy for the cross-layer packet scheduling problem can be computed by formulating the problem as a Markov decision process (MDP) problem. For the MDP problem, when the scheduler takes a particular action in a particular system state it receives a reward or incurs a cost. Note that cost is the negative of reward, so we use them interchangeably in this article.

In this chapter, we consider three costs, namely, transmission power, buffering delay, and packet overflow, which correspond to our objectives to be optimized. The power cost needs to be minimized in green radio networks. For a time-slot n, the power cost $G_P(s^n, u^n)$ is proportional to the actual transmitted power, P_t. Practical applications have varied delay sensitivity and packet overflow tolerance. Therefore, for wireless applications, it is very important to maintain delay and overflow requirements. The delay cost is proportional to the buffer state, and can be computed using Little's law as, $G_D(s^n, u^n) = b^n/a^n$ for time-slot n. The overflow cost, $G_O(s^n, u^n)$ is just the number of packets dropped in time-slot n.

Dynamic programming algorithms are used to find the optimal policy, where the policy that incurs the lowest expected cost is called the optimal policy. Note that our problem is a multi-objective optimization problem, i.e. we want to minimize the transmission power, buffer delay, and packet loss due to buffer overflow with a specified minimum BER. The problem can be handled in two ways: by forming the cost function as a weighted sum of the expected costs for all the objectives and minimizing it, or by minimizing the expected cost of one objective (e.g. power) with bounds on other expected costs for remaining objectives (e.g. delay and overflow loss).

The first formulation is referred to as an unconstrained MDP (UMDP) formulation, where our objective is to find the optimal stationary policy μ^* over all Markov deterministic stationary policies, Π that minimizes the average cost per stage, given by

$$G_T^\mu = \limsup_{H \to \infty} \frac{1}{H} \sum_{n=1}^{H} \mathbb{E}[G_T(s^n, \mu(s^n))]. \quad (12.3)$$

At each decision epoch n, the controller obtains the state s^n of the system. Based on this state, the controller chooses a control action u^n and corresponding transmission rate $\Psi(u^n)$, and incurs a per-stage cost of $G_T(s^n, u^n)$. The weighted sum of all the cost, $G_T(s_i, u)$ can be written as,

$$G_T(s_i, u) = G_P(s_i, u) + \beta_1 G_D(s_i, u) + \beta_2 G_O(s_i, u). \tag{12.4}$$

The weighting factors β_1 and β_2 determine the relative importance of one objective with respect to others. The UMDP formulation can be solved iteratively via the value iteration or policy iteration algorithm using the Bellman equation [13]:

$$\Lambda + h(s_i) = \min_{u \in U} \left[G_T(s_i, u) + \sum_{j=1}^{S} P_{s_i, s_j}(u) h(s_j) \right], \tag{12.5}$$

where Λ is the optimal average cost, and $h(s_i)$ is the differential cost for each state $s_i \in S$. In the relative value iteration algorithm, the relative values of the expected cost for all the states with respect to a reference state, $h(s_i)$, $i = 1, 2, \cdots, S$, are updated iteratively until an equilibrium is reached. That is, until the difference between the maximum and minimum change in updated value is below a given tolerance, ϵ [13]. When equilibrium is achieved the value of the reference state is the optimal average cost Λ and the corresponding policy is the optimal policy. In the policy iteration algorithm, the expected relative cost of each state is calculated for a given policy, and then based on these calculated expected costs a new improved policy is determined that yields the maximum expected relative cost using one-step look-ahead. The algorithm terminates when the policy improvement step yields no change in the expected costs [14]. The policy iteration algorithm for a unichain discrete-time Markov decision process can be described by as follows:

- *Initialization*: Set $k = 0$ and select an arbitrary initial policy $\mu^{(0)}$
- *Policy evaluation*: Select a state s_r as a reference state and put $h^{(k)}(s_r) = 0$. Find $\lambda^{(k)}$, $h^{(k)}(s_i)$, $i = 1, 2, \cdots, S$ by solving the following set of equations:

$$\lambda^{(k)} + h^{(k)}(s_i) = G(s_i, \mu^{(k)}(s_i)) + \sum_{j=1}^{S} P_{s_i, s_j}(\mu^{(k)}(s_i)) h^{(k)}(s_j), \forall s_i \in S. \tag{12.6}$$

- *Policy improvement*: Find new improved policy using

$$G(s_i, \mu^{(k+1)}(s_i)) + \sum_{j=1}^{S} P_{s_i, s_j}(\mu^{(k+1)}(s_i)) h^{(k)}(s_j)$$

$$= \min_{u \in U_{s_i}} \left[G(s_i, u) + \sum_{j=1}^{S} P_{s_i, s_j}(u) h^{(k)}(s_j) \right]. \tag{12.7}$$

- *Termination*: If $\mu^{(k+1)} = \mu^{(k)}$, the algorithm terminates; otherwise, the process is repeated with $\mu^{(k+1)}$ replacing $\mu^{(k)}$.

The second formulation is referred to as a constrained MDP (CMDP) formulation, where the objective of our cross-layer adaptation problem is to find the optimal stationary policy μ^* so that

$$\min_{\mu} G_P^{\mu}, \text{ s.t.: } G_D^{\mu} \leq B_d \text{ and } G_O^{\mu} \leq P_d, \quad (12.8)$$

where B_d and P_d are the maximum allowable average delay and average packet overflow rate. The CMDP problem can be solved using linear programming techniques with a variable $v(s, u)$, which represents the steady-state probability that the system is in state s and action u is applied, and the appropriate constraints as follows [13]:

$$\min_{v} \sum_{s \in S, u \in U_s} G_P(s, u) v(s, u)$$

$$\text{subject to: } \sum_{s \in S, u \in U_s} G_D(s, u) v(s, u) \leq B_d$$

$$\sum_{s \in S, u \in U_s} G_O(s, u) v(s, u) \leq P_d$$

$$\sum_{u \in U_s} v(s', u) = \sum_{s \in S, u \in U_s} v(s, u) p_{s,s'}(u), \forall s' \in S$$

$$\sum_{s \in S, u \in U_s} v(s, u) = 1$$

$$v(s, u) \geq 0, \quad \forall s \in S, \forall u \in U_s, \quad (12.9)$$

whereas the UMDP formulation gives us a deterministic policy, the CMDP formulation provides a simple tool for the determination of optimal randomized scheduling policies. When there exists an optimal solution v^* to the LP problem, then there exists an optimal randomized stationary policy μ^* for the CMDP problem, where μ^* satisfies [13]

$$\theta_{\mu^*(s)}(u) = \frac{v^*(s, u)}{\sum_{u' \in U} v^*(s, u')}, \quad \text{if } \sum_{u' \in U} v^*(s, u') > 0. \quad (12.10)$$

If $\sum_{u' \in U} v^*(s, u') = 0$ for some $s \in S$, an action that drives the system to the recurrent class of states $S_R = \{s \in S : \sum_{u' \in U} v^*(s, u') > 0\}$ is chosen in each state [13]. LP can generally handle a problem with a large number of variables and can give an optimal solution easily and quickly. The equivalent linear program for the CMDP problem above can be solved using modern linear programming techniques (such as interior point methods) and software tools (such as MATLAB optimization toolbox, ILOG CPLEX) very easily and efficiently [15]. There is a one-to-one correspondence between the feasible (and hence optimal) solution of the LP and the feasible (and hence optimal) solution of CMDP. LP is feasible if and only if CMDP is feasible [16].

So far we have concentrated on how to find the optimal policy when the perfect CSI is available at the transmitter. However, there are situations when the perfect CSI is not available before taking the transmission decision. In this case, the problem can be formulated as a partially observable Markov decision process (POMDP). Since the states are hidden (due to hidden channel states), the optimal action cannot be found in a way similar to that for the MDP problem. However, since the observations (in the form of ACK/NAK) for a particular action are available at the end of transmission, a belief state z, which is the probability distribution on the states given the history of actions and observations, can be formed and it can be updated when the prior belief and new observations for the taken action are known. The optimal solution for the POMDP using belief is computationally infeasible. However, it was shown in [17] that there exist some heuristic-based policies that work closer to the fully observable optimal policy. One such heuristic policy is maximum likelihood policy heuristic, where the scheduler finds the most probable state from the current updated belief and the optimal action for the state is applied. In the maximum-likelihood policy heuristic (MLPH) at time-slot n for a particular belief $z^n(s)$, the policy can be represented as

$$\mu_{\text{ML}}(z^n) = \mu^*_{\text{MDP}}(\arg\max_{s} (z^n(s))). \tag{12.11}$$

In this expression, $\mu^*_{\text{MDP}}(s_i)$ is the optimal policy for state s_i of the system. The optimal action of the underlying MDP problem is solved in a similar way as described above assuming that the transition probabilities of the underlying system are known, but the instantaneous state is unknown at the time of transmission.

12.8 Power savings in the cross-layer optimized system

In order to illustrate how much power can be saved in a cross-layer optimized system compared to a single-layer optimized system, we present some numerical results. We also show how the system parameters affect the amount of power saving. The following values for the parameters are assumed: $F = 2$, $C = 4$, $B = 50$, $\lambda_1 = 1$, $A = 15$, time-correlated Rayleigh fading channel, average BER $= 10^{-4}$, $U = 4$ (corresponding transmission rates are 0, 2, 4, and 6 packets per time-slot), and both the channel and traffic states are equiprobable. The transmission power for a particular rate is computed for a given average BER using M-QAM. We use the CMDP formulation to compute the randomized optimal policy. The bound on the average delay is varied to obtain different optimal policies and average transmission powers for the given maximum allowable average packet loss rate of 10^{-4} packets per time-slot.

Figure 12.6 shows that there is a trade-off between the average power and the average delay; the average power decreases when the bounds on the average delay increase. Therefore, for delay-tolerant data traffic, a significant amount of energy can be saved by delaying packet transmissions during bad channel states. As the channel condition improves, it sends those packets using a higher transmission rate (and lower relative average power, because to transmit the same amount of data in a bad channel requires

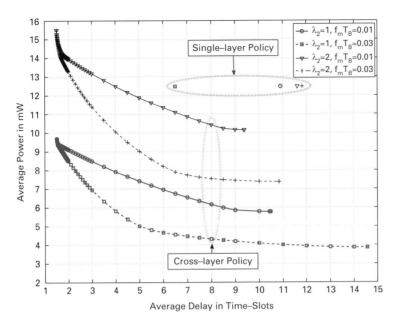

Figure 12.6 Comparison between transmitter power for single-layer and cross-layer packet adaptation policies as a function of packet transmission delay. Whereas the single-layer policy gives a static policy that corresponds to a single point, the cross-layer policies are dynamic since every single point on the cross-layer policy curves corresponds to a unique policy. The effect of fading rates and packet arrivals is shown over Rayleigh fading channels (i.e. for $m = 1$).

more power than in a good channel). For a fast fading channel, the transmission buffer during the bad channel states does not grow as fast as the slow fading channel. As a result, for the same delay, the average transmitter power is less for a fast fading channel. The average power is also higher for a higher packet arrival rate since the transmitter has to send more packets with the same imposed delay bound. The average power for the single-layer policy does not change since it is not buffer dependent. However, the delay and the packet overflow for the single-layer policy increase as the packet arrival rate increases. This single-layer policy is unable to guarantee the delay and loss requirement when the buffer has limited capacity. It is observed that the power requirement for the single-layer policy is significantly higher than that of the cross-layer policy. The power saving is more in a fast fading channel.

We show the effect of a larger buffer on the power vs. delay curves in Figure 12.7. When the size of the buffer is increased from 50 to 100, the scheduler obtains additional space to store packets when the channel condition is poor, therefore even more power can be saved with the larger buffer as can be seen from the plots. As the value of the Nakagami-m severity parameters increases from 1, it gradually approaches an additive white Gaussian noise (AWGN) channel. Therefore, the fluctuations in the channel decrease as $m \to \infty$. When $m = 2, 3, \cdots$, the channel can be approximated as a Rician channel with a line-of-sight component. It can be seen from the figure that by delaying the packets in the buffer, the valuable energy of the device's battery can be saved for a Rician

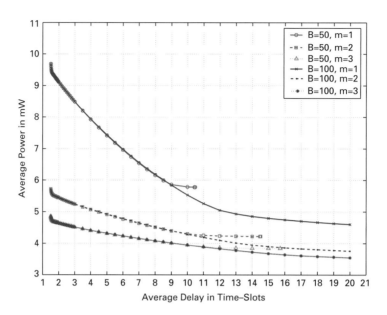

Figure 12.7 The effect of buffer sizes and Nakagami-m severity parameters are shown in terms of Power vs. Delay trade-off curves. For a larger buffer, the scheduler has additional flexibility to store packets in worse channel conditions and send them in better channel conditions. When $m = 2, 3$, the Nakagami-m channel can be considered as a Rician channel with line of sight component. For the same delay, the transmission power is lower in a Rician channel.

channel as well. However, the saving for Rayleigh fading is more than the Rician fading channel.

12.9 Other literature on energy-efficient cross-layer techniques

Energy-efficient cross-layer optimized techniques and designs were a major research attention in the last decade among wireless researchers working in different networks and protocol stacks. In [18], the authors presented an excellent comprehensive overview of cross-layer design for an energy-efficient wireless communications, particularly focusing on a system-based approach towards energy-optimal transmission, resource management across time, frequency and spatial domains, and energy-efficient hardware implementations. For future radio access networks beyond LTE-Advanced, the authors in [19] surveyed various energy-efficient advanced technologies, such as interference mitigation techniques, MIMO, and cooperative communications as well as cross-layer self-organizing networks and the technical challenges. We provide a brief summary on such researches below.

Various energy-efficient techniques have been proposed in physical layers [20],[21]-[23]. A cross-layer energy-efficient cooperative beamforming framework by considering optimal weight design for single-beam beamforming and suboptimal weight designs for multibeam beamforming is discussed in [20]. In [21], the authors propose a cross-layer

12.9 Other literature on energy-efficient cross-layer techniques

approach to switch between multiple-input multiple-output (MIMO) with two transmit antennas and single-input multiple-output (SIMO) to conserve mobile terminals' energy in adaptive MIMO systems. Energy-efficient transmission techniques for Rayleigh fading networks are studied in [22], where the authors show how to map the wireless fading channel to the upper layers parameters for cross-layer design. An energy-efficient cross-layer design for a MIMO downlink SVD channel is given in [23].

The energy efficiency of physical layer transmission using rate and power adaptation is improved through information from the datalink layer in the following works. In [24], energy-efficient operation modes in wireless sensor networks are studied based on cross-layer design techniques over Rayleigh fading channels using a discrete-time queuing model. Using a three-dimensional nonlinear integer programming technique, the authors in [25] have shown that the joint optimization of the physical layer and data link layer parameters (e.g. modulation order, packet size, and retransmission limit) contributes noticeably to the energy saving in energy-constrained wireless networks. In [26], the authors studied the cross-layer trade-off between energy consumption and transmit power within a physical and link layer system model, which jointly considers power allocation, adaptive modulation and coding and ARQ/HARQ retransmission protocols in a downlink multiuser-multicarrier system. The authors in [27] on the other hand studied the energy-efficient cross-layer design of adaptive modulation and coding using a sleep mode to the system.

Game-theoretic approaches are taken in the following works for energy-efficient cross-layer design. The authors in [28], propose an energy-efficient MAC algorithm where each node sets the contention window size with respect to the residual energy, the harvesting energy, and the transmit power using a game-theoretic and cross-layer optimization approach. For CDMA networks, the authors focus on the energy-efficient optimization and cross-layer design problem involving power control and multiuser detection in wireless CDMA data networks via a non-cooperative game-theoretic approach in [29]. In [30], a non-cooperative power control game for maximum energy-efficiency with a fairness constraint on the maximum received powers in cognitive CDMA wireless networks is studied. It is assumed that both the primary and secondary users coexist in the same frequency band. The authors in [31] propose a game-theoretic energy-efficient model in order to study the cross-layer power and rate control problem with quality of service constraints in multiple-access wireless networks.

Cross-layer studies involving multi-hop networks to improve both the energy efficiency and upper layer parameters are considered in the following two works. In [32], the authors propose a cross-layer window control algorithm in the transport layer to improve the throughput and a power control algorithm in the physical layer to reduce power consumption for a wireless multi-hop network, whereas the problem of minimizing the total power consumption in a multi-hop wireless network is studied in [33], and a low-complexity, power-efficient, and distributed algorithm is given.

Energy-efficient cross-layer design is also very important for multimedia networks. In [34], the authors propose and evaluate cross-layer strategies that combine: fountain coding (e.g. Raptor codes at the application layer) and network coding to reduce energy consumption by opportunistically recombining and rebroadcasting the required combinations of packets in a cooperative DVB-H (Digital Video Broadcasting - Handheld) multimedia system for cellular networks. A significance-aware power allocation strategy for wireless multimedia communications is developed in [35], using cross-layer optimization techniques and is shown to reduce power consumption without loss of user-perceivable QoS for MPEG-4 (Moving Picture Expert Group-4) video. For underlay cognitive radio relay networks, a cross-layer design approach to jointly consider optimal relay selection, power allocation, adaptive modulation and coding, and intra-refreshing rate to minimize energy consumption and to improve multimedia transmissions is studied in [36].

A huge volume of literature exists for wireless sensor networks that consider the energy minimization issue using cross-layer approaches. In [37], two heuristics are proposed to compute the routing topology and link data rate in wireless sensor networks that optimize sensor lifetime under both the energy and the bandwidth constraints. A solution to the cross-layer scheduling problem in clustered wireless sensor networks to provide network-wide optimized time division multiple access (TDMA) schedules that can achieve energy efficiency while satisfying a specified reliability objective is proposed in [38]. A cross-layer scheme to optimize the energy along with distortion and encryption performance of video streaming in wireless sensor networks is addressed in [39]. The authors in [40] introduce a cross-layer protocol (XLP) for energy-efficient communication that achieves congestion control, routing, and medium access control in a cross-layer fashion based on the concept of initiative determination. In [41], an energy-constrained system comprising two sensor nodes that avoid interference by exploiting spectrum holes in the time domain is studied. The cross-layer algorithm used for spectrum sensing is designed so as to minimize the average energy required for the successful delivery of a packet accounting for the physical and transport layers parameters. A cross-layer approach in minimizing energy consumption by the receiving and processing circuitry for multi-hop cooperative sensor networks under the end-to-end reliability QoS requirement is investigated in [42] in order to jointly determine the optimized routing, relay selection, and power allocation strategies. An energy-efficient protocol for a wireless sensor network called Breath is implemented and experimentally evaluated on a testbed in [43], which uses a cross-layer interaction to reach maximum efficiency.

In broadcasting and multicasting networks, cross-layer design is found to be energy efficient as well. An energy-efficient real-time data multicasting architecture for ad hoc networks is studied in [44] using a cross-layer design via multicasting through time reservation using adaptive control (MC-TRACE), where the medium access control layer functionality and the network layer functionality are performed by a single integrated layer. In [45], cross-layer jamming detection and mitigation in wireless broadcast networks are studied to optimize power efficiency.

Cross-layer techniques are also studied for different wireless standards, e.g. WiMAX and WLAN. In [46], a cross-layer design involving a physical layer and an application layer for mobile WiMAX is given that achieves reduced power consumption and packet loss along with increased throughput. The authors in [47] present a cross-layer and energy-efficient mechanism for transmitting voice-over-Internet (VoIP) protocol packets over IEEE802.11 WLAN and have shown that it improves the energy-efficiency of a station and WLAN utilization without sacrificing voice quality.

Along with theoretical studies and standards, cross-layer design in electronic devices is shown to be energy optimal. The propagation channel between two half-wavelength dipoles at 2.45 GHz for wireless body area networks is discussed in [48]. The authors propose a semi-empirical path-loss model and then present an application for cross-layer design in order to optimize the energy consumption of different topologies. In [49], the authors design an energy autonomous e-reader device using cross-layer approaches to energy-efficient data transmission based on quality of service requirements and link adaptation. An energy-efficient cross-layer design framework that allows systematic exploration of trade-offs among energy, delay, and quality at the algorithm, architecture, and circuit level of design abstraction for each block of a system considering the interactions between different subblocks of the system is studied in [50]. Experimental results on the power efficiency of the wireless platform on the thin client over an IEEE802.11 link is presented in [51] and it is shown that the cross-layer approach between the application and link layer can save energy in the devices.

12.10 Challenges and future directions

Several challenges exist for designing cross-layer optimized adaptive packet scheduling policies for green radio networks.

- *Metric for energy efficiency*: For green radio systems, appropriate energy-efficiency metrics need to be defined that capture the cost for both communications and computations. The cost metric should be able to realistically abstract the energy-consumption profile of the RF circuitry in the physical layer. The computation cost for the packet scheduler should be able to abstract the power consumption in the processor.
- *Computational complexity and communication overhead*: With the advancement of low-power CMOS VLSI circuits, the speed of computation and the memory are not much of a problem for modern electronic processors. Although the computational complexity increases with the increase in the number of states, however for the problem at hand, the dynamic programming algorithm can find the optimal policy fairly quickly for several hundreds of states with a modern processor at the heart of a device. The linear programming techniques can solve the problem with several thousands of variables very quickly. However, the computation time should be small enough so that optimal cross-layer scheduling can be performed in an online fashion.

  Again, the information required for cross-layer optimized scheduling, such as the channel state and buffer information, should be available at low cost (both computation and communication cost). Designing light weight protocols for information gathering for cross-layer optimization is a big challenge.
- *Cross-layer design for energy-harvesting green networks*: In energy-constrained networks that depend on energy harvested from the environment (e.g. solar-powered networks), cross-layer design of packet transmission and scheduling policies is even

more challenging. This is due to the fact that the energy-generation, storage, and dissipation processes need to be considered in the system model.

- *Application-aware cross-layer design*: Since energy efficiency in green radio networks imposes trade-offs among energy consumption, delay, and throughput performance, the optimal trade-off among these parameters needs to be obtained based on the application requirements. Therefore, application requirements need to be incorporated in the cross-layer optimization framework. Analytical and simulation models will need to be developed to evaluate the trade-off between energy efficiency and application-level QoS performance.
- *Simulation, modeling, and validation tools for cross-layer protocols for green networks*: For the evaluation of cross-layer protocols for green networks, efficient simulation and modeling tools need to be developed. Such a model is expected to be capable of performing a sensitivity analysis of how changes in different system parameters affect the overall system performance in terms of capacity, QoS, and energy efficiency. Again, real-world results from test-beds and field trials will be required to validate the results from simulation and modeling.

12.11 Conclusion

We have shown how to design an intelligent packet transmission scheduler, which takes optimal transmission decisions using cross-layer information, based on Markov decision process formulations. We have discussed how to compute the optimal policies when the channel states are perfectly observable and when they are partially observable. With the help of a simple example, we have shown the benefits of a cross-layer policy over a single-layer policy in terms of energy efficiency. By delaying packet transmissions in an optimal way, a huge amount of power can be saved for delay-tolerant data traffic applications. The amount of saving depends on factors such as the memory of a fading channel and the packet arrival rate. Such a cross-layer optimized packet transmission scheduling method will be a key component in future-generation green wireless networks.

References

[1] J. He et al., "Energy efficient architectures and techniques for green radio access networks," in *Proc. of 5th International ICST Conference on Communications and Networking*, Beijing, China, Aug. 25–27, 2010, pp. 1–6.

[2] A. Dejonghe et al., "Green reconfigurable radio systems," *IEEE Signal Process. Mag.*, vol. 24, no. 3, pp. 91–101, May 2007.

[3] T. Chen, H. Kin, and Y. Yang, "Energy efficiency metrics for green wireless communications," in *Proc. of Wireless Communications and Signal Processing*, Suzhou, Nov. 2010, pp. 1–6.

[4] A. Ephremides, "Energy concerns in wireless networks," *IEEE Wireless Commun. Mag.*, vol. 9, no. 4, pp. 48–59, Aug. 2002.

[5] R. A. Berry and R. G. Gallager, "Communication over fading channels with delay constraints," *IEEE Trans. Inf. Theory*, vol. 48, pp. 1135–1149, May 2002.

[6] D. Rajan, A. Sabharwal, and B. Aazhang, "Outage behavior with delay and CSIT," in *Proc. of IEEE ICC'04*, vol. 1, Paris, France, Jun. 20–24, 2004, pp. 578–582.

[7] Z. K. M. Ho, V. K. N. Lau, and R. S.-K. Cheng, "Cross-layer design of FDD-OFDM systems based on ACK/NAK feedbacks," *IEEE Trans. Inf. Theory*, vol. 55, no. 10, pp. 4568–4584, Oct. 2009.

[8] M. H. Ngo and V. Krishnamurthy, "Optimality of threshold policies for transmission scheduling in correlated fading channels," *IEEE Trans. Commun.*, vol. 57, pp. 2474–2483, Aug. 2009.

[9] E. Uysal-Biyikoglu, A. E. Gamal, and B. Prabhakar, "Energy-efficient packet transmission over a wireless link," *IEEE/ACM Trans. Netw.*, vol. 10, pp. 487–499, Aug. 2002.

[10] A. J. Goldsmith and S. B. Wicker, "Design challenges for energy-constrained ad hoc wireless networks," *IEEE Wireless Commun. Mag.*, vol. 9, no. 4, pp. 8–27, Aug. 2002.

[11] A. Goldsmith and S.-G. Chua, "Variable-rate variable-power MQAM for fading channels," *IEEE Trans. Commun.*, vol. 45, pp. 1218–1230, Oct. 1997.

[12] A. K. Karmokar and V. K. Bhargava, "Performance of cross-layer optimal adaptive transmission techniques over diversity Nakagami-m fading channels," *IEEE Trans. Commun.*, vol. 57, pp. 3640–3652, Dec. 2009.

[13] M. L. Puterman, *Markov Decision Processes: Discrete Stochastic Dynamic Programming*. New York, NY: John Wiley & Sons, 1994.

[14] D. P. Bertsekas, *Dynamic Programming and Optimal Control*. 2nd ed. Belmont, MA: Athena Scientific, 2001.

[15] S. Boyd and L. Vandenberghe, *Convex Optimization*. Cambridge, UK: Cambridge University Press, 2004.

[16] E. Altman, *Constrained Markov Decision Processes: Stochastic Modeling*. London, UK: Chapman and Hall/CRC, 1999.

[17] A. K. Karmokar, D. V. Djonin, and V. K. Bhargava, "POMDP-based coding rate adaptation for type-i hybrid ARQ systems over fading channels with memory," *IEEE Trans. Wireless Commun.*, vol. 5, pp. 3512–3523, Dec. 2006.

[18] G. Miao *et al.*, "Cross-layer optimization for energy-efficient wireless communications: a survey," *Wiley Wireless Communications and Mobile Computing*, vol. 9, no. 4, pp. 529–542, Apr. 2009.

[19] S. Liu *et al.*, "A 25 $Gb/s(/km^2)$ urban wireless network beyond IMT-advanced," *IEEE Commun. Mag.*, vol. 49, no. 2, pp. 122–129, Feb. 2011.

[20] L. Dong, A. P. Petropulu, and H. V. Poor, "Weighted cross-layer cooperative beamforming for wireless networks," *IEEE Trans. Signal Process.*, vol. 57, no. 8, pp. 3240–3252, Sep. 2010.

[21] H. Kim *et al.*, "A cross-layer approach to energy efficiency for adaptive MIMO systems exploiting spare capacity," *IEEE Trans. Wireless Commun.*, vol. 8, no. 8, pp. 4264–4275, Aug. 2009.

[22] G. Li, P. Fan, and K. B. Letaief, "Rayleigh fading networks: a cross-layer way," *IEEE Trans. Commun.*, vol. 57, no. 2, pp. 520–529, Feb. 2009.

[23] D. J. Dechene and A. Shami, "Energy efficient quality of service traffic scheduler for MIMO downlink SVD channels," *IEEE Trans. Wireless Commun.*, vol. 9, no. 12, pp. 3750–3761, Dec. 2010.

[24] X.-H. Lin, Y.-K. Kwok, and H. Wang, "Cross-layer design for energy efficient communication in wireless sensor networks," *Wiley Wireless Communications and Mobile Computing*, vol. 9, no. 2, pp. 251–268, Feb. 2009.

[25] H. Cheng and Y.-D. Yao, "Link optimization for energy-constrained wireless networks with packet retransmissions," *Wiley Wireless Communications and Mobile Computing*, vol. doi: 10.1002/wcm.996.

[26] Q. Bai and J. A. Nossek, "On energy efficient cross-layer assisted resource allocation in multiuser multicarrier systems," in *Proc. of IEEE PIMRC'09*, Sep. 13–16, 2009, pp. 2603–2607.

[27] J. Gong, S. Zhou, and Z. Niu, "Queuing on energy-efficient wireless transmissions with adaptive modulation and coding," in *Proc. of IEEE ICC'11*, Kyoto, Japan, Jun. 2011, pp. 1–5.

[28] H. Kim, H. Lee, and S. Lee, "A cross-layer optimization for energy-efficient MAC protocol with delay and rate constraints," in *Proc. of IEEE International Conference on Acoustics, Speech and Signal Processing (ICASSP)*, 2011, pp. 2336–2339.

[29] S. Buzzi, V. Massaro, and H. V. Poor, "Energy-efficient resource allocation in multipath cdma channels with band-limited waveforms," *IEEE Trans. Signal Process.*, vol. 57, no. 4, pp. 1494–1510, Apr. 2009.

[30] S. Buzzi and D. Saturnino, "A game-theoretic approach to energy-efficient power control and receiver design in cognitive CDMA wireless networks," *IEEE Journal of Selected Topics in Signal Processing*, vol. 5, no. 1, pp. 137–150, Jan. 2011.

[31] F. Meshkati *et al.*, "Energy-efficient resource allocation in wireless networks with quality-of-service constraints," *IEEE Trans. Commun.*, vol. 57, no. 11, pp. 3406–3414, Nov. 2009.

[32] H.-J. Lee and J.-T. Lim, "Cross-layer congestion control for power efficiency over wireless multihop networks," *IEEE Trans. Veh. Technol.*, vol. 58, no. 9, pp. 5274–5278, Sep. 2009.

[33] L. Lin, X. Lin, and N. B. Shroff, "Low-complexity and distributed energy minimization in multihop wireless networks," *IEEE/ACM Trans. Netw.*, vol. 18, no. 2, pp. 501–514, Feb. 2010.

[34] L. Benacem and S. D. Blostein, "Raptor-network coding strategies for energy efficient cooperative DVB-H multimedia communications," in *Proc. of 2010 International Conference on Green Circuits and Systems (ICGCS)*, 2010, pp. 527–532.

[35] S. Hong, Y. Won, and D. I. Kim, "Significance-aware channel power allocation for wireless multimedia streaming," *IEEE Trans. Veh. Technol.*, vol. 59, no. 6, pp. 2861–2873, Jun. 2010.

[36] D. Chen, H. Jiy, and V. C. M. Leung, "Energy-efficient cross-layer enhancement of multimedia transmissions over cognitive radio relay networks," in *Proc. of IEEE WCNC'11*, May 2011, pp. 856–861.

[37] M. Cheng, G. Xuan, and L. Cai, "Joint routing and link rate allocation under bandwidth and energy constraints in sensor networks," *IEEE Trans. Wireless Commun.*, vol. 8, no. 7, pp. 3770–3779, Jul. 2009.

[38] L. Shi and A. O. Fapojuwo, "TDMA scheduling with optimized energy efficiency and minimum delay in clustered wireless sensor networks," *IEEE Trans. Mobile Comput.*, vol. 9, no. 7, pp. 927–940, Jul. 2010.

[39] W. Wang *et al.*, "On energy efficient encryption for video streaming in wireless sensor networks," *IEEE Trans. Multimedia*, vol. 12, no. 5, pp. 417–426, Aug. 2010.

[40] M. C. Vuran and I. F. Akyildiz, "XLP: a cross-layer protocol for efficient communication in wireless sensor networks," *IEEE Trans. Mobile Comput.*, vol. 9, no. 11, pp. 1578–1591, Nov. 2010.

[41] L. Stabellini and J. Zander, "Energy-aware spectrum sensing in cognitive wireless sensor networks: a cross layer approach," in *Proc. of IEEE WCNC'10*, Jul. 2010, pp. 1–6.

[42] Y. Chen, W. Yi, and Y. Yang, "Energy efficient cooperative communication for sensor networks: a cross-layer approach," in *Proc. of IEEE Consumer Communications and Networking Conference (CCNC)*, 2011, pp. 793–797.

[43] P. Pangun *et al.*, "BREATH: an adaptive protocol for industrial control applications using wireless sensor networks," *IEEE Trans. Mobile Comput.*, vol. 10, no. 6, pp. 821–838, Jun. 2011.

[44] B. Tavli and W. Heinzelman, "Energy-efficient real-time multicast routing in mobile ad hoc networks," *IEEE Trans. Comput.*, vol. 60, no. 5, pp. 707–722, May 2011.

[45] J. T. Chiang and Y.-C. Hu, "Cross-layer jamming detection and mitigation in wireless broadcast networks," *IEEE/ACM Trans. Netw.*, vol. 19, no. 1, pp. 286–298, Jan. 2011.

[46] D. Triantafyllopoulou *et al.*, "E-CLEMA: A cross-layer design for improved quality of service in mobile WiMAX networks," *Wiley Wireless Communications and Mobile Computing*, vol. 9, no. 9, pp. 1274–1286, Sep. 2009.

[47] S.-L. Tsao and C.-H. Huang, "An energy-efficient transmission mechanism for VoIP over IEEE 802.11 WLAN," *Wiley Wireless Communications and Mobile Computing*, vol. 9, no. 12, pp. 1629–1644, Dec. 2009.

[48] E. Reusens *et al.*, "Characterization of on-body communication channel and energy efficient topology design for wireless body area networks," *IEEE Transactions on Information Technology in Biomedicine*, vol. 13, no. 6.

[49] C. Isheden, M. Klaus, and G. Fettweis, "Coolreader - an energy autonomous e-reader with broadband wireless connection," in *Proc. of IEEE International Conference on Acoustics, Speech and Signal Processing (ICASSP)*, 2010, pp. 1–5.

[50] G. Karakonstantis, G. Panagopoulos, and K. Roy, "HERQULES: system level cross-layer design exploration for efficient energy-quality trade-offs," in *Proc. of ACM/IEEE Low-Power Electronics and Design (ISLPED)*, 2010, pp. 117–122.

[51] P. Simoens *et al.*, "Characterization of power consumption in thin clients due to protocol data transmission over IEEE 802.11," in *Proc. of 7th International Symposium on Modeling and Optimization in Mobile, Ad Hoc, and Wireless Networks (WiOPT'09)*, 2009, pp. 1–7.

13 Energy-efficient relaying for cooperative cellular wireless networks

Yifei Wei, Mei Song, and F. Richard Yu

13.1 Introduction

The continuously growing demand for ubiquitous network access has led to the rapid development of wireless cellular networks during the last decade. The subscriber number and service traffic in cellular networks have explosively escalated. It is reported that there are now more than 5 billion mobile phone connections worldwide, and more than a billion mobile phone connections have been added globally in just 18 months [1]. The Asia-Pacific region including India and China is the main source of growth, accounting for 47% of global mobile connections at the end of June 2010. The penetration of mobile phones is growing rapidly in developing countries, and the number of subscribers will be astounding, since the world population is expected to reach 9.15 billion in 2050. In parallel with the rapid growth of the number of connections, the service types and traffic load have experienced significant evolvement from voice and short messaging service (SMS) to video and multimedia internet. Such tremendous growth in the information and communication technology (ICT) industry has made it become one of the leading sources of world energy consumption and it is expected to grow dramatically in the future. There are currently more than 4 million base stations (BSs) serving mobile users, each consuming an average of 25 MWh per year. In 2007, four Chinese operators consumed 20 billion KWh, which is equivalent to 8 million tons of coal combustion. ICT already represents around 2% of total carbon emissions, and this is expected to increase from 0.53 billion tonnes (Gt) carbon dioxide equivalent (CO_2e) in 2002 to 1.43 $GtCO_2e$ in 2020 [2]. In addition to the environmental problems that the governments restrict the greenhouse gases and pollution, energy costs also contribute a significant part of a network operator's overall expenditures due to the rapidly growing demand for mobile communication. Therefore, energy saving approaches are urgently required by both governments and network vendors.

Increasingly rigid environmental standards work and rising energy costs in tandem draw attention to the "energy efficiency" aspect of communication technologies. Governments and corporations around the world are tightening energy and emission budgets, thus creating demand for new, energy-aware generations of telecom equipment. This trend has stimulated the interest of researchers in an innovative new research area called green communications technologies, which have a very significant role to play in addressing climate change globally and facilitating efficient and low-carbon development, especially in a rapidly developing country such as China. The traditional objective of network design is to maximize the network capacity or to improve network performance, while green communications technologies change the objective of network design to increase the energy efficiency. The European Commission has recently started some projects within its seventh Framework Programme to address the energy efficiency of communication systems, such as "Towards Real Energy-efficient Network Design (TREND)" [3], "Cognitive Radio and Cooperative strategies for Power saving in multi-standard wireless devices (C2POWER)" [4], and "Energy Aware Radio and NeTwork TecHnologies (EARTH)" [5], which aim at a 50% reduction in the energy consumption of 4th-generation (4G) mobile wireless communication networks within the next 2.5 years. Alcatel-Lucent and 15 major carriers/manufacturers initiated "GreenTouch", with the objective of improving energy efficiency in wired/wireless networks by 1000 times by 2015.

Therefore, minimizing the energy consumption for data transmission has become one of the most important design considerations for future wireless communication networks. Energy consumption can be minimized by many technologies, such as improved transmitter efficiency, fresh aircooling, renewable energy sources and energy saving during low traffic or cell zooming [6], efficient base station (BS) redesign, and energy-efficient wireless architectures and protocols. In [7], the concept of energy-aware cooperative base station power management, which selectively lets BSs go to sleep based on their traffic load while radio coverage can be guaranteed by the remaining active cells, is explored to lead to significant energy savings. The authors in [7] propose an energy-saving approach by switching off BSs and predefine the BS sleeping scheme for cellular networks according to a deterministic traffic variation pattern over time, and [9] extends their work and proposes a BS energy-saving algorithm that can dynamically minimize the number of active BSs according to both time and spatial traffic variation. Moreover, to improve the cellular network performance, two emerging technologies are expected to be deployed in future LTE-Advanced systems; these are coordinated multi-point (CoMP) [9] and wireless relaying [10]. These two transmission technologies can also be used for energy saving, since they can extend BSs' coverage and improve cell-edge user throughput.

Multi-input multi-output (MIMO) techniques based on antenna arrays can dramatically reduce the required transmission power under a certain throughput requirement due to spatial diversity. Since it is difficult to equip hand-held devices with multiple antennas due to size, cost, or hardware limitations [11], the concept of cooperative relaying has been proposed to generate a virtual antenna array [12]. The basic idea of cooperative relaying in wireless networks is that some nodes that overhear the information transmitted from the source node relay it to the destination node, instead of treating it as interference. Since the destination node receives multiple independently faded copies of the transmitted information from the source node and relay nodes, cooperative diversity is achieved and can potentially lead to more energy saving. In this chapter, we analyze the energy-saving performance with a cooperative relay and provide an energy-efficient relaying mechanism based on a selective single relay.

The rest of this chapter is organized as follows. Section 13.2 introduces the techniques for energy savings in wireless networks. The energy-efficient cooperative communication problem is presented in Section 13.3. Section 13.4 presents the system models. We formulate the problem in Section 13.5. Section 13.6 presents the proposed distributed relay selection scheme. Simulation results and discussions are presented in Section 13.7. Finally, we conclude this study in Section 13.8.

13.2 Energy saving in cellular wireless networks

So far, achieving a high data rate has been the primary focus of research in wireless communication systems, without much consideration of energy efficiency. However, many communication technologies significantly increase system complexity and energy consumption. Escalating energy costs and environmental concerns have already created an urgent need for more energy-efficient wireless communication. To address the challenge of increasing power efficiency in wireless networks and thereby maintaining profitability, this chapter will discuss some research issues and challenges and suggest some paradigm-shifting technologies to enable an energy-efficient or green cellular network.

13.2.1 Energy-saving techniques

In cellular networks, BSs dominate the energy consumption of the whole network, and BSs will be more densely deployed to provide higher throughput in the future. Therefore, energy saving from BSs is necessary. The energy consumption of a typical BS can be reduced by improving the BS hardware design and by including power-saving protocols. However, developing more energy-efficient BS hardware has a limited effect, because the energy consumption has a large base level resulting from processing circuits and air conditioning when a BS is switched on. Thus, a system-wide approach by turning off certain BSs when the traffic load is getting low will be more promising and effective. For next-generation cellular networks based on micro-cells, pico-cells, and femto-cells, a static cell deployment is not optimal with fluctuating traffic conditions. Since the operating BS consumes a considerable amount of energy, selectively letting some BSs switch off or go to sleep mode based on their traffic load and letting the remaining active BSs guarantee the radio coverage can lead to a significant amount of energy savings. The authors in [6] present the concept of cell zooming, which lets BSs adjust the cell size according to the network or traffic situation, in order to balance the traffic load, while reducing the energy consumption. Cells with less traffic may go to sleep mode to reduce energy consumption, while the neighboring cells can zoom out and help serve the mobile users cooperatively.

Besides hardware redesign and new system-level features, energy consumption in wireless networks is closely related to their radio resource management schemes. Thus cross-layer design and optimization would be crucial to use energy more efficiently at the system level. For example, to save energy in an ad hoc wireless network, more packets should be transmitted when the channel quality is good, since a high spectral-efficiency modulation scheme can be used and network congestion will be reduced, so that the packets do not need to be retransmitted.

Recently, the research on technologies such as cognitive radio and cooperative relaying has received significant attention by both industry and academia. Cognitive radio was originally proposed to use the radio spectrum in a more efficient manner [20], and it has recently been realized that the intelligent and adaptive allocation of spectrum can reduce power consumption, while maintaining the required quality-of-service (QoS) [21]. From Shannon's capacity formula, we can see that the capacity increases linearly with bandwidth, but only logarithmically with power. This means that in order to reduce power, we should seek more bandwidth using cognitive radio.

Since the wireless signal attenuates nonlinearly and two-hop communication may consume less energy than direct communication, cooperative relaying technologies also enable us to solve the problem of energy efficiency via distributed signal processing. Delivering green communication via cooperative relaying can be achieved either by fixed relays or by user cooperation. Instead of installing new BSs to have a higher BS density, using fixed relays is a flexible and economical solution to provide service to more users using less power compared to systems based on direct transmission. Exploiting users to act as relays not only increases the data rate, but also makes the system more robust and eliminates the cost of installing relay nodes. In [22], it is shown that user cooperation has the potential of simultaneously improving both users' throughput and energy efficiency under different channel conditions. Therefore, this new approach can be a promising technique to increase the system performance in terms of energy efficiency in future wireless mobile networks. However, the increased data rate of one user comes at the price of the energy consumed by another user acting as a relay, and the limited battery life time of mobile users leads to dead users who run out of energy. Thus centralized or distributed algorithms must be designed to dynamically select relays among the users.

13.2.2 Energy-efficiency criteria

To determine whether the technology and architecture is energy efficient or not, the energy-efficiency criteria have to be defined in theory. The energy-efficiency criteria can be defined either by the ratio of output and input power, or by the ratio of some measurable performance to the input energy. While the former is commonly used for circuit systems (e.g. power amplifiers), the latter is a more common metric and can be defined by energy per bit (Joules/bit) or bits per energy (bits/Joule), where information bits are compared with overall input energy. A number of standard efficiency criteria have been defined to optimize various aspects of different components or a wireless system. While the energy-efficiency criteria at the component and equipment level are fairly straightforward to define, it is more challenging to define those metrics at a system or network level. A standardized way of measuring energy efficiency paves the way for forward-looking requirements and goals.

As different equipment vendors develop competing technologies and architectures, this leads to unequal energy consumption between equipments belonging to the same class. Therefore, to determine the winner in energy consumption, it would be sufficient to put two or more network devices under the same load and measure their respective power draw. In order to define an efficiency criteria, equipment vendors normalize energy consumption by effective full-duplex throughput and give us a normalized power efficiency in the form of

$$\rho = \frac{E}{T}, \quad (13.1)$$

where ρ is the normalized power efficiency and is expressed in watts/bps or Joules/bit, E denotes the energy-consumption rate (in watts or Joules per second) and T denotes the effective system throughput (in bits per second), and the values of E and T may come from either internal testing or the vendor's data.

The limited availability of energy on network nodes is one of the most critical issues in distributed battery-powered wireless networks, such as mobile ad hoc networks, wireless sensor networks, and user cooperative relaying networks. Indeed, recharging or replacing the node's battery may be inconvenient, or even impossible in disadvantaged working environments. This implies that the time during which the network functions properly becomes an important performance measure, i.e. network lifetime (L). There are various possible definitions of network lifetime, depending on the network application. For example, it can be considered as the time spanning the instant when the network starts functioning till a certain percentage of nodes run out of energy, or until the network is disconnected [18]. One of the commonly used lifetime definitions is the number of data collections until the number of dead nodes reaches a threshold N_T.

There are two approaches to maximizing network lifetime. One indirect approach aims to minimize energy consumption, which can help extend the network lifetime and does not address the problem precisely, while the other approach directly aims to maximize the network lifetime. This chapter provide the solution that directly aims to maximize the network lifetime.

13.3 Energy-efficient cooperative communication based on selective relay

Relaying technology can be used to increase cell coverage, and improve average user throughput and cell-edge user throughput, and this technology has been considered for the next-generation wireless networks [13]. In fact, early research has shown that relaying techniques can also enable us to solve the problem of energy efficiency via multi-hop transmission and extend the battery life [14]. In particular, multi-hop communication divides a direct path between BS and mobile terminal (MT) into several shorter links, in which wireless channel impairments such as path loss are less destructive, hence lower transmission power can be assigned to the BS and relays. Authors in [15] mention that two-hop communication consumes less energy than direct communication. And finally, it is shown in [16] that using multi-hopping in cellular networks can reduce the average energy consumed per call.

The assumptions for conventional relay networks is that relay stations (RS) can only be used for multi-hop links and there is no cooperation between RS and BS. Recently, cooperative communication techniques have been proposed to create virtual MIMO systems and mitigate the detrimental effects of multipath fading. In cooperative relaying networks, the RS not only forwards the data, but also transmits cooperatively with the BS. Therefore, the signal from the RS and the signal from the direct path between MT and BS can be combined and make a joint decision. Since the destination node receives multiple independently faded copies of the transmitted information from the source node and relay nodes, spatial diversity is achieved. Hence, well-known improvements of MIMO systems including coverage enlarging, capacity enhancement, and energy savings can be achieved via distributed signal processing. Recently, the research on cooperative relaying has received significant attention from both industry and academia. As mentioned earlier, the European Union has already started the C2POWER project with the objective of reducing the power consumption of mobile terminals using cooperative technologies.

Relaying could be implemented using amplify-and-forward (AF), decode-and-forward (DF), or distributed space-time-coded (STC) schemes. As the name implies, in AF mode, the RS receives a noisy version of the signal transmitted by MT and then amplifies and retransmits this noisy version, thus the BS receives two independently faded versions of the signal and make better decisions on the detection of information. In the DF method, which is closest to the idea of a traditional relay, the relay attempts to detect the partner's bits and then retransmits the detected bits. In the distributed STC scheme, a number of RSs transmit the different columns of a space-time coding matrix simultaneously to the destination, which requires all cooperators to be synchronized and co-phased such that the signals from the cooperators can be combined constructively at the destination. The unfixed number of participating antennas and the synchronization difficulties make distributed STC a challenging scheme for implementation [17]. In this chapter, the discussion will focus on the DF relaying scheme since it has advantages in digital processing and avoids noise amplification.

13.3.1 Relay selection schemes

Because of the cooperation overhead, the energy efficiency of the cooperative communication may degrade with the increase of the number of cooperators, i.e. more cooperators may not be more energy efficient [19]. Selective cooperation schemes have been investigated and the concept of *selection diversity* in cooperative systems is proposed in [23], and the authors demonstrated that cooperative relay selection outperforms the distributed STC scheme. Based on a simple relay selection strategy, the authors in [24] investigated the energy efficiency of selective relay cooperation schemes, and showed that the total energy-cost of data transmission with selective relay is lower. In selective single relay cooperation, only one out of a set of potential candidates is chosen to aid the communication process, where the relay selection could be based on distance information or instantaneous channel gains.

Relay selection among available relays is crucial in improving the performance of cooperative relaying [25]–[29]. The authors of [25] present a centralized optimization framework, in which the base station solves the joint relay strategies and resource allocations problem based on the feedback of the receivers' channel estimation, and then informs all users of the appropriate power levels and cooperative strategies. In [26], a semi-distributed relaying algorithm is proposed to jointly optimize the relay selection and power allocation of the system. Distributed relay selection methods based on local instantaneous channel measurements without the topology information are considered in [27]–[29]. The involved nodes exchange their current estimation of channel state information (CSI) and decide the "best" relay for the subsequent frame transmission. The "best" relay is the one that has the maximum instantaneous value of a metric, which is the minimum or the harmonic mean of its source-to-relay (S2R) and relay-to-destination (R2D) channels' gains. In [27], each relay calculates and sends this metric to the source through a feedback channel, then the source uses its source-to-destination (S2D) channel's gain and each relay's metric to determine which relay to cooperate with, and finally the source sends a control signal to the destination and the relays to indicate its decision. In [28], the estimated CSI is gathered at the destination, and the destination chooses the "best" relay based on the metric and then instructs the selected relay to participate in the relaying phase. The authors of [29] deploy timers that are proportional to the channels' gains in each node to help with the relay selection.

Although some work has been done for relay selection in wireless cooperative networks, most previous work uses the current observed channel conditions to make the relay selection decision for the subsequent frame transmission. Specifically, it is assumed that the channel fading is slow enough such that the channel conditions remain in the same state from the current frame to the next. The estimated CSI of the current frame is simply taken as the predicted CSI for the next frame. However, this *memoryless* channel assumption is often not realistic given the time-varying nature of some mobile environments [30]. Finite-state Markov channel (FSMC) models have been widely accepted in the literature as an effective approach to characterize the correlation structure of the fading process, including Rayleigh, Rician, and Nakagami fading channels [31]. In this chapter, the FSMC models are used to predict the upcoming channel state and mitigate error propagation. The adaptive modulation and coding (AMC) scheme is applied to achieve high spectral-efficiency.

In addition, most of the previous work in the literature concentrated on capacity optimization without considering energy consumption. This chapter aims to extend the lifetime of the user cooperation network by relay selection while satisfying certain QoS performance criteria. The distinct features of the proposed scheme include:

- One relay is selected from all available relays according to their states, including residual relay energy and CSI of both S2R and R2D links. A first-order FSMC is used to model the relay state. The source node and relay node can use different modulation and coding schemes (MCS).
- The objectives of the proposed scheme are to mitigate error propagation, increase the spectral efficiency and maximize the network lifetime. We model the relay selection problem as a restless bandit system, which is a powerful stochastic control modeling framework. The optimal relay selection policy can be obtained by solving the restless bandit problem with linear programming (LP) relaxation and primal-dual index heuristic algorithm [32].
- The obtained relay selection policy has an *indexability* property that dramatically reduces the computation and implementation complexity of the relay selection policy.

On-line implementation is further simplified by constructing a lookup table that maps the relay states to the priority indices.
- The proposed scheme is fully distributed and scalable. There is no need for a centralized control point, and relays can join and leave from the set of relay candidates freely.

13.4 System model for the relay selection problem

We consider a distributed cooperative wireless network with peer-to-peer relaying, where each node has the ability to relay data packets for others, as shown in Figure 13.1. Assume that the network uses request-to-send/clear-to-send (RTS/CTS) packet exchanges between the sender and receiver to avoid collision and acquire the current estimation of the signal-to-noise ratio (SNR). The network employs an AMC scheme, thus the source and relay may use different MCSs. The destination can combine the packets delivered by different modulation schemes using digital combining techniques, such as "chase combining" in hybrid automatic repeat request (HARQ) [33]. [34] investigates how to do the combining and detecting when the received data from S and R may be using different modulations.

A receiver-based auto-rate (RBAR) technique is used, in which the decision of the MCS is performed at the receiver side according to the observed channel condition. When the receiver receives the RTS packet, it will determine the MCS for the subsequent frame based on the average received SNR, and inform the sender of the MCS via the CTS packet. Before each time slot, the source broadcasts the RTS, then its destination estimates γ_{S2D} and decides the MCS of S2D (MC_{S2D}), and the relays can estimate γ_{S2R}. After the destination broadcasts the CTS with the MCS information, the source and relays both receive it, and the relays can estimate γ_{R2D} and decide the MCS of R2D (MC_{R2D}). Each MCS has a corresponding spectral efficiency. We assume that there are in total K classes of MCSs with spectral efficiencies denoted as $\eta_0, \eta_1, \ldots, \eta_{K-1}$, and the minimum decoding SNRs for different MCSs are $\gamma_0^*, \gamma_1^*, \ldots, \gamma_{K-1}^*$, respectively.

We assume that there are N available relay candidates that can decode both RTS and CTS between a pair of source and destination, and denote the set of available relays as

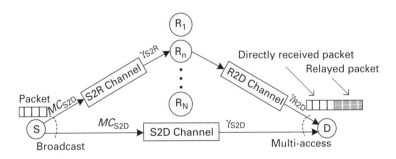

Figure 13.1 Cooperative relaying model.

$N = \{1, 2, \ldots, N\}$. The duration of the whole communication between a pair of source and destination is divided into T time slots which correspond to the time interval between two continuous decisions, and this time interval is also referred to as the *horizon* in this chapter. Let $t \in T = \{0, 1, \ldots, T-1\}$ stand for the time instant when a decision needs to be made. The action of each relay $n \in N$ in time slot t is represented by $a_n(t)$, $a_n(t) \in A = \{0, 1\}$, in which $a_n(t) = 0$ means the relay n is passive (not selected) in time slot t, and $a_n(t) = 1$ means active (selected).

13.4.1 S2R channel

In the first phase of cooperative relaying, the source transmits a data packet to its destination using the modulation and coding scheme MC_{S2D}, and the selected relay overhears it. The FSMC model is widely used in the literature to characterize the wireless channel. First-order FSMC is used in [35] to approximate the channel when designing a decision-feedback maximum-likelihood decoder. The authors in [36] demonstrate that a first-order FSMC is sufficient to model wireless channels, and that the improvements of a second or higher-order FSMC are negligible. Therefore, the first-order FSMC model is used here.

In the FSMC, the channel state is characterized via the received SNR, a parameter that is commonly used to represent the quality of a channel. The range of the average SNR of a received packet is partitioned (quantized) into L levels, and each level is associated with a state of a Markov chain. The channel varies over these states at each time slot according to a set of Markov transition probabilities. That is, the average received SNR at a relay can be modeled as a random variable γ_{S2Rn} evolving according to a finite-state Markov chain, which is characterized by a set of states, $C = \{C_0, C_1, \ldots, C_{L-1}\}$. The S2R channel state realization of γ_{S2Rn} is $\Gamma_n(t)$ for relay n in time slot t. Let $\phi_{g_n h_n}(t)$ denote the probability that γ_{S2Rn} moves from state g_n to state h_n at time t. The $L \times L$ S2R channel state transition probability matrix of relay n is defined as

$$\Phi_n(t) = \left[\phi_{g_n h_n}(t)\right]_{L \times L}, \tag{13.2}$$

where $\phi_{g_n h_n}(t) = \Pr\left(\Gamma_n(t+1) = h_n \mid \Gamma_n(t) = g_n\right)$, and $g_n, h_n \in C$. Relay n can estimate its current channel state $\Gamma_n(t)$ after receiving RTS from the source at time t.

13.4.2 R2D channel

During the second phase (i.e. the relaying phase), the selected relay retransmits the packet to the destination using the modulation and coding scheme MC_{R2D}, and the destination combines both directly received and relayed packets. Given the target bit error rate (BER), the minimum decoding SNRs: $\gamma_0^*, \gamma_1^*, \ldots, \gamma_{K-1}^*$ for different MCSs can be calculated. Since we are concerned with the MCS of the R2D link, the channel state of the R2D link can be divided into discrete K levels as follows:

$$\gamma_{R2D} = \begin{cases} \Upsilon_0, & \text{if } \gamma_0^* \leq \gamma_{R2D} < \gamma_1^* \\ \Upsilon_1, & \text{if } \gamma_1^* \leq \gamma_{R2D} < \gamma_2^* \\ \vdots & \vdots \\ \Upsilon_{K-1}, & \text{if } \gamma_{R2D} \geq \gamma_{K-1}^*. \end{cases}$$

The average received SNR of each packet can be modeled as a random variable γ_{R2Dn} evolving according to a K-state Markov chain, which has a finite-state space denoted as $D = \{D_0, D_1, \ldots, D_{K-1}\}$. The R2D channel state realization of γ_{R2Dn} is $\Upsilon_n(t)$ for relay n in time slot t. Let $\psi_{u_n v_n}(t)$ denote the probability that γ_{R2Dn} moves from state u_n to state v_n at time t. The $K \times K$ R2D channel state transition probability matrix of relay n is defined as

$$\Psi_n(t) = \left[\psi_{u_n v_n}(t) \right]_{K \times K}, \tag{13.3}$$

where $\psi_{u_n v_n}(t) = \Pr(\Upsilon_n(t+1) = v_n \mid \Upsilon_n(t) = u_n)$, and $u_n, v_n \in D$. Relay n can estimate its current channel state $\Upsilon_n(t)$ after receiving CTS from the destination at time t.

13.4.3 Energy model

Since most wireless mobile devices are powered by batteries with limited energy, the battery power should be consumed more carefully and efficiently to maximize the network lifetime. Therefore, we should select the relay with high residual energy to avoid overusing a node. Since the battery energy of a device will decrease due to any application (e.g. a multimedia application or a wireless transmission) run on the device, we don't know exactly the energy state at the next time slot. Therefore, the battery energy can be modeled as a random variable e_n. For simplification, the continuous residual energy realization of e_n can be divided into discrete levels, denoted by $E = \{E_0, E_1, \ldots, E_{H-1}\}$, where H is the number of available energy state levels. Assume the residual energy realization of e_n to be $E_n(t)$ for relay n at time t. The authors of [37] model the transition of the energy levels of nodes in wireless networks as the Markov chain. We adopt this model and define the energy state transition probability matrix of relay n taking action a as

$$\Theta_n^a(t) = \left[\theta_{f_n y_n}^a(t) \right]_{H \times H}, \tag{13.4}$$

where $\theta_{f_n y_n}^a(t) = \Pr(E_n(t+1) = y_n \mid E_n(t) = f_n, a_n(t) = a)$, and $f_n, y_n \in E, a \in A$. The energy model used in some other papers, such as [38], assumes the energy is reduced by a fixed amount after every data transmission action. This model can be considered a special case of the Markov model, where

$$\theta_{f_n y_n}^1(t) = \begin{cases} 1, & \text{if } y_n \text{ is the lower energy state next to } f_n. \\ 0, & \text{otherwise.} \end{cases} \tag{13.5}$$

In real systems, the values in the above transition probability matrices can be obtained from the history observation of the wireless network.

13.4.4 Objectives

We need to find out the optimal relay selection policy, which can set one relay to be active at time slot t according to the relays' states that contain their S2R channel state $\Gamma_n(t) \in C$, R2D channel state $\Upsilon_n(t) \in D$ and residual energy state $E_n(t) \in E$. Here use the following optimization objectives:

- *Maximize network lifetime*: Two approaches are adopted to maximize network lifetime: minimizing the energy consumption required to deliver data packets by selecting the relay with better channel state and balancing energy usage among the relays by selecting the relay with high residual energy.
- *Increase spectral efficiency*: Better channel state $\Upsilon_n(t) \in D$ enables a higher modulation and coding scheme with higher spectral efficiency, and should be reflected in higher reward in our formulation.
- *Mitigate error propagation*: Better channel state $\Gamma_n(t) \in C$ mitigates error propagation with lower BER, and should be reflected in higher reward in our formulation.

13.5 Problem formulation

A restless bandit is used to formulate the stochastic selection problem as follows: Consider a collection of N projects; each project n can be in a state $i_n(t) \in S_n$ in each time slot $t = 0, 1, 2, \ldots$. According to their states, m out of N projects are selected to work, or set to be active ($a_n(t) = 1$), and the other projects are set to be passive ($a_n(t) = 0$). The system reward $R_{i_n(t)}^{a_n(t)}$ is earned when action $a_n(t)$ is taken, and their states change in a Markovian fashion, according to a transition probability matrix (into state $j_n(t+1) \in S_n$ with probability $p_{i_n j_n}^a$). Reward is discounted in time by a discount factor $0 < \beta < 1$. Projects are selected over time under a policy $u \in U$, where U is the set of all Markovian policies (which select the current action as a function, possibly randomized, of the current state and time). The problem is to determine the optimal u that maximizes the total expected discounted reward over the time horizon. The restless bandit problem can be solved according to the indices of the projects, which are calculated by the LP relaxation algorithm. Recent advances in solving the restless bandit problem make it a powerful modeling framework. Thus, the procedure of relay selection can be formulated as a restless bandit problem.

13.5.1 Relay states

The state of an available relay $n \in \{1, 2, \ldots, N\}$ in time slot $t \in \{0, 1, \ldots, T-1\}$ is determined by the realization of the states $\Gamma_n(t), \Upsilon_n(t)$ and $E_n(t)$ of the random variables $\gamma_{S2Rn}, \gamma_{R2Dn}$ and e_n. Consequently, the state of an available relay is the combination of them as follows:

$$i_n(t) = [\Gamma_n(t), \Upsilon_n(t), E_n(t)]. \qquad (13.6)$$

In practice, the change of channel state on S2R, channel state on R2D, and residual energy state are independent of each other, i.e. random variables γ_{S2Rn}, γ_{R2Dn}, and e_n are independent. Therefore, the relay state will change in a Markovian fashion, and the finite-state space of relay n is represented as S_n, $i_n(t) \in S_n$, with the transition probability matrix

$$P_n^a(t) = \left[\phi_{g_n h_n}(t), \psi_{u_n v_n}(t), \theta^a_{f_n y_n}(t)\right]_{G_n \times G_n}, \qquad (13.7)$$

where $\phi_{g_n h_n}(t)$, $\psi_{u_n v_n}(t)$ and $\theta^a_{f_n y_n}(t)$ are defined in (13.2), (13.3) and (13.4), respectively, and $G_n = L \times K \times H$. The element of $P_n^a(t)$ is $p^a_{i_n j_n}(t)$, denoting the transition probability that the state of relay n changes from i_n to j_n, where $i_n, j_n \in S_n$.

13.5.2 System reward

In the restless bandit problem, the system reward represents the optimization objectives. Since the objectives of the proposed scheme is to mitigate error propagation, increase spectral efficiency, and maximize network lifetime, we formulate the system reward to be the function of BER of the S2R link, the spectral efficiency of the R2D link, and the energy consumption of delivering the data packets from source to destination. For the objective of balancing energy usage among the relays, residual energy state $E_n(t)$ should also be one factor of the reward function. The action of a relay determines whether the reward will be gained. Therefore, we define the system reward as:

$$R^{a_n(t)}_{i_n(t)} = a_n(t) R(\omega_p P_b(\Gamma_n(t)), \omega_\eta \eta_k(\Upsilon_n(t)), \omega_J J(P_t, l, r_k), \omega_E E_n(t)), \qquad (13.8)$$

where $|\omega_p| + |\omega_\eta| + |\omega_J| + |\omega_E| = 1$, ω_η, and ω_E are positive weights; ω_p and ω_J are negative weights; P_b is BER function determined by channel state $\Gamma_n(t)$ when a modulation and coding scheme is given; η_k is the spectral efficiency determined by the modulation scheme which adapts to channel state $\Upsilon_n(t)$; J is the energy-consumption function of transmit power, packet length and data rate; P_t is the transmit power of the relay for retransmitting a data packet; l is the length of the packet; and r_k is the data rate.

The instantaneous reward $R^{a_n(t)}_{i_n(t)}$ is earned for relay n in state $i_n(t)$ when it takes action $a_n(t)$ in time slot t. For a stochastic process, a maximum immediate value does not mean the maximum expected long-term accumulated value. Therefore, we need to think about more than just the instantaneous reward that the system can receive. We denote by $u \in U$ the policy, and denote by β the discount factor. Solving the optimal policy for the infinite horizon problems requires the discount factor $0 < \beta < 1$ to ensure that the expected reward is bounded and converges. We assume that the duration of the whole communication is long enough and T is approximately infinite. In the case of undiscounted finite horizon problems (i.e. T is small), we can set $\beta = 1$. The goal of the relay selection is to find a selection policy that maximizes the total expected discounted reward during the whole communication period, and the optimum value is

$$Z^* = \max_{u \in U} E_u \left[\sum_{t=0}^{T-1} (R_{i_1(t)}^{a_1(t)} + R_{i_2(t)}^{a_2(t)} + \cdots + R_{i_N(t)}^{a_N(t)}) \beta^t \right]. \quad (13.9)$$

13.5.3 Solution to the restless bandit problem

To solve the restless bandit problem, a hierarchy of increasingly stronger LP relaxations is developed based on the classical result on LP formulations of Markov decision chains (MDCs) [32].

In order to formulate the restless bandit problem as a linear program, we introduce the following performance measures:

$$x_{i_n}^1(u) = E_u \left[\sum_{t=0}^{T-1} (I_{i_n}^1(t) \beta^t) \right], \quad (13.10)$$

and

$$x_{i_n}^0(u) = E_u \left[\sum_{t=0}^{T-1} (I_{i_n}^0(t) \beta^t) \right], \quad (13.11)$$

where

$$I_{i_n}^1(t) = \begin{cases} 1, & \text{if relay } n \text{ is in state } i_n \text{ and active at time } t \\ 0, & \text{otherwise} \end{cases} \quad (13.12)$$

and

$$I_{i_n}^0(t) = \begin{cases} 1, & \text{if relay } n \text{ is in state } i_n \text{ and passive at time } t \\ 0, & \text{otherwise.} \end{cases} \quad (13.13)$$

The performance measure $x_{i_n}^1(u)$ (resp. $x_{i_n}^0(u)$) represents the total expected discounted time that relay n is in state i_n and active (resp. passive) under selection policy u. We denote by X the corresponding performance region spanned by performance vector $\mathbf{x} = (x_{i_n}^{a_n}(u))_{i_n \in S_n, a_n \in \{0,1\}}$ under all Markovian policies $u \in U$:

$$X = \{\mathbf{x} = (x_{i_n}^{a_n}(u))_{i_n \in S_n, a_n \in \{0,1\}, n \in N} \mid u \in U\}. \quad (13.14)$$

Since the restless bandit problem is naturally formulated as a discounted MDC, the authors of [32] prove that performance region X is a polytope, which they referred to as the *restless bandit polytope* ξ. The restless bandit problem can thus be formulated as the following linear program:

$$\text{(LP)} \quad Z^* = \max_{\mathbf{x} \in X} \sum_{n \in N} \sum_{i_n \in S_n} \sum_{a_n \in \{0,1\}} R_{i_n}^{a_n} x_{i_n}^{a_n}. \quad (13.15)$$

13.5 Problem formulation

The approach developed in [32] is to construct relaxations of polytope X that yield polynomial-size relaxations of the linear program. Denote by $\widehat{X} \supseteq X$ the relaxations, not on the space of the original variables x_i^a, but in a higher-dimensional space that includes new auxiliary variables. Define the polytope $\xi_n^1 = \{x_n = (x_{i_n}^{a_n}(u))_{i_n \in S_n, a_n \in \{0,1\}, n \in N} \mid u \in U\}$, which is precisely the projection of *restless bandit polytope* ξ over the space of the variable $x_{i_n}^{a_n}$ for relay n. Furthermore, ξ_n^1 is also the performance region of the first-order MDC corresponding to relay n. Let α_{i_n} denote the probability that the initial state is i_n, for $i_n \in S_n$, thus the initial state probability vector $\boldsymbol{\alpha} = (\alpha_{i_n})_{i_n \in S_n}$ is given. A complete formulation of ξ_n^1 is given by [32]:

$$\xi_n^1 = \{x_n \in \Re_+^{|S_n \times \{0,1\}|} \mid x_{j_n}^0 + x_{j_n}^1 = \alpha_{j_n} + \beta \sum_{i_n \in S_n} \sum_{a_n \in \{0,1\}} p_{i_n j_n}^{a_n} x_{i_n}^{a_n}, \, j_n \in S_n\}. \quad (13.16)$$

Therefore, the first-order relaxation can be formulated as the following linear program:

$$(\mathrm{LP}^1) \quad Z^1 = \max \sum_{n \in N} \sum_{i_n \in S_n} \sum_{a_n \in \{0,1\}} R_{i_n}^{a_n} x_{i_n}^{a_n}$$

subject to

$$x_n \in \xi_n^1, \quad n \in N,$$

$$\sum_{n \in N} \sum_{i_n \in S_n} x_{i_n}^1 = \frac{m}{1-\beta}. \quad (13.17)$$

In our scheme, only one relay is active, so $m = 1$. We will refer to the feasible space of linear program (LP^1) as the first-order approximation to the *restless bandit polytope* ξ, and denote it by ξ^1. Notice that linear program (LP^1) has $O(N|S_{\max}|)$ variables and constraints, where $|S_{\max}| = \max_{n \in N} |S_n|$, and its size is thus polynomial in the problem dimensions.

The authors of [32] interpret the primal-dual heuristic as a priority-index heuristic under some mixing assumptions on active and passive transition probabilities. Here we use the priority-index rule to select relays. The dual of linear program (LP^1) is

$$(\mathrm{D}^1) \quad Z^1 = \min \sum_{n \in N} \sum_{j_n \in S_n} \alpha_{j_n} \lambda_{j_n} + \frac{m}{1-\beta} \lambda$$

subject to

$$\lambda_{i_n} - \beta \sum_{j_n \in S_n} p_{i_n j_n}^0 \lambda_{j_n} \geq R_{i_n}^0, \quad i_n \in S_n, n \in N,$$

$$\lambda_{i_n} - \beta \sum_{j_n \in S_n} p_{i_n j_n}^1 \lambda_{j_n} + \lambda \geq R_{i_n}^1, \quad i_n \in S_n, n \in N,$$

$$\lambda \geq 0. \quad (13.18)$$

Let $\{\bar{x}_{i_n}^{a_n}\}$ and $\{\bar{\lambda}_{i_n}, \bar{\lambda}\}$ be the optimal primal and dual solution pair to the first-order relaxation (LP^1) and its dual (D^1). Let $\{\bar{\epsilon}_{i_n}^{a_n}\}$ be the corresponding optimal reduced cost coefficients; we have

$$\bar{\epsilon}^0_{i_n} = \bar{\lambda}_{i_n} - \beta \sum_{j_n \in S_n} p^0_{i_n j_n} \bar{\lambda}_{j_n} - R^0_{i_n},$$

$$\bar{\epsilon}^1_{i_n} = \bar{\lambda}_{i_n} - \beta \sum_{j_n \in S_n} p^1_{i_n j_n} \bar{\lambda}_{j_n} + \bar{\lambda} - R^1_{i_n}, \qquad (13.19)$$

which must be nonnegative. The optimal reduced costs $\bar{\epsilon}^0_{i_n}$ and $\bar{\epsilon}^1_{i_n}$ can be interpreted as the rate of decrease in the objective-value of linear program (LP^1) per unit increase in the value of the variable $x^0_{i_n}$ and $x^1_{i_n}$, respectively.

For each available relay $n \in N$, we consider a directed graph that is defined from the passive and active transition probabilities, respectively, as follows: $U_n = (S_n, A_n)$, where $A_n = \{(i_n, j_n) \mid p^0_{i_n j_n} > 0 \text{ and } p^1_{i_n j_n} > 0, \ i_n, j_n \in S_n\}$. Under the mixing assumption that the directed graph U_n is connected for every $n \in N$, every extreme point $\bar{x}$ of polytope ξ^1 has the following properties [32]:

(i) There are at most one relay l and one state $i_l \in S_l$ for which $\bar{x}^1_{i_l} > 0$ and $\bar{x}^0_{i_l} > 0$.
(ii) For all other relays n and all other states either $\bar{x}^1_{i_n} > 0$ or $\bar{x}^0_{i_n} > 0$.

Therefore, starting with an optimal primal solution $\{\bar{x}^{a_n}_{i_n}\}$ and a complementary dual solution pair $\{\bar{\lambda}_{i_n}, \bar{\lambda}\}$, the corresponding reduced cost $\{\bar{\epsilon}^{a_n}_{i_n}\}$ can be computed with (13.18). The index of relay n with current state i_n is defined as

$$\delta_{i_n} = \bar{\epsilon}^1_{i_n} - \bar{\epsilon}^0_{i_n}. \qquad (13.20)$$

The priority-index rule is to set active the relay that has the smallest index.

13.6 Distributed relay selection scheme

In this section, we present the distributed relay selection scheme in wireless cooperative networks. The proposed scheme is based on the RTS/CTS mechanism of collision avoidance, and the current channel state can be observed and the set of available relays can be determined via exchanging RTS/CTS packets.

13.6.1 Available relay candidates

A source node initiates its transmission by sending an RTS packet to its destination node. Its destination and all neighbor nodes will receive this RTS packet, and estimate γ_{S2D} and γ_{S2R}, respectively. The destination decides the MCS of the S2D link (MC_{S2D}) according to the prediction of upcoming channel quality, and replies to the source with a CTS packet containing the MC_{S2D} and γ_{S2D} information. The source and all neighbor nodes of the destination will receive it. Then the source will adjust its MCS to MC_{S2D} for the next time slot. The neighbor nodes also learn which MCS should be adopted to overhear the data packet from the source and get the S2D channel information γ_{S2D}, and they can estimate γ_{R2D} from the CTS packet and decide the MCS of the R2D link (MC_{R2D}) according to the prediction of upcoming channel quality. Here, we assume that the forward and backward channels between the relay and destination are the same, when the transmissions occur on the same frequency band and same coherence interval. The common neighbors of the pair of source and destination that can decode both RTS and CTS could be the set of potential relays. However, in order to provide better relay candidates, we apply the following criterion to examine whether they can be in the set of available relays:

$$\gamma_{S2D} < \min(\gamma_{S2R}, \gamma_{R2D}). \tag{13.21}$$

Those common neighbors that satisfy the criterion that the SNR in direct link is less than the minimum of the SNRs in the relaying path will constitute the set of available relays N. Notice that we use a first-order FSMC model to approximate the time variations of the average SNR in the subsequent time slot, which means that the prediction of upcoming channel quality is the set of K possible states together with the state transition probabilities. Therefore, the decision of which MCS to use in the upcoming time slot is based on the predicted channel state with the highest probability.

13.6.2 Relay selection process

After the handshaking packets, each available relay n gets its current S2R channel state $\Gamma_n(t)$ and R2D channel state $\Upsilon_n(t)$ at time t, and they also can detect their energy state $E_n(t)$. Given the current state $i_n(t)$, each available relay calculates its index δ_{i_n} and broadcasts a candidate index (CI) packet containing its index together with MC_{R2D} information. In order to reduce the probability of collision between CI packets, relays can contend for the channel first using a standard carrier-sense multiple access (CSMA) splitting scheme. In CSMA, a node first senses the channel to make sure that it is idle before transmitting. Adopting other channel reservation mechanisms may solve the "hidden node" and "exposed node" problem, but will impose additional communication overhead and complexity. When other available relays receive this CI packet, they will compare the received index with their own index and broadcast their own CI packet only if their own index is smaller than the received index; otherwise they will not broadcast their own CI packet and will keep silent without listening to the source transmission, which can be energy efficient from a network sense. We can simply set a timer on the source node and destination node to stop receiving the delayed CI packets. After the timer expires, the source receives all the CI packets or the sub-set of all CI packets due to the collision between them, and it will select the relay with the smallest index to cooperate with. After receiving all the CI packets or the sub-set of all CI packets, the destination adopts the (MC_{R2D}) information from the relay with the smallest index to receive the retransmission from this relay.

We would like to emphasize that feeding back the index to the source and making the decision at the source can solve the hidden terminal problem that the relays may be hidden from each other and cannot hear the CI broadcast from each other. Since the priority indices can be computed and stored in a table off-line before transmission, the relay selection process can be divided into the off-line stage and the on-line stage:

- *Off-line computation*: Before data transmission, the set of indices $\{\delta_{i_n}\}$ can be computed with the input information of the state transition probability $p_{i_n j_n}^{a_n}$, the reward $R_{i_n}^{a_n}$, the discount factor β, and the initial state probability vector $\boldsymbol{\alpha}$. Relays store these indices and the corresponding $p_{i_n j_n}^{a_n}$, $R_{i_n}^{a_n}$, and $\boldsymbol{\alpha}$ in a table.
- *On-line selection*: After each handshaking between the pair of source and destination, each available relay $n \in N$ looks up its index table to find out the index δ_{i_n} corresponding to its current state i_n, and then broadcasts its CI packet if it has not received any CI packet or its own index is smaller than the received indices. The relay selection operates in a distributed manner, so there is no need for a centralized control point. If a relay leaves the set of available relays during data transmission, it does not send the CI packet, thus it will not be selected. Another relay with the smallest index will be selected. Therefore, relay nodes can leave and join the relay candidate set freely.

13.6.3 Cost evaluation

Although some communication overhead is added due to the necessary information exchange between the available relays, the cost is light compared with the user data transmission. The relays can share the parameters used for computing the priority index by adding their information to the existing broadcast packets in the routing protocol or other multicast algorithm. In our simulations, the indices can be computed in 1 second using a normal PC. These actions take effect only once before real-time data transmission, and result in little cost.

During data transmission, a relay just looks up the index table according to its current state and sends its current index by a rather small packet only if its own index is smaller than the received indices. Therefore, there will be at most $(N-1)$ CI broadcast packets. Because the CI packet only contains the node ID and its index, the length of the packet is very short and the communication overhead does not have much effect on the system performance.

13.7 Simulation results and discussions

This section illustrates the performance of the proposed scheme by simulations. We compare the proposed scheme with an existing *memoryless* selection method and the random selection method. We use M-ary phase-shift keying (i.e. BPSK, QPSK, and 8PSK) modulations as the available MCSs. We assume that the state of the S2R channel can be bad (s0) or good (s1), and the state of the R2D channel can be good-for-BPSK (d0), good-for-QPSK (d1), or good-for-8PSK (d2), and the state of residual energy can be dead (e0), low (e1), or high (e2). We set the state transition probability matrices of

13.7 Simulation results and discussions

the S2R channel and R2D channel as follows:

$$\Phi = \begin{pmatrix} 0.7 & 0.3 \\ 0.3 & 0.7 \end{pmatrix}, \quad \Psi = \begin{pmatrix} 0.6 & 0.3 & 0.1 \\ 0.2 & 0.6 & 0.2 \\ 0.1 & 0.3 & 0.6 \end{pmatrix}.$$

Let the residual energy state transition probability matrix be Θ^0 when the relay is passive, and Θ^1 when the relay is active.

$$\Theta^0 = \begin{pmatrix} 1.00 & 0.00 & 0.00 \\ 0.01 & 0.99 & 0.00 \\ 0.00 & 0.01 & 0.99 \end{pmatrix}, \Theta^1 = \begin{pmatrix} 1.00 & 0.00 & 0.00 \\ 0.08 & 0.92 & 0.00 \\ 0.00 & 0.08 & 0.92 \end{pmatrix}.$$

The relay state transition probability matrices P^0 and P^1 can be easily acquired according to (13.7). Assume that the BER of a bad channel is about $P_b = 10^{-3}$, and the BER of a good channel is about $P_b = 10^{-5}$. We take $\lg(P_b/10^{-3})$ as the first component of reward function weighted by negative ω_p. The spectral efficiencies of BPSK, QPSK, and 8PSK are 1, 2, and 3, respectively. In general, the better the channel state, the lower the transmission energy requirement [38]. Here we assume that the transmit power is fixed and the energy consumption for retransmitting a data packet is inversely proportional to the data rate or spectral efficiency. Let the energy consumption for d0 channel, d1 channel, and d2 channel be 1, 1/2, and 1/3, respectively. We set the reward to be 0, ω_E, and $2\omega_E$ for energy states e0, e1, and e2, respectively. The reward function $R_{i_n}^{a_n}$ will be zero if the relay is passive ($a_n = 0$) or the energy state is dead (e0). We set the discount factor $\beta = 0.8$ in the simulations. Each simulation lasts 2000 seconds, and there are $N = 20$ available relays at the beginning.

13.7.1 System reward

Firstly, we illustrate the performance improvement of the proposed scheme by system reward. The four design criteria are set to be $\omega_J = -0.1$, $\omega_E = 0.2$, $\omega_p = -0.3$, and $\omega_\eta = 0.4$. We obtain the system reward of different schemes, as shown in Figure 13.2. It can be seen that the proposed scheme has better performance during most of the simulation time. The system reward will decline with time, because more and more relays run out of energy after transmission for some time. There are hardly any available relays at 2000 s, so the system reward drops to near zero in all three schemes. If we change the weights with different values, we can obtain similar results.

13.7.2 Error propagation mitigation

Figure 13.3 shows the average BER of different schemes. We can see that the proposed scheme obtains the average BER of about 10^{-5} when there are many available relays, whereas the existing memoryless selection method gets an average BER of about 10^{-4}, since it selects a relay for the subsequent frame according to the current states which may change in the subsequent frame. We can also observe that the average BER of the

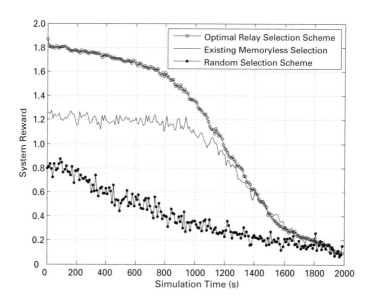

Figure 13.2 System performance of integrated reward.

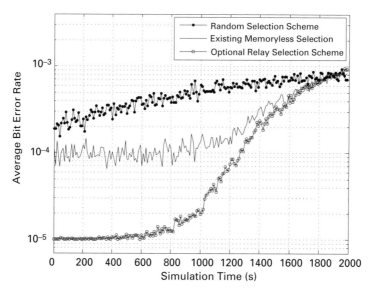

Figure 13.3 Average BER of different relay selection schemes.

random selection scheme increases from the beginning of the simulation, but the average BER of the proposed scheme does not increase until 800 s. This is because the energy is consumed more carefully and efficiently in the proposed scheme, which prolongs the network lifetime. The two approaches of maximizing network lifetime are also integrated into the memoryless selection method in the simulation; therefore, the average BERs of both the memoryless selection method and the proposed scheme increase after 800 s.

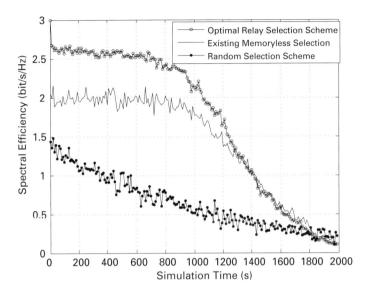

Figure 13.4 Spectral efficiency of different relay selection schemes.

13.7.3 Spectral efficiency improvement

Figure 13.4 is the spectral efficiency of different schemes. It can be seen that the spectral efficiency will decline with time, which is similar to the performance of the system reward. More and more relays run out of energy, and there are hardly any live relays at 2000 s, so the spectral efficiency drops to near zero in all three schemes. We can also observe that the spectral efficiency of the random selection scheme declines at the beginning of the simulation, but the spectral efficiencies of both the proposed scheme and the memoryless selection method decline after 800 s. The proposed scheme outperforms the other two schemes during most of the simulation time.

13.7.4 Network lifetime

Here we use the lifetime definitions that the number of dead nodes reaches a threshold N_{th}, when the network can no longer achieve the target performance. When the network lifetime is the most important issue, we set the two related weights as $\omega_E = 0.5$ and $\omega_J = -0.5$, and assign 0 for the other weights. Therefore, the residual energy and energy consumption will be the system reward. We set the initial state of each relay as high (e2) energy, and the energy consumption for delivering a packet in good-for-8PSK (d2) channel as a fixed amount, and the energy consumption in good-for-BPSK (d0) channel is 3 times the amount, and the energy consumption in good-for-QPSK (d1) channel is 2 times the amount. The number of dead relays will increase with transmission time. Figure 13.5 compares the network lifetime of different schemes with varying N_{th}. As expected, the lifetime of all schemes increases with N_{th}, and the proposed scheme always has a longer lifetime. As can be seen from the figure, the first dead relay appears round about 250 s in the random selection scheme, and the first dead relay appears

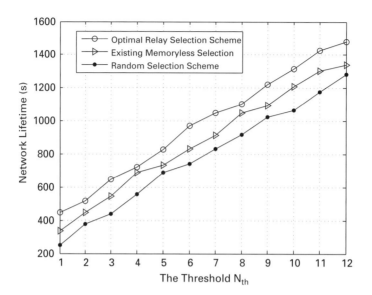

Figure 13.5 Network lifetime comparison under different thresholds.

round about 330 s and 450 s in the memoryless method and our proposed scheme, respectively.

13.8 Conclusion

We have presented a brief survey of methods to achieve energy efficiency in wireless networks. Relaying-based cooperative user communication has emerged as a promising technique to increase cell coverage and improve user throughput. However, increased performance of one user comes at the price of the energy consumed by another user acting as a relay, and the limited battery lifetime of mobile users leads to dead users who run out of energy. Thus the centralized or distributed algorithms must be designed to dynamically select relays among the users. Most previous work concentrates on capacity optimization without considering energy consumption. This chapter aims to extend the lifetime of the user cooperation network by relay selection while satisfying certain QoS performance criterion.

We have presented a distributed relay selection scheme considering finite-state Markov channels, adaptive modulation and coding, as well as residual relay energy. Two approaches are adopted to maximize network lifetime: minimizing the energy consumption required to deliver data packets by selecting the relay with better channel state, and balancing energy usage among the relays by selecting the relay with high residual energy. We have formulated the relay selection problem as a stochastic restless multi-armed bandit system. The solution can be derived by linear programming relaxation and primal-dual index heuristic algorithm. The obtained relay selection policy has an indexability property, and the priority indices can be computed and stored in a table off-line. For the on-line relay selection, each relay just needs to look up its index table to find out the index corresponding to its current state, which dramatically reduces the on-line computation and implementation complexity. Simulation results have been presented to illustrate that the proposed relay selection scheme significantly outperforms the memoryless method and random selection method.

References

[1] BBC News, "Over 5 billion mobile phone connections worldwide," [Online] Available: www.bbc.co.uk/news/10569081

[2] The climate group, "Enabling the low carbon economy in the information age," SMART 2020 Report.

[3] TREND, "Towards real energy-efficient network design," [Online] Available: www.fp7-trend.eu/

[4] C2POWER, "Cognitive radio and cooperative strategies for power saving in multi-standard wireless devices," [Online] Available: www.ict-c2power.eu/

[5] EARTH, "Energy aware radio and network technologies," [Online] Available: https://www.ict-earth.eu/

[6] Z. Niu et al., "Cell zooming for cost-efficient green cellular networks," *IEEE Commun. Magazine*, vol. 48, no. 11, pp. 74–79, Nov. 2010.

[7] M. A Marsan et al., "Optimal energy savings in cellular access networks," in *IEEE International Conference on Communications Workshops*, June 2009, pp. 1–5.

[8] S. Zhou et al., "Green mobile access network with dynamic base station energy saving," in *Proc. of MobiCom'09 (Poster)*, Sep. 2009.

[9] H. Huang et al., "Increasing downlink cellular throughput with limited network MIMO coordination," *IEEE Trans. Wireless Commun.*, vol. 8, no. 6, pp. 2983–2989, June 2009.

[10] Y. Song et al., "Relay station shared by multiple base stations for intercell interference mitigation," in *IEEE C802.16m-08/1436rl*, Nov. 2008.

[11] A. Sendonaris, E. Erkip, and B. Aazhang, "User cooperation diversity part I and part II," *IEEE Trans. Commun.*, vol. 51, no. 11, pp. 1927–1948, Nov. 2003.

[12] A. Nosratinia, T. E. Hunter, and A. Hedayat, "Cooperative communication in wireless networks," *IEEE Commun. Magazine*, vol. 42, no. 10, pp. 74–80, Oct. 2004.

[13] 3GPP TR 36.814, "Further advancements for E-UTRA physical layer aspects (Release 9)," [Online] Available: www.3gpp.org/ftp/Specs/html-info/36814.htm

[14] J. N Laneman and G. W Wornell, "Energy-efficient antenna sharing and relaying for wireless networks," in *Proc. of IEEE WCNC'00*, Mar. 2000.

[15] J. Y. Song, H. Lee, and D. H. Cho, "Power consumption reduction by multi-hop transmission in cellular networks," in *Proc. of IEEE VTC-F'04*, Sept. 2004.

[16] A. Radwan and H. S Hassanein, "NXG04-3: does multi-hop communication extend the battery life of mobile terminals?," in *Proc. of IEEE GLOBECOM'06*, Nov. 2006.

[17] M. M. Fareed and M. Uysal, "A novel relay selection method for decode-and-forward relaying," in *Proc. of Canadian Conference on Electrical and Computer Engineering (CCECE'08)*, Niagara Falls, ON. May 2008.

[18] Q. Dong, "Maximizing system lifetime in wireless sensor networks," in *Proc. of 4th Int. Symp. Information Processing in Sensor Networks*, Apr. 2005.

[19] Z. Zhou et al., "Energy-efficient cooperative communication in clustered wireless sensor networks," in *Proc.of IEEE Milcom'06*, Oct. 2006.

[20] F. R. Yu, M. Huang, and H. Tang, "Biologically inspired consensus-based spectrum sensing in mobile ad hoc networks with cognitive radios," *IEEE Network*, pp. 26–30, June 2010.

[21] A. He et al., "Minimizing energy consumption using cognitive radio," in *Proc. of IEEE International Performance, Computing and Communications Conference (IPCCC'08)*, Dec. 2008.

[22] M. Nokleby and B. Aazhang, "User cooperation for energy-efficient cellular communications," in *Proc. of IEEE ICC'10*, May 2010.

[23] E. Beres and R. Adve, "Selection cooperation in multi-source cooperative networks," *IEEE Trans. Wireless Commun.*, vol. 7, no. 1, pp. 118–127, Jan. 2008.

[24] R. Madan et al., "Energy-efficient cooperative relaying over fading channels with simple relay selection," *IEEE Trans. Wireless Commun.*, vol. 7, no. 8, pp. 3013–3025, Aug. 2008.

[25] T. C.-Y. Ng and W. Yu, "Joint optimization of relay strategies and resource allocations in cooperative cellular networks," *IEEE Journal on Selected Areas in Commun.*, vol. 25, no. 2, pp. 328–339, Feb. 2007.

[26] J. Cai et al., "Semi-distributed user relaying algorithm for amplify-and-forward wireless relay networks," *IEEE Trans. Wireless Commun.*, vol. 7, no. 4, pp. 1348–1357, Apr. 2008.

[27] A. S. Ibrahim et al., "Relay selection in multi-node cooperative communications: when to cooperate and whom to cooperate with?," in *Proc. of IEEE GLOBECOM'06*, Nov. 2006.

[28] M. M. Fareed and M. Uysal, "A novel relay selection method for decode-and-forward relaying," in *Proc. of Canadian Conference on Electrical and Computer Engineering (CCECE'08)*, Niagara Falls, ON, May 2008.

[29] A. Bletsas et al., "A simple cooperative diversity method based on network path selection," *IEEE Journal on Selected Areas in Commun.*, vol. 24, no. 3, pp. 659–672, Mar. 2006.

[30] J. Yang, A. K. Khandani, and N. Tin, "Statistical decision making in adaptive modulation and coding for 3G wireless systems," *IEEE Trans. Veh. Tech.*, vol. 54, no. 6, pp. 2066–2073, Nov. 2005.

[31] Y. Wei, F. R. Yu, and M. Song, "Distributed optimal relay selection in wireless cooperative networks with finite state Markov channels," *IEEE Trans. Veh. Tech.*, vol. 59, no. 5, pp. 2149–2158, June 2010.

[32] D. Berstimas and J. Niño-Mora, "Restless bandits, linear programming relaxations, and a primal dual index heuristic," *Operations Research*, vol. 48, no. 1, pp. 80–90, 2000.

[33] J.-F. Cheng, "Coding performance of hybrid ARQ schemes," *IEEE Trans. Commun.*, vol. 54, no. 6, pp. 1017–1029, June 2006.

[34] Y. Zhang, Y. Ma, and R. Tafazolli, "Modulation-adaptive cooperation schemes for wireless networks," in *Proc. of IEEE VTC-S'08*, May. 2008.

[35] L. Li and A. J. Goldsmith, "Low-complexity maximum-likelihood detection of coded signals sent over finite-state Markov channels," *IEEE Trans. Commun.*, vol. 50, no. 4, pp. 524–531, Apr. 2002.

[36] H. S. Wang and P.-C. Chang, "On verifying the first-order Markovian assumption for a Rayleigh fading channel model," *IEEE Trans. Veh. Tech.*, vol. 45, no. 2, pp. 353–357, May 1996.

[37] P. Hu et al., "The HMM-based modeling for the energy level prediction in wireless sensor networks," in *Proc. of IEEE 2nd Conf. on Industrial Electronics and Applications*, Harbin, P. R. China, May 2007.

[38] Y. Chen et al., "Transmission scheduling for optimizing sensor network lifetime: a stochastic shortest path approach," *IEEE Trans. Signal Proc.*, vol. 55, no. 5, pp. 2294–2309, 2007.

14 Energy performance in TDD-CDMA multi-hop cellular networks

Hoang Thanh Long, Xue Jun Li, and Peter Han Joo Chong

14.1 Introduction

In traditional cellular communication systems, there is only one path between a mobile station (MS) and a base station (BS). In a multi-hop cellular network (MCN) [1, 2] the direct communication path between the MS and BS is divided into several shorter transmission links. Since the signal attenuation exponentially increases with the distance, shorter links in terms of distance require much lower transmission power. Consequently, lower transmission power introduces lower interference, which is an important factor affecting the quality of calls. All these factors, low signal attenuation, low transmission power, and low interference will have a great impact on the system capacity and total transmission power. In [3], the authors propose a virtual cellular network (VCN) and investigate the effect of multi-hop relaying on transmission power. The losses considered in the system include the path loss, shadowing loss, and Rayleigh fading. The average transmission power decreases with the number of multiple hops. However, as stated in [4], the transmission power should include both transmission energy and the hardware consumption energy. Therefore, a more accurate model taking the transmitter and receiver processing energy into account should be considered. In a multi-hop system, the total energy consumed may exceed the direct transmission link [5] according to the model used in [4]. In [6], the energy consumed by every bit of an MCN system is analyzed in detail. From the results, we can see that the energy consumed per bit in the multi-hop system is much lower than the single-hop system.

In this chapter, we develop a more practical model including the inter-cell and intra-cell interferences to study the performances for different services in terms of various data rates for a multi-hop TDD-CDMA system [7]. This is different from previous studies [3, 5, 6], which consider only propagation loss, path loss, and shadowing fading. The energy consumption including the hardware processing energy for two single-hop schemes and two multi-hop schemes are studied. Furthermore, in order to find the optimal relay station structure for the multi-link, the energy-consumption profiles for different relay architectures are compared.

14.2 Structure of relay stations and power consumption

In this section, we provide an analysis of how multi-hop transmission results in lower transmission power for each node and for the whole system. We consider four cases of

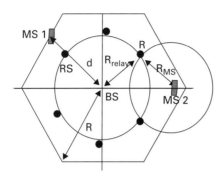

Figure 14.1 Two-hop FRS structure in MCNs.

relay station architectures: one fixed and three random relay station structures. In the fixed relay station (FRS) structure, the locations of the RSs are determined in advance based on a certain algorithm. We assume that each cell in the MCN has six relay stations located at an optimal location, which is R_{relay} away from the BS, as shown in Figure 14.1. The optimal location of relay stations achieves the best network performance and it will be investigated foremost in this chapter.

For the single-hop transmission in uplink, the transmission power, P_t, of the transmitter (MS) is proportional to the distance from its receiver (BS) as given by

$$P_t = P_r d^\alpha, \quad (14.1)$$

where P_r is the received power at BS (receiver), d is the distance between MS and BS, and α is the path-loss exponent. The worst case for single-hop is that the MS is located at the corner of a cell for $d = R$, where R is the cell radius.

In two-hop transmission in uplink, the total transmission power of two-hop transmission is usually lower than that of single-hop transmission because the communication link is split into two shorter paths. Consider an MS located at the cell edge: a special feature of two hop is that an RS is found between the BS and the MS or in other words, the BS, the RS, and the MS are located linearly as shown in Figure 14.1 for MS 1. In this case, the total minimum transmission power, P_{min_tot}, can be achieved. Assuming the distance between the MS and the RS is d and the distance between the RS and the BS is R_{relay}, then the P_{min_tot} is given by

$$P_{min\_tot} = P_r \left(d^\alpha + R_{relay}^\alpha \right). \quad (14.2)$$

Referring to MS 2 in Figure 14.1, the total transmission power, P_{t_tot}, of MS-RS and RS-BS links becomes larger and it is obtained by

$$P_{t\_tot} = P_r \left(R_{MS}^\alpha + R_{relay}^\alpha \right). \quad (14.3)$$

In FRS, if the distance between the MS and the BS is greater than R_{relay}, a relay transmission should be used. Then, the MS should select the closest RS for relay transmission.

14.2 Structure of relay stations and power consumption

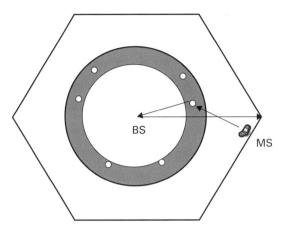

Figure 14.2 RSs located in an area from 0.4R to 0.55R.

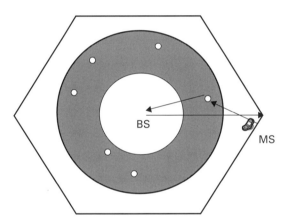

Figure 14.3 RSs located in an area from 0.2R to 0.7R.

14.2.1 Random relay station (RRS) structure

In the RRS structure, the RSs can be randomly placed around the BS without any predetermination. Based on our previous study, the optimal location of the relay station has been found to be about $0.4R$ for different network environments. In the RRS, we study the random distribution of RSs around this optimal value. Three different cases are considered. The relay stations are located randomly in the range of $0.3R$ to $0.55R$, $0.2R$ to $0.7R$, and in the whole cell as shown in Figures 14.2–14.4.

Despite the different relay structures, a factor for an MS to select a proper RS to forward its data to the BS is entirely based on power consumption. For example, if an MS selects an RS to forward its data to the BS, P_{t_tot} is proportional to $R^{\alpha}_{MS-RS} + R^{\alpha}_{RS-BS}$. Otherwise, if the MS selects a direct transmission to the BS, P_{t_tot} is proportional to R^{α}_{MS-BS}. In order to decide whether relay transmission should be used, based on the

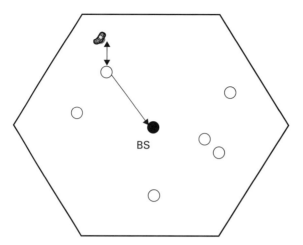

Figure 14.4 RSSs located in the entire cell area.

consideration of power consumption, the minimum value of the set in (14.4) should be identified and given as

$$\{R^{\alpha}_{MS-RS}, \left(R^{\alpha}_{MS-RS_i} + R^{\alpha}_{RS_i-BS}\right)\Big|_{i\in\{1,2,...,n\}}\}, \qquad (14.4)$$

where RS_i refer to the ith RS and n is the total number of RSs in each cell. Since the path-loss, α, can be varied from time to time and place to place depending on a number of conditions, it puts an engineering burden on the network provider to find a closed-form expression. For simplicity, we consider a free-space path loss for which $\alpha = 2$; then, (14.4) becomes

$$\{R^{2}_{MS-RS}, \left(R^{2}_{MS-RS_i} + R^{2}_{RS_i-BS}\right)\Big|_{i\in\{1,2,...,n\}}\}, \qquad (14.5)$$

which can be simply represented visually in Figure 14.5.

In order to find an RS for an MS, firstly, a circle is drawn between an MS and BS with a diameter, d, as shown in Figure 14.5. If there are any RSs located inside that circular area, relay transmission should be used and the closest RS to the center of the circle should be selected. This RS selection will give the minimum value of R_{MS-RS} in (14.5). Otherwise, a direct link between the MS and the BS is in use as R^{2}_{BS-MS} is the smallest value in (14.5). We have found that this circling approach to select an RS is effective for other α as well.

14.3 Time-slot allocation schemes

In our study, four different time-slot allocation schemes are used for the TDD-CDMA system. For the single-hop system [10], they are fixed time-slot allocation (FTSA) and

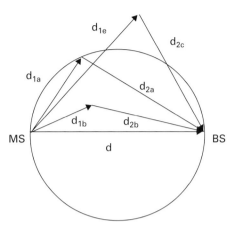

Figure 14.5 RSSs located in the entire cell area.

dynamic time-slot allocation (DTSA). In addition, the two schemes for MCN are multi-link fixed time-slot allocation (ML-FTSA) and multi-link dynamic time-slot allocation (ML-DTSA). The energy-consumption model in the previous discussion will be used in all four different methods.

14.3.1 Fixed time-slot allocation (FTSA)

It is a single-hop transmission scheme. A fixed switching point is used to separate the time frame. The 15 time slots then are divided into two parts; for TS 1 to TS 7, they are used for the uplink transmission. The remaining eight time slots are used for the downlink transmission. When a new call initiates in a given cell, the algorithm begins searching for a TS between TS 1 and TS 7 to support the uplink connection. TS 1 will first go through the call admission process (which will be described later). If the call admission process is successful for that MS in TS 1, TS 1 is selected for uplink connection. Otherwise, the algorithm continues testing the next uplink TS until TS 7. If the call admission process was not successful in any TS between TS 1 and TS 7, the new call will be blocked. After selecting the uplink TS, the TSs between TS 8 and TS 15 are tested for downlink connection. Downlink TS selection follows the same procedure, except that searching for downlink TS starts from the rightmost TS, TS 15, in the frame and continues to TS 8. Similarly, if the call admission process was not successful in any TS between TS 8 and TS 15, the new call will be blocked. A new call is accepted when TSs are successfully allocated to both uplink and downlink. As shown in Figure 14.6, a TS can support multiple MSs as long as their data rate and SIR requirements are satisfied in the TS.

14.3.2 Dynamic time-slot allocation (DTSA)

This is also a single-hop scheme, but there is no switching point in the time frame. When a new call initiates in a given cell, the algorithm starts from TS 1 to look for an empty TS or an uplink TS (which already supports uplink transmission) to allocate the uplink

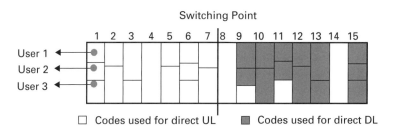

Figure 14.6 An example of time-slot allocation using the FTSA scheme.

Figure 14.7 An example of time-slot allocation using the DTSA scheme.

connection. If TS x is an empty TS or an uplink TS, it will go through the call admission process. If call admission succeeds for that MS in TS x, TS x is selected for uplink connection. Otherwise, the algorithm continues testing the next empty or uplink TS until it reaches TS 15. If the call admission process does not succeed for any TS, the new call will be blocked. The same procedure is followed for downlink TS selection, except that searching for an empty or a downlink TS starts from TS 15. If the call admission process does not succeed in any TS, the new call will be blocked. A new call is accepted when TSs are allocated to both uplink and downlink. Removing the switching point enables the algorithm to fully utilize all free TSs of the frame and increases the chance of accepting a newly arriving user. Figure 14.7 shows an example of time-slot allocation using the DTSA scheme. It can be seen that any TS can be used for uplink or downlink transmission. However, since each cell independently arranges the TSs of a frame for uplink and downlink, crossed-time slot interference may appear between adjacent cells. CTS interference happens when a TS is used for uplink in one cell and for downlink in a neighboring cell.

14.3.3 Multi-link fixed time-slot allocation (ML-FTSA)

This is a multi-hop scheme; it is developed based on the FTSA, and there is also a switching point between time slot 7 and time slot 8. The left part is for uplink transmission and the right part is for downlink transmission, as shown in Figure 14.8. In addition, a relay station cannot transmit and receive the signal in one time slot simultaneously, which will reduce the self-interference at relay stations. Also, due to this separation of uplink and downlink, there is no self-interference at the base station. For ML-FTSA, a 2-hop is accepted if TS allocations for uplinks, MS-to-RS and RS-to-BS links, and

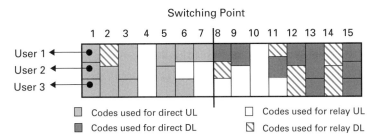

Figure 14.8 An example of time-slot allocation using the ML-FTSA scheme.

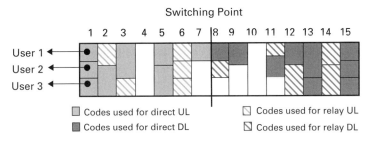

Figure 14.9 An example of time-slot allocation using the ML-DTSA scheme.

downlinks, BS-to-RS and RS-to-MS links, are all successful. An example of time-slot allocation for ML-FTSA is shown in Figure 14.8. It can be seen that each TS can support two links, a direct link and a relay link.

14.3.4 Multi-link dynamic time-slot allocation (ML-DTSA)

This time-slot allocation scheme is based on DTSA with multi-hop capability. Therefore, in this case, there is no switching point to separate the uplink and downlink transmissions. Consequently, for one time slot, it can be used as uplink and downlink simultaneously. The time slot can be fully used to accept more users and increase the data capacity. The time-slot allocation is shown in Figure 14.9.

For the single-hop transmission, we only consider the energy consumed for each MS since the hardware power of a cell phone is common to both single-hop and multi-hop; to simplify the model it is not included in the calculation. For the multi-hop cases, two different kinds of energies are considered. One is energy consumption just for the mobile user; the other is the energy for the whole uplink process. The RS hardware energy consumption is also included for energy per link.

14.4 System model

A cluster of 19 cells with two-tier configuration is used as shown in Figure 14.10(a). A two-hop MCN is considered in this study and the RSs can be placed one hop from the

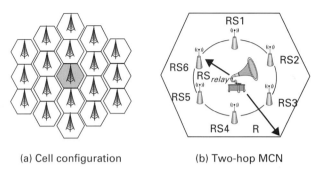

(a) Cell configuration   (b) Two-hop MCN

Figure 14.10 (a) Simulated 19-cell network, (b) two-hop MCN RS structure.

BS using FRS or RRS as shown in Figure 14.10(b). Users are uniformly distributed over these 19 cells. When a new call is generated in the system based on Poisson process, it is randomly assigned to a cell and the duration is exponentially distributed. After the connection is established, the user is assumed to be stationary, which makes the model simple by not considering the hand-off issue. For TDD-CDMA single-hop transmission, the time allocation schemes FTSA and DTSA will be used. For ML-FTSA and ML-DTSA, MSs can use multi-hop transmission to save energy. In addition, we assume that the system is under perfect power control. For the TDD-CDMA scheme in our model, a frame structure is used. For each frame, the time duration is 10 ms, which also is subdivided into 15 time slots. For single-hop transmission, each user can occupy one time slot for uplink transmission. On the other hand, if a relay station is used for a two-hop transmission, two time slots are assigned to that user. The maximum spreading factor is 16 and the maximum data rate that can be supported is 256 kbps. Therefore, the data rate can range from 256 to 16 kbps (256/16). In addition, for each time slot, it can support multi-users.

In order to compare energy consumption in different TDD-CDMA time-slot allocation schemes, a model including the hardware energy is built to help the analysis. One commonly used model assumes that the transceiver power is proportional to the transmission power. In that case, if we need to reduce the energy consumption, we have to reduce the data rate, which is not reasonable for the high demand for the increasing high data rate requirement. Another model [9], which is more complicated and practical, uses the same assumption when the data rate is larger than 100 kbps. When the data rate is lower than 100 kbps, it is assumed that the power used by hardware is almost constant, as shown in Figure 14.11.

Therefore, the total energy consumption in our model is the summation of transmission energy and hardware energy as given by

$$E = (P_T + P_H)T, \qquad (14.6)$$

where P_T is the transmitter power used for transmission and it changes according to the different call scenario to maintain the quality of service. P_H is the transmitter processing power; its value varies with the transmission data rates according to Figure 14.11. T is

Table 14.1. Traffic classes

Service	R_{UL}	R_{DL}	Target E_b/N_0	Duty factor (δ)	Average holding time
Class A	16 kpbs	32 kbps	5.0 dB	0.4	120 s
Class B	32 kpbs	64 kbps	7.0 dB	1.0	120 s

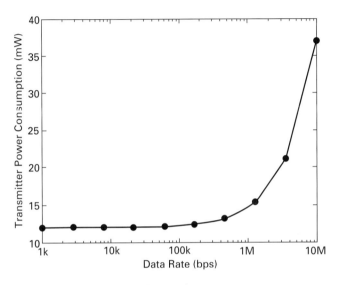

Figure 14.11 Transmitter power consumption vs. data rate.

the time for transmission. For TDD-CDMA systems, each user only occupies a certain time slot, hence the data transmission is not continuous; therefore, T can be derived as

$$T = \frac{\Delta t}{t_{frame}} t_{slot}, \quad (14.7)$$

where frame duration of t_{frame}=10 ms is used, and there are 15 time slots in each frame; hence for each time slot, the time duration is t_{slot}=10 ms/15 = 6.7×10^{-4} s. Δt is the duration of each call. Two traffic loads are considered in our system. One is low traffic, Class A, which is used for the voice service and the other is high traffic, Class B, which is used for low-quality video service. The parameters for each traffic class are shown in Table 14.1 [11, 12].

14.5 Simulation results and discussions

In order to verify the improvement from single-hop to multi-hop in terms of the energy consumption, and the blocking and dropping probability, and to investigate the optimal structure of relay stations, we have simulated the four time-slot allocation schemes using

Table 14.2. Simulation parameters

Number of cells	19
Cell radius, R	0.5 km
Maximum MS/RS power, P_{MS}/P_{RS}	21 dBm
Path-loss exponent, α	3.52
Path-loss constant, K	128 dB
Background noise, η	-103 dBm
Log-normal shadowing (μ, σ),	(0,8) dB

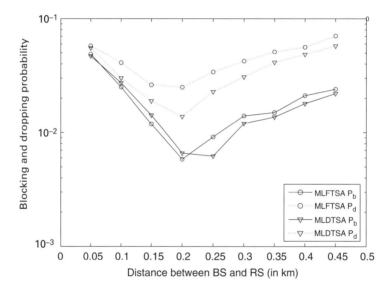

Figure 14.12 Blocking and dropping probabilities vs. location of relays stations.

C++. The parameters used for the simulation are listed in Table 14.2. The propagation model for micro-cellular has been well studied in [13]. We assume a log-normal fading model to simplify the problem.

First, we attempt to find the optimal locations of RSs in FRS architecture. In this model, only six relays are considered and they are distributed uniformly on a circle of radius at the relay from the base station. The dependences of blocking and dropping probability, uplink energy per mobile, uplink energy per call, and downlink energy per call with different relays are studied in order to find the optimal RS location.

As shown in Figure 14.12, the blocking and dropping probabilities reach their minimum values when the separation between the base station and relay stations are from 0.2 to 0.25 km, which is about 40% to 50% of the cell radius, R. In Figure 14.13, uplink energy per mobile is minimum when R_{relay} is 0.25 km. Uplink energy per call is smallest at $R_{relay} \sim 0.15$–0.2 km. Similarly, refering to Figure 14.14, the optimal

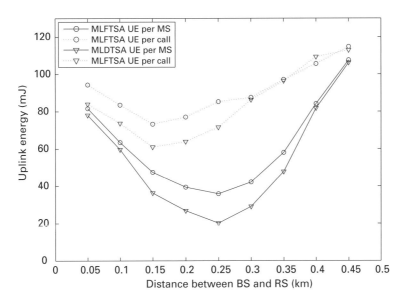

Figure 14.13 Uplink energy per mobile and per call vs. location of relay stations.

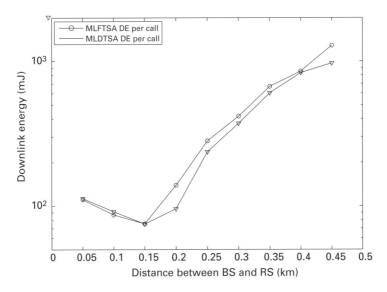

Figure 14.14 Downlink energy per mobile and per call vs. location of relay stations.

downlink energy is around $R_{relay} \sim 0.15\text{–}0.2$ km. It can be seen that although each of the above criteria, blocking and dropping probabilities, uplink energy per mobile/call, and downlink energy per call, have their different optimal RS locations, these optimal locations fall in the ranges of 0.15 to 0.25 km, which is equivalent to 30%–50% of the cell radius. Furthermore, when the locations of the relay stations vary in this range, the values of those criteria change insignificantly around their bottoms. Thus, without loss of generality, we set the optimal RS location for FRS at $R_{relay} = 0.2$ km.

Since for the multi-hop scenario, there are relay stations used which consume extra hardware energy, we are also interested in studying whether multi-hop is better in terms of the total transmission power. In addition, we also examine the differences between FRS and RRS to find the best RS structure.

14.5.1 Blocking and dropping probabilities for high and low data rate traffic

Blocking probability, P_b, is the statistical probability that a telephone connection cannot be established due to insufficient transmission resources in the network. Usually it is expressed as a percentage or decimal equivalent of calls blocked by network congestion during the busy hour [14].

In our simulation, we define a call admission process to decide whether to accept a new user. This call admission process is to check whether the selected time slot can support the user's data rate with a reasonably quality of service. One of the most important criteria for the new user to meet is to keep a received signal-to-noise ratio larger than the target E_b/N_0 with a certain margin, which is 1 dB in our simulation, while the transmission power is less than the maximum permissible transmission power. The target E_b/N_0 depends on the traffic type, as shown in Table 14.1.

After a new call meets the call admission criteria for uplink and downlink requirements, these new downlink and uplink transmissions will cause interference to the existing calls in the system. Therefore, there is a possibility that a call in neighboring cells using the same time-slot will be dropped if it does not meet the target E_b/N_0, that makes the dropping probability, P_d.

Figures 14.15 and 14.16 show the blocking and dropping probability for four time-slot allocation schemes for high and low traffic. In both figures, it can be seen that the increase of P_b is very small for all four schemes because as traffic increases, the

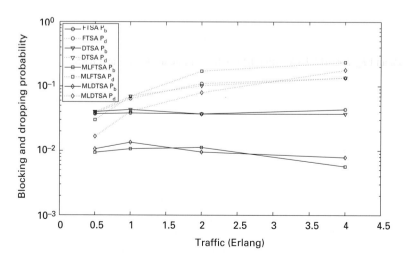

Figure 14.15 Comparison of time-slot allocation schemes in high-data-rate traffic.

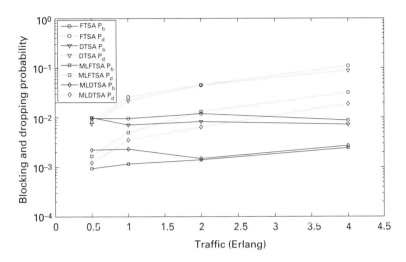

Figure 14.16 Comparison of time-slot allocation schemes in low-data-rate traffic.

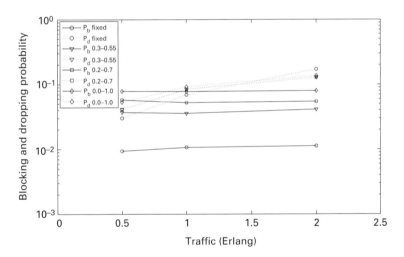

Figure 14.17 P_b and P_d for MLFTSA with 6 RSs in high traffic using FRS and RRS.

P_d will also increase. Then, more time slots will be available for the new connections and thus, the blocking probability is kept almost constant. The results show that multi-hop schemes perform better in terms of these two probabilities, since for the multi-hop schemes, the transmission link is divided into two shorter paths, which provide a better signal quality with higher E_b/N_0, for each path. Therefore, more users can be accepted. In the meantime, small transmission power is needed for shorter paths, so that smaller interferences are generated to other existing users. As a result, a good quality of service is still being maintained for existing users.

Figure 14.17 shows P_b and P_d for MLFTSA with six RSs in high traffic using FRS and RRS structures. It can be seen that FRS has lower P_b and P_d than RRS because when the RSs are randomly distributed, there is a higher chance of not using an RS. As

Table 14.3. Consumed energy (mJ) of four schemes with $P_H = 0$

Traffic type	ρ (Erlangs/cell)	FTSA (UL/DL)	DTSA (UL/DL)	MLFTSA (UL/MS)
Class A	2	79.77/910.23	82.14/1006.82	24.49
	4	103.75/1558.95	105.94/1727.35	39.64
Class B	0.5	121.69/1633.5	132.01/1298.05	46.23
	2	187.13/3806.91	176.78/3653.18	57.18
Traffic type	ρ (Erlangs/cell)	MLFTSA (UL/DL)	MLDTSA (UL/MS)	MLDTSA (UL/DL)
Class A	2	46.64/65.35	15.44	36.96/53.18
	4	71.24/59.61	19.9	52.67/66.58
Class B	0.5	93.14/197.47	30.02	78.42/117.03
	2	121.44/539.59	49.02	114.6/299.74

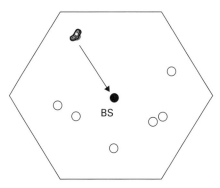

Figure 14.18 A poor scenario of random relay station distribution in RRS structure.

shown in Figure 14.18, it is very difficult for users in the upper part of the cell to find an RS under the RRS structure. Thus, this call might be blocked easily.

14.5.2 Energy consumption for single-hop and multi-hop transmission using FRS

Firstly, we analyze energy consumption without considering hardware power-consumption. Table 14.3 shows the energy consumption for four time-slot allocation schemes for six RS using FRS. From Table 14.3, we can see that, for the two single-hop schemes, the energy consumption is almost the same with the same traffic type and traffic load. This is because the mobile users are uniformly distributed in the system, which will cause the average distance between the base station and the mobile station to be the same for both single-hop schemes. However, when the traffic load increases, there will be more interference. In order to maintain the quality of service, the transmitter will increase the transmission power and thus, consume more energy.

In Table 14.3, uplink energy per mobile (UL/MS) is defined as the average energy consumed by an MS to transmit data to a base station or a relay. Uplink energy per call (UL/call) is defined as the average energy consumed per call. The energy of a call can be the energy consumed by an MS only if it is a single-hop call from MS to BS directly. If it is a two-hop call, the energy of a call is the sum of the energy consumed by both MS and RS. Uplink energy per mobile is a useful indicator to evaluate whether the network

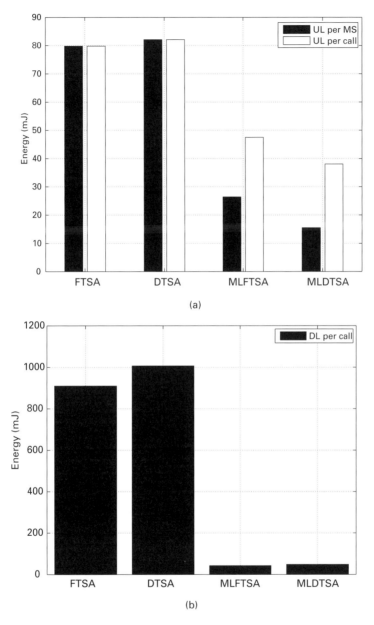

Figure 14.19 Energy consumption for Class A traffic with $P_H = 0$ at $\rho = 2$ Erlangs/cell: (a) for uplink, and (b) for downlink.

structure is energy friendly to the end user. In other words, lower energy consumption makes mobile phones last for a longer time. On the other hand, energy per call indicates the amount of energy the system has consumed for the whole operation and indicates whether the network structure is environmentally friendly.

Figures 14.19 and 14.20 show the energy consumption of all four schemes in low-data-rate traffic and high-data-rate traffic, respectively, at $\rho = 2$ Erlangs/cell and $P_H = 0$.

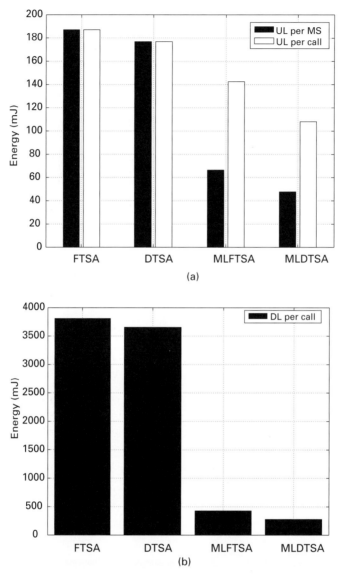

Figure 14.20 Energy consumption for Class B traffic with $P_H = 0$ at $\rho = 2$ Erlangs/cell: (a) for uplink, and (b) for downlink.

From these two figures, it can be seen that the performance of the two single-hop schemes is almost the same. As expected, the energy consumption for high data rate traffic is much higher than that for low data rate traffic. For multi-hop schemes, ML-FTSA and ML-DTSA consume much less energy for both high-data-rate and low-data-rate traffic. The uplink energy per MS of multi-hop schemes is about 20–30% of that for single-hop schemes in both traffics. It means that the cell phone battery can last for a much longer time, which is a great benefit for the end user. Uplink energy per call in multi-hop

14.5 Simulation results and discussions

schemes is about 50% of that in single-hop scheme. Downlink energy per call in multi-hop schemes is only 10% of that in single-hop scheme. This is because MCN uses much shorter transmission links than a single-hop network, which, in turn, results in much smaller transmission power. Thus, even for the total transmission energy, which is the sum of uplink and downlink energy per call, MCN needs much less energy. Another observation is that among two multi-hop allocation schemes, MLDTSA performs slightly better. This is due to the asymmetric data rates of uplink and downlink and a better algorithm in utilizing channels of the dynamic time-slot allocation scheme in MLDTSA.

Next, we add hardware processing energy, P_H, for comparison in order to get a whole picture of the transmission energy. From the model used in [9], it is suggested that when the data rate is 16 kbps and 32 kbps, the hardware power is almost the same, around 12 mW. Thus, we use $P_H = 12$ mW for our study.

Figures 14.21 and 14.22 show the energy consumption in low-data-rate traffic and high-data-rate traffic, respectively, with $P_H = 12$ mW. Including hardware power-consumption, the energy consumption for multi-hop schemes is increased significantly as compared to no hardware power. However, it is still lower than single-hop schemes. Uplink energy per mobile in multi-hop schemes is about half of its values in single-hop schemes, while downlink energy per call for multi-hop schemes is about 20–25% of that for single-hop schemes.

In our model, we consider the average call duration is 120 seconds and thus, the energy consumed by the hardware is

$$E_H = 12 \text{ mW} \times \frac{120 \text{ s}}{10 \text{ ms}} \times 0.667 \text{ ms} = 96 \text{ mJ}. \tag{14.8}$$

A significant amount of energy is added to all profiles. In multi-hop networks, more hardware equipment is required and used so that pulls up the energy level consumed in this type of network structure. However, savings from better network structure using in multi-hop network still surpass this additional amount of energy caused by hardware power-consumption. Therefore, by using the multi-hop scheme for voice service, not only the blocking and dropping probabilities can be decreased by 10%, but also the total energy-consumption can be decreased significantly.

14.5.3 Energy consumption for RRS structure

In this section, we study the energy consumption of an RRS. Figures 14.23 and 14.24 show the energy consumption of an FRS and various ranges of RRS for MLDTSA with six RSs with $P_H = 0$.

From Figure 14.23, it can be seen that the uplink energy per mobile increases with the degree of randomness of the relay stations. FRS has the lowest uplink energy per mobile, whereas allowing the RSs to be randomly placed within the whole cell in RRS, i.e. 0-R, gives the highest uplink energy per mobile. If the relay stations are randomly distributed in the cell, in some areas there will be more than enough RRSs and in other areas there will be no relay station at all. Hence, it is very difficult for some users to find the RS to build the two-hop connection. When the random area gets larger, as in Figure 14.24,

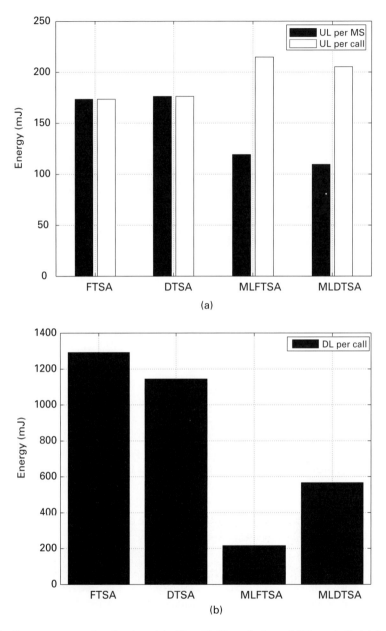

Figure 14.21 Energy consumption for class A traffic with $P_H = 0$ at $\rho = 2$ Erlangs/cell: (a) for uplink and (b) for downlink.

this effect becomes more obvious. Therefore, the energy reduction is not as dramatic as the fixed relay station. For uplink energy per call, as shown in Figure 14.24(a), RRS uses slightly more energy compared to FRS. As shown in Figure 14.24(b), for downlink energy per call, RRS uses much more energy especially for high-data-rate traffic, because

14.5 Simulation results and discussions

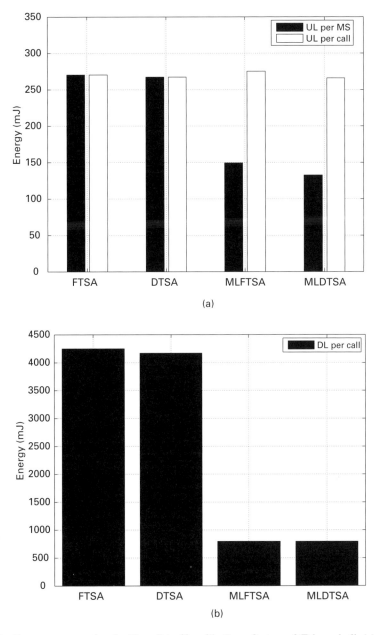

Figure 14.22 Energy consumption for Class B traffic with $P_H = 0$ at $\rho = 2$ Erlangs/cell: (a) for uplink and (b) for downlink.

of much stronger interferences from existing calls. In order to maintain those single-hop calls that are far away, the BS needs to use high power.

Even though RRS performs worse than FRS in terms of blocking and dropping probabilities and energy consumptions, in some cases FRS may not be implementable due to

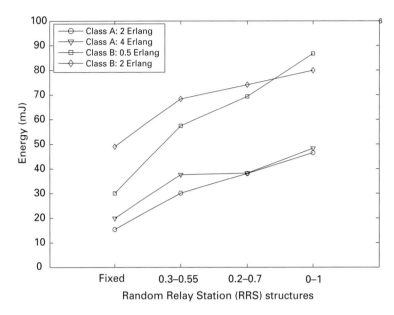

Figure 14.23 Uplink energy consumption per MS for MLDTSA with 6 RSs using RRS.

physical limitations and an unstable optimal RS location. RRS may need to be used to provide a flexible choice of implementation. This is why we are still interested in looking at the RS selection and its performance.

14.6 Conclusion

In this chapter, we have studied energy consumption for various time-slot allocation schemes for TDD-CDMA cellular networks. We have compared the performances of two conventional single-hop schemes, namely, FTSA and DTSA, with multi-hop schemes, ML-FTSA and ML-DTSA, for multi-hop cellular networks. In these multi-hop schemes each time slot supports several calls with different links as long as the transceiver does not send and receive data at any given time. In addition, the main feature of ML-FTSA or ML-DTSA is that it allows allocating one time slot to support both uplink and downlink connections. In addition, a random relay structure is proposed to examine the energy consumption compared to a fixed relay structure.

We have also studied and compared the energy of these schemes using two traffic scenarios – low-data-rate traffic and high-data-rate traffic. The results show that in every traffic scenario multi-hop schemes significantly reduced call dropping and call blocking probabilities and hence, increased system capacity. The results indicate that energy consumption decreased dramatically for multi-hop links both in the cases of high traffic and low traffic. We have further studied the dependence of energy consumption on the degree of randomness of relay stations in multi-hop networks and found that the more structured the relay network is, the better the network saves energy, such as with the fixed relay

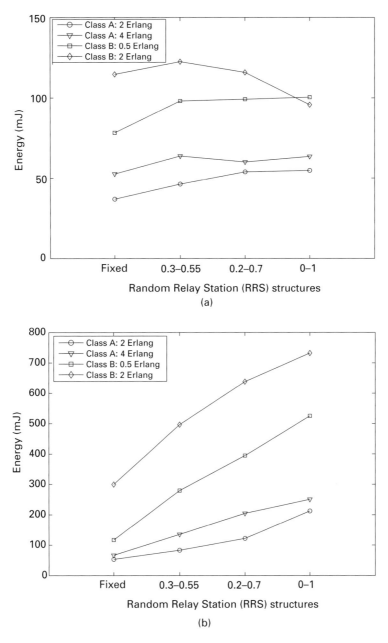

Figure 14.24 Energy consumption per link for MLDTSA with 6 RSs using RRS: (a) for uplink and (b) for downlink.

station structure. Increasing the degree of randomness of relay stations made choices more flexible at the cost of network performance. However, even with the maximum level of randomness, a multi-hop structure is observed to be superior to conventional single-hop networks.

For future study, we suggest investigating energy reduction if the number of multi-hop links is increased to three or more. In addition, investigation to find the optimal number of random relay stations is also important.

References

[1] *3rd Generation Partnership Project: Opportunity Driven Multiple Access*, 3GPP, Tech. Rep. Tech. Spec. 3G TR 25.924 version 1.0.0 (1999–12), 1999.

[2] P. Lin, W.-R. Lai, and C.-H. Gan, "Modeling opportunity driven multiple access in UMTS," *IEEE Transactions on Wireless Communications*, vol. 3, no. 5, pp. 1669–1677, September 2004.

[3] E. Kudoh and F. Adachi, "Transmit power efficiency of a multi-hop virtual cellular system," in *Proc. of IEEE Vehicular Technology Conference Fall-2003*, Orlando, Florida, vol. 5, pp. 2910–2914, October 2003.

[4] R. Min and A. Chandrakasan, "Top five myths about the energy consumption of wireless communication," *Mobile Computing and Communications Review*, vol. 7, no. 1, pp. 65–67, 2003.

[5] J. Y. Song, H. Lee, and D. H. Cho, "Power consumption reduction by multi-hop transmission in cellular networks," in *Proc. of IEEE Vehicular Technology Conference*, Los Angeles, CA, vol. 5, pp. 3120–3124, September 2004.

[6] A. Radwan and H. S. Hassanein, "Does multi-hop communication extend the battery life of mobile terminals?" in *Proc. of IEEE Global Telecommunications Conference*, San Francisco, CA, pp. 1–5, November 2006.

[7] Y. Ming and P. H. J. Chong, "Time slot allocation schemes for multi-hop TDD-CDMA cellular systems," in *Proc. of IEEE Wireless Communications and Networking Conference*, pp. 3099–3104, March 2007.

[8] M. Arabshahi and P. H. J. Chong, "High-speed multimedia services for TDD-CDMA multi-hop cellular networks," in *Proc. of IEEE IWCMC 2008*, Crete, Greece, pp. 482–487, August 2008, (invited).

[9] A. Wang and C. Sodini, "On the energy efficiency of wireless transceivers," in *Proc. of IEEE Global Telecomm. Conf.*, pp. 3205–3209, November 2005.

[10] L. S. El Alami, E. Kudoh, and F. Adachi, "Blocking probability of a DS-CDMA multi-hop virtual cellular network," *IEICE Transactions on Fundamentals of Electronics Communications and Computer Sciences*, vol. E89A, pp. 1875–1883, July 2006.

[11] S. J. Lee, H. W. Lee, and D. K. Sung, "Capacities of single-code and muticode DS-CDMA systems accommodating multiclass services," *IEEE Transactions on Vehicular Technology*, vol. 48, no. 2, pp. 376–384, March 1999.

[12] M. Lindstrom and P. Lungaro, "Resource delegation and rewards to stimulate forwarding in multi-hop cellular networks," in *Proc. of IEEE Vehicular Technology Conference-Spring 2005*, Stockholm, Sweden, vol. 4, pp. 2152–2156, 2005.

[13] S. Y. Tan and H. S. Tan, "A theory for propagation path loss characteristics in a city street-grid," *IEEE Transactions on Electromagnetic Compatibility*, vol. 37, no. 3, pp. 333–342, Aug. 1995.

[14] Blocking probability, March 22, 2010 [Online]. Available: www.javvin.com/wireless/BlockingProbability.html

15 Resource allocation for green communication in relay-based cellular networks

Umesh Phuyal, Satish C. Jha, and Vijay K. Bhargava

15.1 Introduction

The energy efficiency of wireless communication systems and their impact on the environment have been largely ignored in the past during the design and implementation of existing wireless networks. Increasing energy-consumption in these networks has been recently identified as a global problem due to its adverse effects on the environment and increasing cost of operation [1]–[5]. Cellular systems constitute a major part of wireless communications and their use in daily life is increasing more than ever [6]. Mobile devices are expected to surpass personal computers as the main web-accessing devices in the near future [7]. Therefore, mobile communications can contribute up to 15–20% of the overall energy-consumption in information and communication technologies (ICT) [7], which can no longer be disregarded.

The total energy-consumption of ICT itself is difficult to estimate because studies vary depending on the definition of ICT, the methodology used to generate the estimates, and the proportion of a device's energy consumption that is attributed to ICT [8, 9]. Several studies have suggested that the fraction of overall electricity consumption due to ICT infrastructure corresponded to around 7.8% in the European Union in 2005 [10], which is expected to rise to 10.9% by 2020. Around 3% of the world's electricity consumption is attributed to ICT, contributing to about 2% of worldwide CO_2 emission [1, 9, 11, 12]. It is, therefore, apparent that there is an urgent need to design a sustainable cellular wireless communication system by developing energy-efficient (green) technologies.

Green communication is a vast research area that covers energy saving at all layers in the protocol stack of wireless access networks, as well as in the design of wireless architectures and techniques of future-generation cellular systems [13]. In this chapter, we mainly focus on reducing the transmit energy-consumption in cellular wireless networks.

The main focus of this chapter is to explore the energy-efficient resource allocation strategies in a cellular system using a relay-based dual-hop transmission approach. Various approaches to achieve green communication in a cellular network are first discussed. Then, the advantages of and implementation issues for a relay-based cooperative transmission approach are analyzed in the context of its potential to enable green communication in cellular networks. Some performance metrics that can provide a quantitative measure of the green performance of a cellular network are also discussed.

A literature survey of notable resource allocation schemes for a relay-based cooperative cellular network (CCN) is provided to facilitate the discussion on the advantages and disadvantages of existing schemes in the context of their energy-efficiency performance. A novel resource allocation scheme is then introduced in order to address some of the identified shortcomings of the existing schemes and to optimize the energy efficiency or "green performance" (J/bit) of CCN. Performances of different schemes are evaluated and analyzed. Since the improvement of green performance comes at the cost of degraded overall system throughput in general, a trade-off between system throughput and green performance is also analyzed using multi-objective optimization approach.

15.2 Enabling green communication in cellular wireless networks

Energy saving can be achieved at various levels in cellular wireless networks. These levels can be broadly classified in three categories: component level, equipment level, and network level.

15.2.1 Component level

Each piece of equipment in a cellular system, e.g. BS and MS, consists of several components such as antennas, power amplifier, baseband processor, power supply, and other supporting components. The energy efficiency of these components has a huge impact on the overall energy-efficiency of the system [2, 7]. Various *green performance metrics* can be used to measure the energy efficiency of these components. For example, the ratio of radiated power to input power can be used as a green metric for antennas, while the ratio of effective output power to input power can be used as a green metric for power amplifiers and power supply units [7]. A significant fraction of energy is wasted as heat in the very essential components such as power amplifiers. It goes without saying that green components form the basis for achieving green communication.

15.2.2 Equipment level

BSs and MSs are the most used equipment in cellular systems. Since up to 80% of total power is consumed at the BS, its energy efficiency is the most critical in achieving green communication [1, 11]. Generally, power consumption per unit effective system throughput is the green metric used to calculate the energy efficiency of equipment such as BSs. Here, the effective throughput generally includes the frame overhead from the physical and link layers. Similarly, power consumption should also count in the energy consumption in any auxiliary devices such as a cooling unit that is needed to operate the equipment.

15.2.3 Network level

Energy efficiency at the network level depends on various factors such as deployment topology, network layout, transmission management, and resource management strategies. Since most of the existing cellular networks are optimized for high load, their performance in terms of energy efficiency is poor at low load. Network-level performance parameters may include one or more of total system throughput, total coverage area, coverage area per BS, number of subscribers served per BS, etc. Therefore, green metrics used for network-level energy efficiency can be, for example, coverage area per unit power-consumption, number of subscribers served per unit power-consumption, and useful throughput (excluding frame overhead) per unit power-consumption. Coverage area per unit power-consumption seems to be a green metric better suited for networks in rural areas, while number of subscribers served per unit power-consumption may be better suited for urban-area networks.

15.2.4 Computational complexity versus transmit-power-saving

When energy-efficient algorithms to save transmit power are implemented at the equipment level such as in the BS and MS, the computational complexity introduced by the algorithm may demand an increased energy requirement for signal processing as well as encoding and decoding, resulting in decreased network-level energy-efficiency. Therefore, the trade-off analysis of computational complexity versus saving in transmit-power requirement is an important issue. Emphasis should be given to designing low-complexity schemes resulting in easy implementation as well as a gain in overall network-level energy efficiency.

15.3 Relay-based green CCN

A number of approaches at the network level that may reduce the energy consumption and thus enable green communication in cellular networks have been under discussion. For example, discontinuous transmission by the BS where some of the hardware components are switched off, multi-hop transmission approaches, and cooperative multipoint architectures are being discussed as energy-saving measures [7]. One of the most popular approaches is relay-based cooperative transmission, which can reduce the overall energy-consumption without requiring many changes in the existing infrastructure of cellular systems [14, 15].

In a traditional cellular network shown in Figure 15.1(a), each BS serves a number of MSs within its service area. On the other hand, in a relay-based CCN, one or more relay stations (RSs) provide an alternative path to transmit data from BS to MS. In such a network, the BS and RSs cooperate to serve the MSs within their service area as illustrated in Figure 15.1(b).

The method of relaying was first introduced by Van der Meulen in 1971 [16], and was later studied from an information theoretic point of view by Cover and El Gamal in 1979 [17]. Relay paths can offer better communication channels when the direct path (BS–MS) is not reliable due to, for example, dead spots created by adverse channel conditions (e.g. blockage by an obstacle and high attenuation at cell edge), or a hotspot created by higher traffic. Cooperative transmission also offers new frontiers for research, particularly in the area of resource allocation. For example, resources now need to be smartly shared between the BS and RSs in order to achieve optimal utilization.

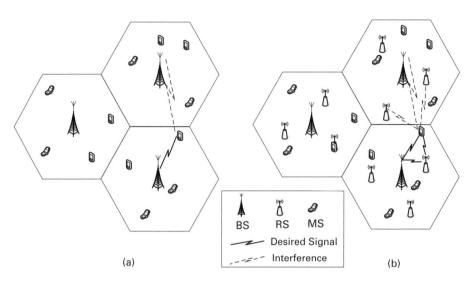

Figure 15.1 Cellular networks with ideal hexagonal cells: (a) traditional, and (b) relay-based CCN. Relays not only offer alternative communication paths, but also cause more intercell interference to near-cell-edge users.

A relay-based cooperative transmission approach seems to be cost-effective and more practical to enable green communication in cellular networks compared to other approaches such as micro-cell, pico-cell, or femto-cell and cell zooming (CZ). Micro-, pico-, and femto-cells require the installation of new BSs and reliable wired backhaul connection. Femto-cells provide an efficient way of improving performance for indoor users; however, they may not improve the performance if only outdoor BSs are deployed because the major bottleneck is the path loss from outdoor BSs to indoor MSs [18]. CZ suffers from several implementation issues in practice [19]. First, CZ is possible only with a very close coordination among neighboring cells, which usually requires an extra entity called a CZ server connected to all these cells. Second, the effectiveness of a CZ entirely depends on the knowledge of traffic-load fluctuation in the network, which is difficult to trace and feed back to the CZ server concisely and timely in practice. Third, it can also cause problems like increased inter-cell interference and coverage holes.

15.3.1 Implementation issues and challenges

Some of the major implementation issues and research challenges in enabling green communication by the cooperative relay-based transmission approach are as follows:

1. *Relay deployment cost versus green performance enhancement*: There is always the question: will using relay(s) increase performance for the current scenario? The answer depends on various factors. There may be cases when the channels between BS–MS, BS–RS, and RS–MS are of similar quality. It is obvious that a relay does not help much in such a scenario. Relays are useful only when they can offer better channel conditions compared to the direct (BS–MS) path.

Enabling relaying in a cellular network also requires a software (and possibly hardware) upgrade at the BS and MSs. Moreover, while relay-based cooperative transmission offers several advantages, the deployment of RSs causes more inter-cell interference to near-cell-edge users, as illustrated in Figure 15.1. It is necessary for the operator that performance enhancement due to relaying is worthwhile relative to the investment in RSs. Therefore, efficient use of relaying is necessary to improve the performance of the system in terms of green metrics (which will be briefly discussed later), network coverage, and system capacity in order to justify the investment.

2. *Protocol design for relaying*: When the answer to the above question is "yes", the next question is about the choice of relaying protocol: whether to use dumb relays, which simply amplify-and-forward (AF) the received signal; smart ones, which decode the received signal and retransmits after re-encoding and re-modulating (i.e. decode-and-forward); or something in between these two, which demodulates the received signal, removes the noise and retransmits after re-modulation without decoding (regenerate-and-forward). Since AF relays are transparent to source/destination coding and modulation schemes and have low complexity, they seem to be more practical compared with other types.

3. *Relay mobility*: In a CCN, cooperation among the nodes can be realized in various ways such as by installing fixed RSs at predefined locations, by enabling relay functionality in MSs, or by doing both. Because of the power supply constraint (battery life) and other hardware limitations of the MS, it may be difficult to enable relay functionality in the MS. Moreover, MSs need to have sufficient incentive to cooperate, for example, prioritized resource allocation in upcoming time-slots. Where there is no such incentive, there is a risk of MSs reporting a false channel state or battery life to avoid relaying. In contrast, fixed relays deployed by the wireless operator do not typically have these constraints. Network planning and routing design also become easier with such relays [20]. While installing such fixed relays, their positions are usually selected to optimize the system performance. Moreover, installation of fixed RSs is a trade-off point between the following two options: i) very costly hardware changes by installing new BSs and using more spectrum, and ii) low-cost but not very reliable ad hoc relaying by the MSs. Therefore, in this chapter, we mainly focus on CCN with pre-installed fixed AF relays to achieve green communication in a more realistic way.

4. *Relay positioning and selection*: The performance of a relay-based cooperative communication system depends highly on the location of the relays. Therefore, for a fixed relay–based system, relays should be installed such that the system performance is optimized. If the source decides to use relay-based transmission when there are multiple potential RSs present, then smart relay selection to optimize system performance is a crucial step.

Efficient relay positioning and selection strategies are necessary to harness the potential benefits of cooperative transmission in order to enhance the energy efficiency of CCN such as by improving frequency- and/or code-reuse in a cellular network. To illustrate this, let us take an example network scenario, shown in Figure 15.3. MS_1 is outside the BS transmission range. When RS_1 and RS_2 are positioned such that they are not within the transmission range of each other, the BS can transmit data to RS_2 using the same frequency band (or codeword) used in the RS_1–MS_1 transmission. Note, however, that if the data from the BS is intended for MS_2 via RS_2, then MS_2 should not be inside the transmission range of RS_1 and the BS at the same time.

5. *Resource distribution between BS and RS*: Since the total transmit power is limited in a cellular system due to regulatory constraints, it must be smartly distributed among the BS and RSs. Therefore, the design of an efficient resource allocation scheme can be one of the major ways to enable green communication in CCN. At low-traffic demand, temporarily switching off network elements is a state-of-the-art solution to save energy [7]. However, it may not be effective in modern cellular networks where the traffic load remains consistently higher. Energy-efficient resource allocation is a potential alternative to save energy in such networks. Most of the work in the literature so far focuses on optimizing the spectral efficiency and transmission reliability. Developing a resource allocation strategy to maximize the green performance of CCN is an open research challenge which demands more research.

6. *Synchronization among BS, RS, and MS*: Synchronization among the operation of different elements is essential in CCN. In this relay-based system, an MS may receive multiple copies of a signal via multiple (direct and relay) paths. Frequency synchronization is required for the successful detection and demodulation of those copies, whereas time synchronization is required for successful combining. *Code synchronization* is essential for decoding the received copies of a signal if *network coding* is implemented by the relays. A centralized synchronization mechanism may be implemented because the BS can coordinate all activities within a cell. Synchronization in cellular networks is a well-explored area in the literature and is out of the scope of this chapter.

7. *End-user node complexity*: From the above discussions, it is clear that the implementation of a relay-based cooperative approach increases the complexity of BS and/or MS nodes, as they need to decide whether to transmit via relay or not. If they decide to use relaying, they need to be able to smartly select relay(s). If MSs are also allowed to function as relays, they need to be upgraded with such a capability. Finally, at the receiver end, the MS needs to be able to make good use of multiple copies of data obtained via direct and/or relay paths and effectively combine them in a useful manner. Because of limited computational capacity and energy resources available at end-user MSs, these requirements may significantly increase the MS node's complexity.

15.3.2 Advantages of fixed relay-based CCN

The main advantage of a fixed relay-based approach is that it requires little to no hardware modification in the existing cellular architecture. Hence it is more practical as only software upgrades at the BS and MSs may be sufficient. In addition, the deployment of fixed relays is cheaper and includes less planning than installing new BSs. Therefore, the cooperative approach is more economical and feasible compared to micro-, pico-, and femto-cell approaches to implement a green cellular system [2]. Moreover, the cooperative technique is an easier way to implement CZ [19].

The relay-based approach shortens the propagation distance between nodes, reducing the required transmission power, and hence providing the potential to save energy. In addition, RS installation does not need expensive backhaul links which are necessary for BSs.

As discussed earlier, frequency- and/or code-reuse can be improved by using relay-based communication. Other inherent benefits of fixed relay-based cooperative transmission include increases in system throughput, transmission reliability, and network coverage [21]. Due to these advantages of the fixed relay-based cooperative technique, it has been recently envisioned as one of the potential candidates for future-generation green cellular networks.

15.3.3 Green performance metrics for resource allocation

Performance metrics are indicators of the efficiency and effectiveness of a scheme or design approach to achieve an objective. Understanding and selection of these metrics are vital for the quantitative evaluation of the scheme or design approach. In the context of the green performance of cellular networks, the energy-efficiency metrics can be defined at component, device, and system/network levels [2]. The metrics at the component and device levels indicate the energy-efficiency performance of individual components and devices, and have been well established [7]. In this chapter, we are interested in the energy-efficiency metrics for resource allocation at the system/network level.

System performance per unit energy or power-consumption is a vital green metric at network level. For example, power consumption per unit coverage area of a cell (W/m^2), power consumed per unit achieved system throughput (J/bit), and the fraction of total power saved by an energy-aware scheme, are green metrics at the network level. A detailed discussion on green metrics can be found in [2, 7], and the references therein. We consider the power consumption required for unit system throughput (J/bit) as a basis for optimizing the resource allocation. Minimizing J/bit enhances green performance; however it may result in overall poor system throughput. Therefore, this green metric combined with a quality-of-service (QoS) metric which guarantees a predefined minimum data rate or signal-to-noise ratio (SNR) to each MS seems to be a more practical basis for resource allocation.

15.4 Resource-allocation schemes for CCN: a brief survey

In this section, we review the most notable state-of-the-art resource-allocation schemes for the relay-based CCN in the literature. We then discuss the approaches and objectives of these schemes and analyze their effectiveness towards energy-efficiency performance. Resource allocation in such networks can be optimized to achieve various objectives such as system throughput maximization, transmission power minimization, QoS provisioning, and energy efficiency. Generally, a trade-off between these objectives is needed in practice. To facilitate the discussion, we start with the general classification of these schemes as shown in Figure 15.2, where they are broadly classified into three major

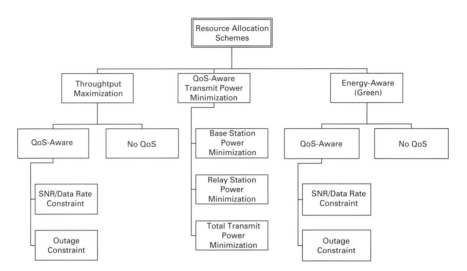

Figure 15.2 Classification of resource allocation schemes for CCN.

categories: throughput maximization, QoS-aware transmit power minimization, and energy-aware green schemes.

Although a similar classification is possible for the allocation of other network resources such as frequency bands, time slots, and codewords, we mainly focus on the allocation of transmit power in the following discussion.

15.4.1 Throughput maximization schemes

Modern cellular networks consistently face high traffic demand. As a result, capacity maximization is of prime importance for these networks. These schemes maximize the overall system throughput and generally use all of the available transmit power [22]. They are very effective in fulfilling the high traffic demand in the system. However, such schemes favor MSs with better direct and/or relay channels and hence may not be fair to all users in terms of data rate and outage. To combat this problem, throughput maximization schemes with QoS provisioning exist in the literature (such as in [23]) which maximize the system throughput and at the same time guarantee either a predefined data rate (or SNR) for each user or a predefined system outage. The schemes with QoS provisioning, however, have a lower overall system throughput than those without QoS provisioning. Neither type of scheme considers the energy efficiency of cellular networks. Therefore, in terms of the green performance metric (i.e. J/bit), their performance can be unsatisfactory.

15.4.2 QoS-aware transmit power minimization schemes

The main objective of schemes in this category is to reduce power consumption in the system rather than maximizing the system capacity. They minimize the transmit power (BS, RS, or total power) providing a QoS guarantee (data rate or SNR) at each MS.

Jointly minimizing the sum of BS and RS powers (such as in [24]) provides better results compared to minimizing BS or RS power alone, at a cost of computational complexity. These schemes perform better than throughput maximization schemes in terms of the green performance metric. However, the overall system throughput in this case is usually lower than that of throughput maximization schemes.

15.4.3 Energy-aware green schemes

The energy efficiency of a cellular system can be best achieved by directly optimizing the green performance metric (J/bit) during the resource-allocation process. Energy-aware green resource allocation is a relatively new research concept. Only a little work on saving energy in a CCN has been reported in the literature (e.g. [14, 15], and [25]). In [14], MSs are partitioned into groups. Only one RS is selected for the transmission in the next time-slot among those RSs for which at least one user in their group has BS–RS SNR higher than a threshold value (minimum end-to-end SNR required at the MS). The idea is to switch off those RSs whose first-hop transmission cannot guarantee the end-to-end SNR, and save transmit power without increasing outage probability. However, since at most one RS is active at a time, the overall system throughput and fraction of users served at a time may be very low in practice.

In [15] and [25], the authors propose schemes to maximize the network life by minimizing the energy consumption. Based on the channel conditions and power requirement, [15] propose a scheme that selects between direct non-cooperative transmission or relay-based cooperative transmission, or decides to postpone the transmission for possible future better channel conditions. The authors in [15] show that cooperation can reduce the energy consumption in the network, while satisfying the QoS performance in terms of maximum allowable system outage. On the other hand, [25] proposes a scheme that exploits the existence of a BS–MS path. However, with the proposed scheme, the increase in lifetime and energy saving comes at the cost of increased outage. In addition, contrary to the assumption in [25], relay-based communication is required for users at the cell-edge that do not have a strong direct path in practical scenarios. When the direct path is in the deep fade, the users are hidden from the BS, giving the opportunity for the BS to transmit data to other users in the second time-slot reusing the resources, as discussed in Section 15.3. None of these existing schemes optimizes the green performance metric (J/bit). Moreover, the effect of energy saving on the overall system throughput has not been discussed. Therefore, more research is required to develop an efficient green resource-allocation scheme for CCN.

15.5 Design of a green power allocation scheme

In this section, we introduce a novel energy-efficient (green) power allocation scheme for a relay-based CCN. This scheme minimizes the required transmit power per unit achievable throughput (i.e. J/bit) to optimize the green performance of the system and at the same time it guarantees a QoS in terms of end-to-end data rate required by each MS [26].

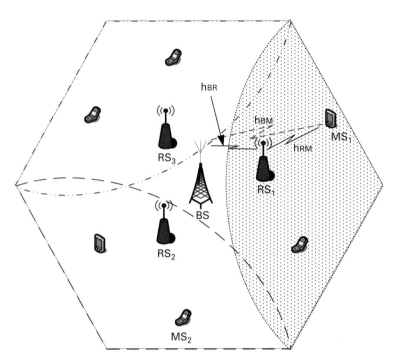

Figure 15.3 An ideal hexagonal cell of a relay-based CCN. The shaded region shows the service area (transmission region) of fixed relay RS_1 which contains the nodes of our interest.

15.5.1 System model

Let us consider a CCN where relays help a source node (i.e. a BS) in transmitting its data to an intended MS. More specifically, let us study a cellular scenario where the BS is fixed and a cell-edge MS is served by one of the multiple fixed RSs deployed by the operator, as shown in Figure 15.3. Multiple fixed RSs with sectorial antennas may be installed by an operator at predetermined locations in order to physically isolate the service areas of individual RSs. In contrast, there may be multiple RSs that can serve a particular BS–MS communication; however, for a less complex and practical system implementation, assume that only one RS is selected to assist any MS at a particular time [27].

Although selecting multiple relays may provide better communication opportunities, there may be a price to be paid in terms of system complexity, including the requirement of complex operations such as RSs synchronization, coordination, and joint resource allocation; in addition, the MS may need to be able to simultaneously detect and combine signals from multiple sources. Moreover, since there are a limited number of fixed RSs deployed by the operator in practice, it is reasonable to say that generally each MS finds one of the RSs much better than the others. There may be several ways to choose the "best" RS for an MS. The BS selects such an RS using the implemented relay selection strategy, which is essentially a routing issue [28]. Since relay selection is not within the main scope of this chapter, we assume that the BS chooses an RS that can potentially maximize the achievable end-to-end SNR of the MS. Interested readers are referred to

[28] and references therein for a description of the various relay selection strategies in relay-based CCNs.

Consider the downlink transmission of a two-hop two-time-slot-based half-duplex system with AF relaying. The complete transmission takes place in two stages: i) the BS broadcasts the signal towards the RS and the MS in the first time-slot, and ii) the RS amplifies the signal received in the first time-slot and retransmits it to the MS in the second time-slot. However, we assume that the direct link between the BS and the MS is in deep fade, which necessitates transmission via an RS to guarantee the QoS.

The signal received by the RS in the first time-slot is given by

$$y_R = \sqrt{P_B} h_{BR} X_B + \eta_R, \tag{15.1}$$

where P_B is the BS transmit power, h_{BR} is the channel gain of the BS–RS path; X_B is the symbol transmitted by the BS and drawn from a unit variance constellation; and η_R is the receiver noise at the RS, which is modeled as zero-mean complex Gaussian random variable with variance σ_R^2.

In the second time-slot, the RS amplifies the received signal y_R such that its transmit power becomes P_R, and forwards it to the destination MS. Therefore, the signal received by the MS in the second time-slot is given by

$$y_M = \sqrt{P_R} h_{RM} X_R + \eta_M, \tag{15.2}$$

where h_{RM} is the gain of the RS–MS channel, $X_R = y_R/|y_R|$ is the unit energy symbol transmitted by the RS, and η_M is the receiver noise at the MS modeled as a zero-mean complex Gaussian RV with variance σ_M^2.

Using (15.1) and (15.2),

$$y_M = \frac{\sqrt{P_B}\sqrt{P_R} h_{BR} h_{RM}}{\sqrt{P_B |h_{BR}|^2 + \sigma_R^2}} X_B + \eta'_M, \tag{15.3}$$

where

$$\eta'_M = \frac{\sqrt{P_R} h_{RM}}{\sqrt{P_B |h_{BR}|^2 + \sigma_R^2}} \eta_R + \eta_M \tag{15.4}$$

is the total noise at the MS receiver and has a zero-mean complex Gaussian distribution. From (15.3), the SNR at the MS is given by

$$\gamma = \frac{P_B |h_{BR}|^2 P_R |h_{RM}|^2}{P_B \sigma_M^2 |h_{BR}|^2 + P_R \sigma_R^2 |h_{RM}|^2 + \sigma_R^2 \sigma_M^2}$$

$$= \frac{\beta_1 \beta_2 P_B P_R}{\beta_1 P_B + \beta_2 P_R + 1}, \tag{15.5}$$

where $\beta_1 = |h_{BR}|^2/\sigma_R^2$ and $\beta_2 = |h_{RM}|^2/\sigma_M^2$ are non-negative constants whose values depend on the channel conditions and noise variances. It is also worthwhile noting that, for any communication link to be possible between the BS and the MS, β_1 and β_2 cannot be zero.

Let us assume that the RS and the MS estimate the channels with the help of a pilot signal transmitted by the BS and the RS or using blind channel estimation methods and then send this information to the BS using a feedback channel, either directly or via an RS. Based on these assumptions, a green power allocation (GPA) scheme [26] is described below.

15.5.2 Green power allocation scheme

In order to find the energy-efficient power allocation between the BS and the RS, consider a scenario where the total transmit power is constrained to P_{max}. We are interested in allocating the available power between the BS and the RS such that the minimum power is spent per unit achievable system throughput and the predefined QoS is guaranteed. Therefore, the optimization problem for GPA can be formulated as

$$\underset{\{P_B, P_R\}}{\text{minimize}} \quad f(P_B, P_R) = \frac{P_B + P_R}{\log(1 + \gamma)} \quad (15.6)$$

$$\left. \begin{aligned} \text{subject to} \quad & P_B + P_R \leq P_{max} \\ & \frac{1}{2}\log(1 + \gamma) \geq R_{min} \\ & P_B > 0, \quad P_R > 0, \end{aligned} \right\} \quad (15.7)$$

where $R_{min} > 0$ is the minimum end-to-end data rate needed to fulfill the QoS requirement of the user. The factor $\frac{1}{2}$ in the second constraint accounts for the fact that the BS–MS data transmission via the RS is achieved over two time-slots. Note, however, that this factor is not included in the objective function as it simply scales the value of the objective function but does not affect the solution point of the optimization problem. When P_B or P_R is zero, from (15.5), γ will be zero. In such a case, the second constraint cannot be satisfied. Therefore, it is clear that both P_B and P_R must have positive values as indicated by the last two constraints.

Let us briefly discuss the characteristics of the objective function $f(P_B, P_R)$. Determining the convexity (or concavity) of $f(P_B, P_R)$ doesn't seem to be straightforward since γ given by (15.5) is neither convex nor concave in (P_B, P_R). This can be shown by calculating the determinant of the Hessian matrix of γ:

$$\begin{aligned} |\mathbf{H}_\gamma| &= \begin{vmatrix} \frac{\partial^2 \gamma}{\partial P_B^2} & \frac{\partial^2 \gamma}{\partial P_B \partial P_R} \\ \frac{\partial^2 \gamma}{\partial P_R \partial P_B} & \frac{\partial^2 \gamma}{\partial P_R^2} \end{vmatrix} \\ &= \begin{vmatrix} -\frac{2\beta_1^2 \beta_2 P_R (\beta_2 P_R + 1)}{(\beta_1 P_B + \beta_2 P_R + 1)^3} & \frac{\beta_1 \beta_2 (\beta_1 P_B + \beta_2 P_R + 2\beta_1 \beta_2 P_B P_R + 1)}{(\beta_1 P_B + \beta_2 P_R + 1)^3} \\ \frac{\beta_1 \beta_2 (\beta_1 P_B + \beta_2 P_R + 2\beta_1 \beta_2 P_B P_R + 1)}{(\beta_1 P_B + \beta_2 P_R + 1)^3} & -\frac{2\beta_1 \beta_2^2 P_B (\beta_1 P_B + 1)}{(\beta_1 P_B + \beta_2 P_R + 1)^3} \end{vmatrix} \\ &= -\frac{\beta_1^2 \beta_2^2}{(\beta_1 P_B + \beta_2 P_R + 1)^4}, \end{aligned} \quad (15.8)$$

which turns out to be negative. However, the Bordered Hessians [29, §3.4] of γ are given by

$$D_1 = \begin{vmatrix} 0 & \dfrac{\partial \gamma}{\partial P_B} \\ \dfrac{\partial \gamma}{\partial P_B} & \dfrac{\partial^2 \gamma}{\partial P_B \partial P_R} \end{vmatrix} = -\dfrac{\beta_1^2 \beta_2^2 P_R^2 (\beta_2 P_R + 1)^2}{(\beta_1 P_B + \beta_2 P_R + 1)^4} \qquad (15.9)$$

and

$$D_2 = \begin{vmatrix} 0 & \dfrac{\partial \gamma}{\partial P_B} & \dfrac{\partial \gamma}{\partial P_R} \\ \dfrac{\partial \gamma}{\partial P_B} & \dfrac{\partial^2 \gamma}{\partial P_B^2} & \dfrac{\partial^2 \gamma}{\partial P_B \partial P_R} \\ \dfrac{\partial \gamma}{\partial P_R} & \dfrac{\partial^2 \gamma}{\partial P_B \partial P_R} & \dfrac{\partial^2 \gamma}{\partial P_R^2} \end{vmatrix} = \dfrac{2 \beta_1^3 \beta_2^3 P_B P_R (\beta_1 P_B + 1)(\beta_2 P_R + 1)}{(\beta_1 P_B + \beta_2 P_R + 1)^5}. \qquad (15.10)$$

Since P_B, P_R, β_1, and β_2 can all have only positive values, we see that $D_1 < 0$ and $D_2 > 0$. Therefore, γ is quasi-concave in (P_B, P_R).

Since $\log(\cdot)$ is a monotonically increasing function for a positive argument, the function $l(P_B, P_R) = \log(1 + \gamma)$, where γ is given by (15.5), must be quasi-concave in (P_B, P_R) [30]. Moreover, as described below, it can be proved that $l(P_B, P_R)$ is strictly concave in (P_B, P_R) when $\gamma \geq 0.5$.

The Hessian matrix of $l(P_B, P_R)$ is given by

$$\mathbf{H}_l = \begin{bmatrix} \dfrac{\partial^2 l}{\partial P_B^2} & \dfrac{\partial^2 l}{\partial P_B \partial P_R} \\ \dfrac{\partial^2 l}{\partial P_R \partial P_B} & \dfrac{\partial^2 l}{\partial P_R^2} \end{bmatrix}$$

$$= \begin{bmatrix} \dfrac{\beta_1^2}{(\beta_1 P_B + \beta_2 P_R + 1)^2} - \dfrac{\beta_1^2}{(\beta_1 P_B + 1)^2} & \dfrac{\beta_1 \beta_2}{(\beta_1 P_B + \beta_2 P_R + 1)^2} \\ \dfrac{\beta_1 \beta_2}{(\beta_1 P_B + \beta_2 P_R + 1)^2} & \dfrac{\beta_2^2}{(\beta_1 P_B + \beta_2 P_R + 1)^2} - \dfrac{\beta_2^2}{(\beta_2 P_R + 1)^2} \end{bmatrix}.$$

(15.11)

The first-order leading principal minor of the matrix $\mathbf{H}_l$ (which is the top-left element) is always negative because both β_2 and P_R are positive, which results in the first term being smaller than the second term. The second-order leading principal minor is the determinant of $\mathbf{H}_l$, which can be written as

$$|\mathbf{H}_l| = \dfrac{\beta_1^2 \beta_2^2 (1 - \psi)}{(\beta_1 P_B + 1)^2 (\beta_2 P_R + 1)^2}, \qquad (15.12)$$

where

$$\psi = \frac{(\beta_1 P_B + 1)^2 + (\beta_2 P_R + 1)^2}{(\beta_1 P_B + \beta_2 P_R + 1)^2}. \tag{15.13}$$

Therefore, we need $\psi < 1$ for $|\mathbf{H}_l|$ to be positive, which translates to the condition: $2\beta_1 \beta_2 P_B P_R > 1$. This condition is satisfied when $\gamma \geq 0.5$. In such a case, the Hessian becomes negative-definite [29, §3.2.2], and thus $l(P_B, P_R)$ will be strictly concave in (P_B, P_R) [31, §3.1.4].

In any practical system, SNR must be much higher than 0.5 for successful communication. Furthermore, the objective function in (15.6) is a positive quasi-convex function and the feasible region defined by (15.7) is a convex set. Standard algorithms exist in the literature to solve such an optimization problem efficiently, for example, see [31, §4.2.5]. Note that when the optimization problem (15.6)–(15.7) is infeasible, such an event contributes to QoS outage since the QoS guarantee cannot be provided by the available power budget.

15.5.3 Performance analysis of GPA scheme

Let us evaluate the performance of the GPA scheme described above and compare it with that of other selected power-allocation schemes in the literature. For performance evaluation, consider the following two metrics: i) *green performance metric* defined as total power required to achieve unit end-to-end data rate (J/bit), and ii) *QoS outage* defined as the probability that the predefined throughput requirement for QoS guarantee cannot be achieved in a two-time-slot transmission cycle. Obviously, lower values of both of these metrics are desirable to enable green communication and to ensure better QoS. For comparison, we consider the following power-allocation schemes:

1. *Throughput maximization power allocation (TMPA)*: Power allocation to maximize the achievable system throughput has been extensively studied in the literature. In such a scheme, the optimal power allocation between BS and RS is generally found by maximizing $\log(1 + \gamma)$ subject to the same constraints as (15.7). Therefore, the throughput maximization power-allocation solution can be found by solving a convex optimization problem.
2. *Uniform power allocation (UPA)*: This is one of the easy-to-implement schemes widely used to calibrate the performance of other schemes. It is also known as equal power allocation because all the transmitting nodes and/or frequency channels are provided with equal transmit power [32]. For our scenario, this means that both BS and RS transmit powers will be $P_{max}/2$.
3. *GPA with no QoS provisioning (GPANQ)*: Let us also consider the GPA without QoS guarantee for comparison. The optimization problem for this case will be similar to

(15.6)–(15.7) but without the data rate constraint, i.e.

$$\underset{\{P_B, P_R\}}{\text{minimize}} \quad f(P_B, P_R) = \frac{P_B + P_R}{\log(1+\gamma)} \qquad (15.14)$$

$$\left.\begin{array}{ll}\text{subject to} & P_B + P_R \leq P_{max} \\ & P_B > 0, \quad P_R > 0.\end{array}\right\} \qquad (15.15)$$

Since a constraint is relaxed in GPANQ compared to GPA, it is easy to see that the energy efficiency obtained by GPANQ serves as a bound for the energy efficiency of the GPA scheme. In other words, J/bit for GPA in general will be equal to or higher than J/bit for GPANQ.

15.5.4 Adaptive interrupted transmission

Since GPA, TMPA, and UPA possess a QoS constraint in terms of the minimum guaranteed throughput, it seems reasonable for these schemes to interrupt and suspend the transmission in situations when the QoS requirement cannot be guaranteed. By doing so, these schemes can further improve their energy efficiency without introducing added outage. Therefore, let us assume that GPA, TMPA, and UPA dynamically decide whether or not to carry out relay-based transmission based on their instantaneous channel condition. It will soon be clear that adapting the dynamic transmission interruption, these power allocation schemes can provide better energy efficiency compared to GPANQ.

15.5.5 Simulation results

Now, we present some simulation results demonstrating the performance of the different power-allocation schemes discussed above. The channel gains h_{BR} and h_{RM} are modeled as zero mean i.i.d. Rayleigh variables assumed to be constant over the period of two time-slots. The effect of log-distance-based path loss [33, §4.9.1] is taken into account by changing the variance of the channel gains. The list of various simulation parameters and their corresponding values is given in Table 15.1. The numerical solutions of the power-allocation optimization problems are obtained using the interior point method.

First, we study the effect of the position of the RS on the performance of the considered power-allocation schemes. For this purpose, we simulate scenarios for different BS–RS distances keeping the BS–MS distance fixed at 1 km. We move the RS along the path from BS to MS and plot the corresponding results in Figures 15.4–15.6. It can be observed in Figure 15.4 that the sensitivity of green performance metric[1] to the change in relay position is less for GPA, TMPA, and GPANQ. In contrast, this metric for UPA significantly depends on the relay position, and is better when the RS is near the middle of the distance between BS and MS. This is because GPA, TMPA, and GPANQ possess the ability to adapt their power allocation to the channel gains of BS–RS and RS–MS links. This ability is lacking in UPA because UPA assigns a static transmit power regardless of the channel conditions. Note that the green performance metric for GPA is even better

---

[1] Note that in all the simulation results, throughput is normalized by bandwidth and, therefore, the unit of this metric is J/(bit/Hz).

Table 15.1. Simulation parameters

Parameter		Value	
		Range	Default
Cell radius (cell-edge BS–MS distance)			1 km
BS–RS distance		250–750 m	500 m
Path-loss exponent			3.0
Carrier frequency			2 GHz
Channel bandwidth			5 MHz
Receiver noise figure			5 dB
QoS data-rate requirement	R_{min}	0.75–2.0 bit/s/Hz	1.5 bit/s/Hz
Total transmit power constraint	P_{max}	4.5–7.5 W	6.0 W

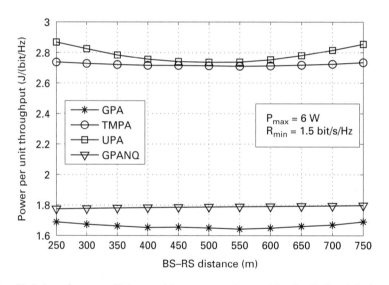

Figure 15.4 Variation of energy efficiency with respect to relay position for QoS satisfied users.

than that of GPANQ, thanks to the adaptive interrupted transmission method described in Section 15.5.4.

For a similar reason, it is obvious that the QoS outage must be higher for UPA, which is verified by Figure 15.5. The QoS outage for UPA rises sharply when the RS is moved away in either direction from halfway between the BS and MS. In contrast, GPA and TMPA are robust against the change in RS position and the outage does not change much for a wider range of relay positions. Note that system outages for GPA are observed to be equal to that of TMPA. This is because as long as TMPA can provide the QoS guarantee within the prescribed power budget, GPA can also do it. GPA raises its transmit power to provide the QoS guarantee even at the cost of energy efficiency if it is possible to do so without violating the power budget constraint. Note also that the QoS outage for

15.5 Design of a green power allocation scheme

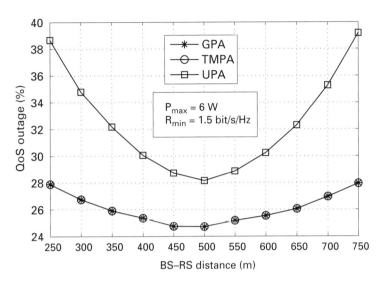

Figure 15.5 Variation of QoS outage with respect to relay position.

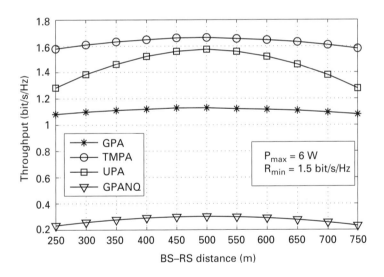

Figure 15.6 Variation of average system throughput with respect to relay position.

GPANQ is not shown in the plot. This is because there is no QoS guarantee in GPANQ and therefore QoS outage is not defined for this scheme.

From the above results, energy efficiency is observed to be much better for GPA compared to other schemes for all relay positions without an added penalty in the QoS outage. As discussed earlier, due to the absence of the QoS constraint, better energy efficiency may be achieved by GPANQ in the absence of adaptive interruption in transmission. However, GPANQ suffers from decreased system throughput as illustrated by Figure 15.6, where the variation of average system throughput provided by different

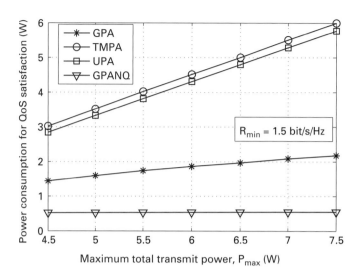

Figure 15.7 Total transmit power-consumption for QoS satisfaction by different power-allocation schemes.

power-allocation schemes is plotted. It is clear that other schemes perform much better than GPANQ in providing system throughput. Therefore, GPANQ may not be a good choice for commercial systems. Note that the average system throughput is less sensitive towards the position of relay. It is also worthwhile to note that average system throughput is less than R_{min} because of the existence of QoS outage events where transmission is suspended and no throughput is achieved. Figure 15.6 further shows that average throughput is better for all schemes when RS is about midway between BS and MS.

It may be argued that throughput is higher for TMPA and UPA; however, the total power-consumption is also much higher for those schemes. This is demonstrated in Figure 15.7 where total power-consumption by various power-allocation schemes is plotted against various values of power budget, P_{max}. Since all of the power-allocation schemes perform better when the relay is positioned near the middle of the distance between BS and MS, for fair comparison we take the BS–RS distance (equal to the RS–MS distance) of 500 m for this and subsequent simulations. A linear increase in total power-consumption with P_{max} is observed in Figure 15.7 for GPA and TMPA. Although the power consumption of GPA also increases almost linearly with P_{max}, the rate of increase for GPA and TMPA is almost four times higher compared to GPA. This results in a higher performance gap for higher P_{max}. The power consumption of GPANQ is the least among the schemes considered and is almost invariant with P_{max}.

Figure 15.8 shows the variation of the energy-efficiency metric (J/bit) against P_{max} for different schemes. The figure further confirms that GPA uses less power to deliver unit throughput compared to TMPA and UPA for all values of P_{max}. For an increase in P_{max}, the performance gap between GPA and TMPA increases. On the other hand, the gap between GPA and GPANQ decreases with an increase in P_{max}. This is because the QoS outage decreases with the increase in P_{max} resulting in less events of suspended transmission, as depicted in Figure 15.9. As shown by the plot in Figure 15.9, the outage

15.5 Design of a green power allocation scheme

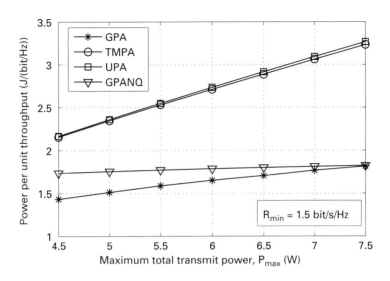

Figure 15.8 Average power per unit throughput for different P_{max}.

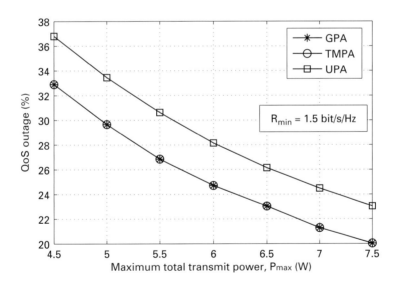

Figure 15.9 QoS outage for different P_{max}.

for GPA is less than that for UPA for all values of P_{max}. This reconfirms that GPA provides better energy efficiency without introducing any outage penalty.

The variation of the performance parameters with a change in the QoS constraint (i.e. the required minimum throughput) for a fixed P_{max} is depicted in Figures 15.10–15.12. As expected, a more stringent QoS constraint results in an increased outage probability, which is confirmed by Figure 15.10. However, GPA outperforms other schemes even with a significant increase in the throughput requirement.

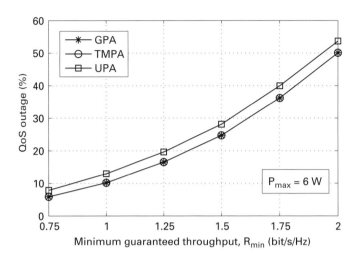

Figure 15.10 Variation of QoS outage for different R_{min}.

Figure 15.10 further shows that the outage probability increases more rapidly for a higher throughput requirement. The results are plotted without making any assumptions on the acceptable outage limit. This figure also presents the trade-off between the two QoS parameters, namely, throughput requirement and acceptable outage probability of the system, as described in the following example.

Example 1: Figure 15.10 can be used to find the upper bound on the throughput guarantee that can be satisfied for a certain outage limit and vice versa for a given P_{max}. For example, for $P_{max} = 6$ W and a maximum outage probability of 10%, UPA can handle the throughput requirement of up to 0.85 bit/s/Hz, whereas GPA can satisfy the requirements of as much as 1 bit/s/Hz in the simulation scenario considered.

For a fixed P_{max}, the average throughput of UPA and TMPA monotonically decreases with an increase in R_{min} because of the increase in the QoS outage. This is depicted in Figure 15.11. In contrast, the average throughput for GPA first increases with an increase in R_{min}, attains a peak, and then starts decreasing. For lower R_{min}, the throughput offered by GPA is lower in order to offer a significant enhancement in green performance compared to UPA and TMPA. For higher R_{min}, the average throughput for GPA also decreases because of the outage. Therefore, it is observed in Figure 15.12 that more power per unit throughput is required by GPA to satisfy a more stringent QoS constraint. The performance gap between GPA and TMPA decreases with an increase in throughput requirement because the energy efficiency of TMPA and UPA improves for an increased throughput requirement at the cost of rising outage and falling average throughput.

As illustrated in Figures 15.11 and 15.12, the throughput and energy efficiency offered by GPANQ are independent of R_{min}. The throughput of GPANQ is lowest among the considered power-allocation schemes, whereas the energy efficiency is better than TMPA and UPA but mediocre compared to GPA.

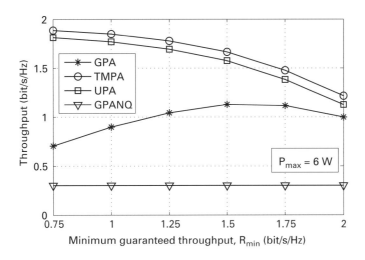

Figure 15.11 Average throughput for different R_{min}.

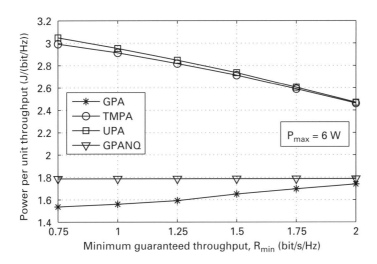

Figure 15.12 Variation of energy efficiency for different R_{min}.

15.6 Green performance versus system capacity

In Section 15.5, we have focused on the green performance of a relay-based CCN and formulated the problem to minimize J/bit. As discussed and observed in previous sections, minimization of J/bit generally degrades the throughput capacity of the network. Due to ever-growing traffic demand, network capacity has been a prime focus for most of the cellular operators. Achieving green communication at the cost of degraded overall system throughput may not be reasonable in practice. As a result, a trade-off between system capacity and energy efficiency is inevitable.

In this section, we discuss the multi-objective problem formulation method [34] that can jointly optimize the energy efficiency and system capacity. This can be achieved by minimizing the total transmit power $(P_B + P_R)$ while maximizing throughput (minimizing its reciprocal) at the same time.

Let us formulate a multi-objective optimization problem as

$$\underset{\{P_B,\, P_R\}}{\text{minimize}} \quad f_m(P_B, P_R) = \alpha\, \bar{f}_{pwr}(P_B, P_R) + (1-\alpha)\, \bar{f}_{thr}(P_B, P_R) \tag{15.16}$$

subject to (15.7),

where the objective function consists of a weighted sum of two separate functions: $\bar{f}_{pwr}(P_B, P_R)$ accounting for power consumption, and $\bar{f}_{thr}(P_B, P_R)$ accounting for achievable throughput. $0 \le \alpha \le 1$ is a trade-off parameter that decides the relative priorities given to green performance and system throughput. A value of α higher than 0.5 assigns higher priority to green transmission.

In multi-objective optimization, it is essential to transfer the individual objective functions to a dimensionless entity and normalize their values [34]. Since $0 \le (P_B + P_R) \le P_{max}$, the objective function $\bar{f}_{pwr}(P_B, P_R)$ can be formed by normalizing $(P_B + P_R)$ by P_{max}. Similarly, due to the QoS constraint, we are interested in the throughput values such that $\frac{2}{\log_2(1+\gamma)} \le \frac{1}{R_{min}}$. Therefore, $\bar{f}_{thr}(P_B, P_R)$ can be formed by first normalizing $\frac{1}{2}\log_2(1+\gamma)$ by R_{min} and then taking its reciprocal to get a fair multi-objective optimization. Hence, the multi-objective function becomes

$$f_m(P_B, P_R) = \alpha \frac{(P_B + P_R)}{P_{max}} + (1-\alpha) \frac{2 R_{min}}{\log(1+\gamma)}. \tag{15.17}$$

As discussed in Section 15.5.2, $\log(1+\gamma)$ is strictly concave for $\gamma \ge 0.5$. Since the reciprocal of a positive concave function is convex, $1/\log(1+\gamma)$ is a convex function in (P_B, P_R) for $\gamma \ge 0.5$. As α, P_{max}, and R_{min} are constants, the objective function $f_m(P_B, P_R)$ given by (15.17) is convex in (P_B, P_R). Therefore, (15.16) is a convex optimization problem.

15.6.1 Performance analysis

For the same system model and simulation environment as described in Section 15.5, Figures 15.13–15.15 demonstrate how the multi-objective optimization-based power allocation results in a trade-off between the two performance parameters – power consumption and achievable throughput – for various values of α.

It is observed in Figure 15.13 that the average power-consumption decreases with increase in the trade-off parameter α, $\alpha = 1$ corresponding to the power minimization scheme. This is also clear from (15.17). Consequently, the average system throughput decreases with increase in α and is highest for $\alpha = 0$, which corresponds to the throughput maximization scheme. This is demonstrated in Figure 15.14.

Comparing the two plots in Figures 15.13 and 15.14, it can be observed that the power-consumption curve declines sharply with an increase in α and quickly approaches its

15.6 Green performance versus system capacity

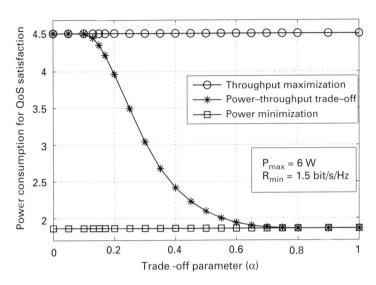

Figure 15.13 Variation of average power-consumption with respect to trade-off parameter α.

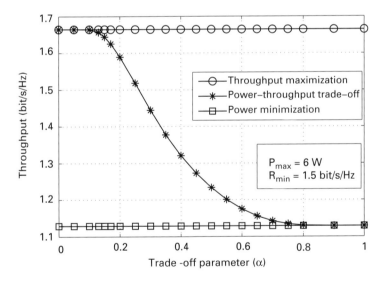

Figure 15.14 Variation of average throughput with respect to trade-off parameter α.

lower bound, whereas the throughput curve declines with a comparatively slower rate. This means that the energy-efficiency metric, J/bit, can be expected to decrease for an increase in values of α. However, as observed in Figure 15.15, this is true for an increase in α only up to a limit. Within this limit, enhanced green performance may be achieved at the cost of decreased throughput. In contrast, the energy efficiency does not improve with an increase in α beyond this limit. Therefore, only minimizing the total transmit power may not be the ultimate energy-efficient solution. Finding the right balance

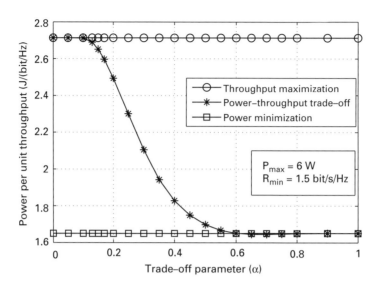

Figure 15.15 Variation of energy efficiency with respect to trade-off parameter α.

between throughput and power consumption during the power allocation can significantly improve the throughput performance of a relay-based cooperative system while maintaining the J/bit metric, which is illustrated in the following example. It should be noted that system outage is independent of α.

Example 2: For $P_{max} = 6$ W and $R_{min} = 1.5$ bit/s/Hz, Figure 15.15 shows that, though only by a small margin, the energy-efficiency metric is minimum at $\alpha = 0.7$, meaning that the system is more energy efficient than the power minimization scheme for this choice of α in this particular scenario. The energy-efficiency metric does not change much for values of α above 0.6. Compared to $\alpha = 1$ (power minimization), $\alpha = 0.6$ provides a 4.07% higher throughput (see Figure 15.14) without any penalty on the energy-efficiency metric, while $\alpha = 0.5$ provides a 12.42% increase in throughput with only a 2.92% deterioration in the energy-efficiency metric.

15.7 Conclusion

In this chapter, we have argued that implementation of relay-based cooperative transmission in cellular networks can be an economical and easier approach to enable green communication which requires minimal modification in existing cellular infrastructure. We have analyzed the benefits and implementation challenges of this approach. Transmit power required per unit achievable throughput has been considered as the main green performance metric. Then we have investigated the potential of resource allocation in enhancing green performance and studied an energy-efficient resource allocation scheme for downlink transmission in fixed relay-based cooperative cellular networks. This green power allocation (GPA) scheme minimizes the power per unit throughput while fulfilling the QoS in terms of minimum throughput guarantee. Simulations have showed that GPA outperforms other schemes such as throughput maximization and uniform power allocation, and helps in enabling green communication with a QoS guarantee.

It has also been observed that maximizing green performance generally degrades the overall system throughput. Since higher system capacity is a crucial factor in practical cellular networks, a balance between green performance and system capacity is inevitable. Therefore, trade-off between these two parameters has been analyzed by introducing a design parameter (trade-off factor) in a multi-objective optimization-based resource allocation scheme. The trade-off factor provides extra flexibility to wireless operators while designing a resource allocation mechanism to maintain a balance between green performance and system capacity.

References

[1] E. Oh *et al.*, "Toward dynamic energy-efficient operation of cellular network infrastructure," *IEEE Commun. Mag.*, vol. 49, no. 6, pp. 56–61, Jun. 2011.

[2] L. Correia *et al.*, "Challenges and enabling technologies for energy aware mobile radio networks," *IEEE Commun. Mag.*, vol. 48, no. 11, pp. 66–72, Nov. 2010.

[3] J. Hoydis, M. Kobayashi, and M. Debbah, "Green small-cell networks," *IEEE Veh. Technol. Mag.*, vol. 6, no. 1, pp. 37–43, Mar. 2011.

[4] C. Han *et al.*, "Green radio: radio techniques to enable energy-efficient wireless networks," *IEEE Commun. Mag.*, vol. 49, no. 6, pp. 46–54, Jun. 2011.

[5] R. Devarajan *et al.*, "Energy-aware user selection and power allocation for cooperative communication system with guaranteed quality-of-service," in *Proc. of IEEE 12th Canadian Workshop on Information Theory (CWIT)*, May 2011, pp. 216–220.

[6] J. Gozalvez, "Green radio technologies," *IEEE Veh. Technol. Mag.*, vol. 5, no. 1, pp. 9–14, Mar. 2010.

[7] T. Chen, H. Kim, and Y. Yang, "Energy efficiency metrics for green wireless communications," in *Proc. WCSP'10*, Oct. 2010, pp. 1–6.

[8] R. Bolla *et al.*, "Energy efficiency in the future internet: a survey of existing approaches and trends in energy-aware fixed network infrastructures," *Commun. Surveys Tuts.*, vol. 13, no. 2, pp. 223–244, 2011.

[9] C. Despins *et al.*, "Leveraging green communications for carbon emission reductions: Techniques, testbeds, and emerging carbon footprint standards," *IEEE Commun. Mag.*, vol. 49, no. 8, pp. 101–109, Aug. 2011.

[10] European Commission DG INFSO, *Impacts of Information and Communication Technologies on Energy Efficiency, Final Report*, Sep. 2008.

[11] W. Vereecken *et al.*, "Power consumption in telecommunication networks: overview and reduction strategies," *IEEE Commun. Mag.*, vol. 49, no. 6, pp. 62–69, Jun. 2011.

[12] H. Bogucka and A. Conti, "Degrees of freedom for energy savings in practical adaptive wireless systems," *IEEE Commun. Mag.*, vol. 49, no. 6, pp. 38–45, Jun. 2011.

[13] Y. Chen *et al.*, "Fundamental trade-offs on green wireless networks," *IEEE Commun. Mag.*, vol. 49, no. 6, pp. 30–37, Jun. 2011.

[14] V.-A. Le *et al.*, "Green cooperative communication using threshold-based relay selection protocols," in *Proc. of ICGCS'10*, Jun. 2010, pp. 521–526.

[15] I. Krikidis, J. Thompson, and P. Grant, "Cooperative relaying with feedback for lifetime maximization," in *Proc. of IEEE ICC'10*, May 2010, pp. 1–6.

[16] E. C. Van der Meulen, "Three-terminal communication channels," *Adv. Appl. Prob.*, vol. 3, no. 1, pp. 120–154, 1971.

[17] T. Cover and A. El Gamal, "Capacity theorems for the relay channel," *IEEE Trans. Inform. Theory*, vol. 25, no. 5, pp. 572–584, Sep. 1979.

[18] G. Fettweis *et al.*, "Field trial results for LTE-advanced concepts," in *Proc. of IEEE International Conference on Acoustics Speech and Signal Processing (ICASSP)*, Mar. 2010, pp. 5606–5609.

[19] Z. Niu *et al.*, "Cell zooming for cost-efficient green cellular networks," *IEEE Commun. Mag.*, vol. 48, no. 11, pp. 74–79, Nov. 2010.

[20] X. J. Li, B.-C. Seet, and P. H. J. Chong, "Multihop cellular networks: technology and economics," *Computer Networks*, vol. 52, no. 9, pp. 1825–1837, 2008.

[21] A. Nosratinia, T. Hunter, and A. Hedayat, "Cooperative communication in wireless networks," *IEEE Commun. Mag.*, vol. 42, no. 10, pp. 74–80, Oct. 2004.

[22] S. Ren and M. van der Schaar, "Distributed power allocation in multi-user multi-channel cellular relay networks," *IEEE Trans. Wireless Commun.*, vol. 9, no. 6, pp. 1952–1964, Jun. 2010.

[23] S. Kadloor and R. Adve, "Relay selection and power allocation in cooperative cellular networks," *IEEE Trans. Wireless Commun.*, vol. 9, no. 5, pp. 1676–1685, May 2010.

[24] R. Zhang, C. C. Chai, and Y.-C. Liang, "Joint beamforming and power control for multi-antenna relay broadcast channel with QoS constraints," *IEEE Trans. Signal Process.*, vol. 57, no. 2, pp. 726–737, Feb. 2009.

[25] S. Mousavifar, T. Khattab, and C. Leung, "Lifetime maximization with predictive power management in selective relay networks," in *Proc. of IEEE PIMRC'09*, Sep. 2009, pp. 340–344.

[26] U. Phuyal, S. C. Jha, and V. K. Bhargava, "Green resource allocation with QoS provisioning for cooperative cellular network," in *Proc. of IEEE 12th Canadian Workshop on Information Theory (CWIT)*, May 2011, pp. 206–210.

[27] R. Pabst *et al.*, "Relay-based deployment concepts for wireless and mobile broadband radio," *IEEE Commun. Mag.*, vol. 42, no. 9, pp. 80–89, Sep. 2004.

[28] V. Sreng, H. Yanikomeroglu, and D. Falconer, "Relay selection strategies in cellular networks with peer-to-peer relaying," in *Proc. IEEE VTC 2003-Fall*, vol. 3, Oct. 2003, pp. 1949–1953.

[29] M. J. Osborne, *Mathematical methods for economic theory: a tutorial*, 2007. [Online]. Available: www.economics.utoronto.ca/osborne/MathTutorial/index.html

[30] H. J. Greenberg and W. P. Pierskalla, "A review of quasi-convex functions," *Operations Research*, vol. 19, no. 7, pp. 1553–1570, 1971.

[31] S. Boyd and L. Vandenberghe, *Convex Optimization*. Cambridge University Press, Mar. 2004.

[32] U. Phuyal *et al.*, "Power loading for multicarrier cognitive radio with MIMO antennas," in *Proc. IEEE WCNC'09*, Apr. 2009, pp. 1–5.

[33] T. S. Rappaport, *Wireless Communications: Principles and Practice*. 2nd ed. NJ: Prentice Hall PTR, 2002.

[34] R. T. Marler and J. S. Arora, "Survey of multi-objective optimization methods for engineering," *Struct. and Multidisciplinary Optim.*, vol. 26, no. 6, pp. 369–395, 2004.

Part V

Green radio test-bed, experimental results, and standardization activities

16 How much energy is needed to run a wireless network?

Gunther Auer, Vito Giannini, István Gódor, Oliver Blume,
Albrecht Fehske, Jose Alonso Rubio, Pål Frenger, Magnus Olsson,
Dario Sabella, Manuel J. Gonzalez, Muhammad Ali Imran, and
Claude Desset

16.1 Introduction

The global mobile communication industry is growing rapidly. Today there are already more than 4 billion mobile phone subscribers worldwide [1], more than half the entire population of the planet. Obviously, this growth is accompanied by an increased energy-consumption of mobile networks. Global warming and heightened concerns for the environment of the planet require a special focus on the energy efficiency of these systems [2].

Many approaches to wireless energy-efficiency are limited to the power consumption of single nodes, e.g. a base station [3]–[5]. This scope is comparably easy to specify and to measure, but it fails to capture the network performance aspects (e.g. system throughput) implied by coverage and interference issues. Other methodologies are very broad, capturing the ICT industry in total [6]. Recently an assessment framework for the power consumption of deployed wireless networks has been published, the mobile energy-efficiency (MEE) network benchmarking service [7], based on metering all components of a network. However, for the energy efficiency it is not possible to directly compare, e.g. an Indian network with a Scandinavian network, therefore MEE has to introduce correction terms for the climate, for the number of base stations operated off-grid, and for the generations of equipment in the field.

However, the above approaches do not give insight into which parts of a network are most energy intensive or which provide the highest energy-saving potentials. There is a need for a simulation tool studying theoretically the effect of improvements in hardware, deployment strategies, and network management. Such a tool can be used to guide energy-efficiency research activities and it will enable network operators to design more efficient networks. The EARTH[1] project [8] is a concerted research effort to substantially reduce the energy consumption of cellular networks, and as part of its objectives, EARTH devised a holistic framework to evaluate and compare the energy efficiency of arbitrary embodiments of wireless communication networks.

In order to quantify potential energy savings, the power consumption of the entire network needs to be captured, complemented by adequate performance measures including

---

[1] EU funded research project EARTH (Energy-Aware Radio and neTwork tecHnologies), FP7-ICT-2009-4-247733-EARTH, Jan. 2010 to June 2012. https://www.ict-earth.eu.

network capacity, quality-of-service (QoS), and coverage. The EARTH energy-efficiency evaluation framework (E^3F) presented in Section 16.2 comprises methodologies and metrics that allow for a fair comparison between different networks, e.g. between a state-of-the-art 3GPP LTE cellular network, and an enhanced system with improved energy-efficiency. The E^3F provides the key criteria for assessing the overall energy-efficiency over a large area. It captures various deployments from dense urban to rural areas, as well as including busy and off-peak hours. The E^3F primarily builds on well-established methodology for radio network performance evaluation developed in 3GPP; the most important addendums, introduced in Sections 16.3 and 16.4, are to add a sophisticated power model of the base stations (BSs) as well as a large-scale long-term traffic model extension to existing 3GPP traffic scenarios. Then, using the metrics defined in Section 16.5, in Section 16.6 the E^3F is applied in order to provide an assessment of the BS energy efficiency of a 3GPP LTE network deployed within an average European country. The energy efficiency of LTE compared to that of already deployed networks is discussed in Section 16.7, and targets for the energy efficiency of future wireless networks are given.

16.2 Energy-efficiency evaluation framework (E^3F)

The widely accepted state-of-the-art method to evaluate the performance of a wireless network is to simulate the relevant aspects of the radio access network (RAN) at the system level. The computed results are, e.g. the system throughput measured in bit/s, QoS metrics, and fairness in terms of cell-edge user throughput. In order to ensure that the results generated by different RAN system simulation tools are comparable, well-defined reference systems and scenarios are specified. This is an outcome of extensive consensus work from standardization bodies, such as 3GPP [9], and international research projects, such as the EU project Wireless World Initiative New Radio (WINNER) [10], with partners from academia as well as from industry. The most recent example is the global effort in ITU to evaluate system proposals for compliance with IMT-Advanced requirements [11]. Towards that end, the EARTH E^3F builds on the 3GPP evaluation framework for LTE [9].

Figure 16.1 shows the necessary enhancements over existing performance evaluation frameworks, such that the energy efficiency of the entire network, comprising component, node, and network level, over an extended time frame can be quantified. The EARTH E^3F illustrated in Figure 16.1, identifies the essential building blocks that are necessary for an accurate holistic assessment of energy-efficiency enhancements. Although the specific realization of a system-level simulation tool largely depends on the specific problem at hand, as well as the chosen software implementation, it is envisaged that for the assessment of combinations of energy-efficiency enhancements integrated into one holistic system concept, the E^3F should capture the following aspects:

- A sophisticated power model (specified in Section 16.3), that maps the RF output power radiated at the antenna elements to the total supply power of a BS site. The

16.2 Energy-efficiency evaluation framework (E³F)

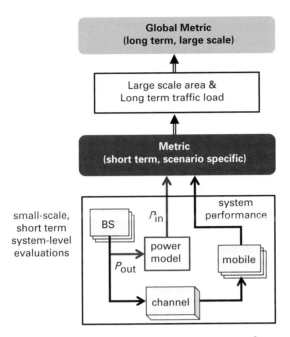

Figure 16.1 EARTH energy-efficiency evaluation framework (E³F).

power model maps the gains on the component level (e.g. an improvement in the energy efficiency of the power amplifiers) to energy savings on the entire network.
- Long-term traffic models (established in Section 16.4), that describe load fluctuations over a day and complement the statistical short-term traffic models.
- Large-scale deployment models (developed in Section 16.4) of large geographical areas are considered to extend the existing small-scale deployment scenarios.
- Energy-efficiency metrics (discussed in Section 16.5), that relate the required energy to the system performance, e.g. to the covered area or to the transmitted traffic volume.

16.2.1 Small-scale, short-term system-level evaluations

Statistical traffic models (e.g. FTP file download or VoIP calls), specific small-scale deployment scenarios (e.g. urban macro-cell consisting of 57 hexagonal cells with uniformly distributed users), and power models that quantify the power consumption of components within a node, constitute small-scale, short-term system-level evaluations (bottom block in Figure 16.1). The small-scale, short-term system-level evaluations are carried out by a system-level simulation platform, augmented by a model capturing the BS power consumption.

16.2.2 Global E³F

Modeling the power consumption of a wireless network on a countrywide or even global scale comprises different deployment areas (such as dense urban and rural areas, which are characterized by different population densities), and daily traffic variations. In order

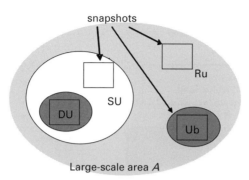

Figure 16.2 Representing a large network by typical deployment specific small-scale areas (snapshots).

to extend small-scale, short-term evaluations to a global scale, covering countrywide geographical areas and ranging over a full day or week, long-term traffic models and large-scale deployment maps are to be integrated into the E^3F. As it is infeasible to compute the performance or power consumption on this large scale, representative deployment-specific scenarios (snapshots) are simulated, as illustrated in Figure 16.2. The global assessment of network energy-efficiency comprises of the following steps:

1. Small-scale, short-term evaluations are conducted for four typical small-scale deployment scenarios, namely, dense urban (DU), urban (Ub), suburban (SU), and rural (Ru).
2. For each deployment the performance is evaluated for a representative set of traffic loads, which captures the range between the minimum and the maximum load observed in that deployment.
3. The system-level evaluations provide energy consumption and other performance metrics (e.g. throughput and QoS) for each small-scale deployment and a certain traffic load.
4. Given the daily/weekly traffic profile of each deployment, the power consumption over a day/week is generated by weighted summing of the short-term evaluations.
5. Finally, the mix of deployment scenarios that quantify the area covered by cities, suburbs, highways, and villages, yields the global set of the large-scale system energy-consumption.

The simulation-based E^3F, obtained by weighted summing over all representative "snapshot" scenarios, may be compared to the measurement-based approach of GSMA MEE [7]. Moreover, the simulation-based E^3F provides an added value of anticipating energy-saving effects *before* deployment, and may thus serve to identify the potentials of improved components, resource management algorithms, and deployment strategies for future green wireless networks. The E^3F furthermore allows us to assess the interactions of individual solutions; while some combinations may cancel each other out, others may provide synergy that yields additional gains.

16.3 Power model

It is important to note that for each deployment (dense urban, urban, suburban, and rural) different deployment strategies and hardware improvements may be beneficial. A technique that provides good gains in dense urban scenarios may not be attractive for rural areas. Likewise, for different times of day (busy vs. off-peak hours) energy-efficient network management may switch between different configurations, e.g. switching off cells and/or carriers or reducing the bandwidth of base stations. The E^3F is an enabler to derive best-practice design rules for integrated solutions that are tailored for each scenario.

16.3 Power model

This section provides a power model for various types of LTE base station. The power model constitutes the interface between component and system level, which allows quantifying how energy savings on specific components enhance the energy efficiency at the node and network level.

16.3.1 Base station power-consumption breakdown

Figure 16.3 shows a simplified block diagram of a complete BS that can be generalized to all BS types, including macro, micro, pico, and femto BSs. A BS consists of multiple transceivers (TRXs), each of which is serving one transmit antenna element. A TRX comprises a power amplifier (PA), a radio frequency (RF) small-signal transceiver section, a baseband (BB) interface including a receiver (uplink) and transmitter (downlink) section, a DC-DC power supply, an active cooling system, and an AC-DC unit (mains supply) for connection to the electrical power grid. In the following the various TRX parts are analyzed.

Antenna interface: The influence of the antenna type on power efficiency is modeled by a certain amount of losses, including the feeder, antenna band-pass filters, duplexers, and matching components. Since macro BS sites are often situated at different physical locations from the antennas, a feeder loss of about $\sigma_{feed} = 3$ dB needs to be added. The feeder loss of a macro BS may be mitigated by introducing a remote radio head (RRH), where the PA is mounted at the same physical location as the transmit antenna. Likewise, feeder losses for smaller BS types are typically not relevant.

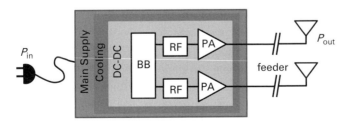

Figure 16.3 Block diagram of a base station transceiver.

Power amplifier (PA): Typically, the most efficient PA operating point is close to the maximum output power (near saturation). Unfortunately, non-linear effects and OFDM modulation with non-constant envelope signals force the power amplifier to operate in a more linear region, i.e. 6 to 12 dB below saturation [12]. This prevents adjacent channel interference (ACI) due to non-linear distortions, and therefore avoids performance degradation at the receiver. However, this high operating back-off gives rise to poor power-efficiency η_{PA}, which translates to a high power-consumption P_{PA}. Digital techniques such as clipping and digital pre-distortion [13, 14] in combination with Doherty PAs [12] improve the power efficiency and linearize the PA, while keeping ACI under control, but require an extra feedback for pre-distortion and significant additional signal processing [14]. While these techniques are necessary in macro and micro BSs, they are omitted in smaller BSs at the expense of an increased PAPR; the fact that the PA power consumption for small BS types accounts for a smaller percentage of the power breakdown allows for a higher operating back-off.

The *small-signal RF transceiver (RF-TRX)* comprises a receiver and a transmitter for uplink (UL) and downlink (DL) communication. The linearity and blocking requirements of the RF-TRX may differ significantly depending on the BS type, and so its architecture. Typically, low-IF (Intermediate-Frequency) or super-heterodyne architectures are the preferred choice for macro/micro BSs, whereas a simpler zero-IF architecture is sufficient for pico/femto BSs [15]. Parameters with highest impact on the RF-TRX energy consumption, P_{RF}, are the required bandwidth, the allowable signal-to-noise and distortion ratio (SiNAD), the resolution of the analog-to-digital conversion, and the number of antenna elements for transmission and/or reception.

Baseband (BB) interface: The baseband engine (performing digital signal processing) carries out digital up/down-conversion, including filtering, FFT/IFFT for OFDM, modulation/demodulation, digital-pre-distortion (only in DL and for large BSs), signal detection (synchronization, channel estimation, equalization, compensation of RF non-idealities), and channel coding/decoding. For large BSs the digital baseband also includes the power consumed by the serial link to the backbone network. Finally, platform control and medium access control (MAC) operation add a further power consumer (control processor).

The silicon technology significantly affects the power consumption P_{BB} of the BB interface. This technology scaling is incorporated into the power model by extrapolating on the international technology roadmap for semiconductors (ITRS). The ITRS anticipates that silicon technology is replaced by a new generation every two years, each time doubling the active power-efficiency but multiplying the leakage by three [16]. The increasing leakage puts a limit on the power reduction that can be achieved through technology scaling. Apart from the technology, the main parameters that affect the BB power consumption are related to the signal bandwidth, number of antennas and the applied signal processing algorithms. While the consumed power scales linearly with the bandwidth; MIMO signal detection scales more than linearly with the number of antennas.

Power supply and cooling: Losses incurred by DC-DC power supply, mains supply, and active cooling scale linearly with the power consumption of the other components,

16.3 Power model

Table 16.1. Base station power consumption at maximum load of a LTE system with 2 × 2 MIMO for different BS types as of 2010

			Macro	RRH	Micro	Pico	Femto
BS	Max Tx power	[dBm]	43.0	43.0	38.0	21.0	17.0
	(average) P_{max}	[W]	20.0	20.0	6.3	0.13	0.05
	Feeder loss σ_{feed}	[dB]	3	0	0	0	0
PA	Back-off	[dB]	8.0	8.0	8.0	12.0	12.0
	Max PA out (peak)	[dBm]	54.0	51.0	46.0	33.0	29.0
	PA eff. η_{PA}	[%]	31.1	31.1	22.8	6.7	4.4
	Total PA, $\frac{P_{max}}{\eta_{PA}\cdot(1-\sigma_{feed})}$	[W]	**128.2**	**64.4**	**27.7**	**1.9**	**1.1**
RF	P_{TX}	[W]	6.8	6.8	3.4	0.4	0.2
	P_{RX}	[W]	6.1	6.1	3.1	0.4	0.3
	Total RF, P_{RF}	[W]	**12.9**	**12.9**	**6.5**	**1.0**	**0.6**
BB	Radio (inner Rx/Tx)	[W]	10.8	10.8	9.1	1.2	1.0
	Turbo code (outer Rx/Tx)	[W]	8.8	8.8	8.1	1.4	1.2
	Processors	[W]	10.0	10.0	10.0	0.4	0.3
	Total BB, P_{BB}	[W]	**29.6**	**29.6**	**27.3**	**3.0**	**2.5**
DC-DC, σ_{DC}		[%]	7.5	7.5	7.5	9.0	9.0
Cooling, σ_{cool}		[%]	10.0	0.0	0.0	0.0	0.0
Mains Supply, σ_{MS}		[%]	9.0	9.0	9.0	11.0	11.0
Total per TRX chain		[W]	**225.0**	**125.8**	**72.3**	**7.3**	**5.2**
# Sectors		#	3	3	1	1	1
# Antennas		#	2	2	2	2	2
# Carriers		#	1	1	1	1	1
Total N_{TRX} chains, P_{in}		[W]	**1350.0**	**754.8**	**144.6**	**14.7**	**10.4**

and may be approximated by the loss factors σ_{DC}, σ_{MS}, and σ_{cool}, respectively. Note that active cooling is only applicable to macro BSs, and is omitted in smaller BS types. Moreover, for RRHs active cooling is also obsolete, since the PA is cooled by natural air circulation, and the removal of feeder losses σ_{feed} allow for a lower PA power consumption, $P_{PA} = \frac{P_{out}}{\eta_{PA}\cdot(1-\sigma_{feed})}$, where η_{PA} denotes the PA power efficiency.

Assuming that the BS power consumption grows proportionally with the number of transceiver chains N_{TRX}, the breakdown of the BS power consumption at maximum load, $P_{out}=P_{max}$, amounts to

$$P_{in} = N_{TRX} \cdot \frac{\frac{P_{max}}{\eta_{PA}\cdot(1-\sigma_{feed})} + P_{RF} + P_{BB}}{(1-\sigma_{DC})(1-\sigma_{MS})(1-\sigma_{cool})}. \quad (16.1)$$

The *efficiency* is defined by $\eta = P_{out}/P_{in}$, whereas the *loss* factor is defined by $\sigma = 1-\eta$. Note that the maximum RF output power per transmit antenna, P_{max}, is measured at the input of the antenna element, so that losses due to the antenna interface (other than feeder losses) are *not* included in the power breakdown.

Table 16.1 summarizes the state-of-the-art power consumption of various LTE BS types as of the year 2010. By introducing RRHs in macro BS sites, so that feeder

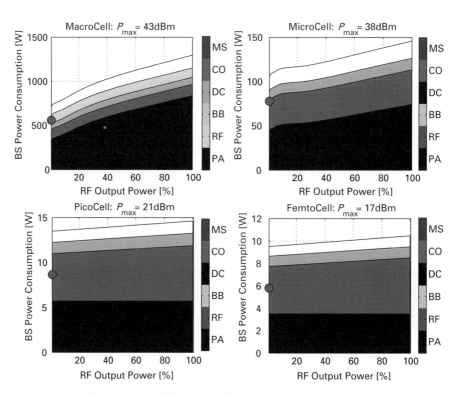

Figure 16.4 Power consumption for various BS types as a function of the RF output power. An LTE system with 10 MHz system bandwidth and 2×2 MIMO configuration is considered. Macro BSs employ 3 sectors per site. Legend: PA: power amplifier, RF: small-signal RF transceiver, BB: baseband processor, DC: DC-DC converters, CO: active cooling (only applicable to macro BS), MS: mains power supply.

losses σ_{feed} and active cooling are avoided by mounting the PA close to the transmit antenna, the power savings exceed 40%.

16.3.2 BS power consumption at variable load

In a conventional BS, the power consumption depends on the traffic load; it is mainly the PA power consumption that scales down due to reduced traffic load. This mainly happens when, e.g. the number of occupied subcarriers is reduced in idle mode operation, and/or there are subframes not carrying data. Naturally this scaling over signal load largely depends on the BS type; for macro BSs the PA accounts for 55–60% of the overall power-consumption at full load, whereas for low power nodes the PA power consumption amounts to less than 30% of the total.

Figure 16.4 shows BS power-consumption curves for an LTE system with 10 MHz bandwidth and 2×2 MIMO configuration. Three sectors are considered for macro BSs, whereas omni-directional antennas are used for the smaller BS types. While the power consumption P_{in} is load dependent for macro BSs, and to a lesser extent for micro BSs, there is a negligible load dependency for pico and femto BSs. The reason is that for

Table 16.2. Power model parameters for different BS types

BS type	N_{TRX}	P_{max} [W]	P_0 [W]	Δ_p	P_{sleep} [W]
Macro	6	20.0	118.7	5.32	93.0
RRH	6	20.0	100.0	3.4	93.0
Micro	2	6.3	53.0	3.1	39.0
Pico	2	0.13	6.8	4.0	4.3
Femto	2	0.05	4.8	7.5	2.9

low-power BSs, the impact of the PA is diminishing. Other components hardly scale with the load in a state-of-the-art implementation; although some more innovative designs could lead to an improved power scaling at low loads. As can be seen in Figure 16.4, the relations between relative RF output power, P_{out}, and BS power consumption, P_{in}, are nearly linear. Hence, the following linear approximation of the power model is justified:

$$P_{in} = \begin{cases} N_{TRX} \cdot (P_0 + \Delta_p \, P_{out}), & 0 < P_{out} \leq P_{max}, \\ N_{TRX} \cdot P_{sleep}, & P_{out} = 0, \end{cases} \quad (16.2)$$

where P_{max} denotes the maximum RF output power at maximum load, P_0 is the power consumption calculated at the minimum possible output power, assumed to be 0.1% of P_{max}, and Δ_p is the slope of the load-dependent power consumption.

Also indicated in Figure 16.4 and Table 16.2, is a sleep mode power-consumption, P_{sleep}. In future base stations, fast deactivation of components, i.e. to put them into sleep when there is nothing to transmit, is believed to be an important solution to save energy. The sleep mode power-consumption is introduced here to capture such solutions.

16.4 Traffic model

In order to provide a realistic analysis of the energy efficiency of wireless networks, it is essential to know the traffic demand to be served by the network. Thus, it is important to identify the spatial and temporal variation of the traffic demand both on large as well as on small scale.

16.4.1 Deployment areas of Europe

The geographical distribution as well as the population densities are fairly similar for most European countries; however, the Nordic countries (Finland, Norway, and Sweden) and Russia substantially deviate from the European average. Let a_d, $0 \leq a_d \leq 1$, denote the share of the area covered by deployment scenario d normalized to the total area of a given country or region. The European average of the geographical distribution a_d of the considered deployment areas and the corresponding population densities p_d in citizen/km^2 shown in Table 16.3 therefore excludes the Nordic countries and Russia. Note that in central districts of a metropolis, the population density may exceed $p_d = 20,000$ citizen/km^2, but these are omitted due to their negligible covered area a_d.

Table 16.3. Deployment areas in Europe (excluding Nordic and Russia)

Deployment d	Population density, p_d [citizen/km²]	Covered area, a_d	Relative pop. density, $p_d/\bar{p}$
Dense urban	3000	1%	12
Urban	1000	2%	4
Suburban	500	4%	2
Rural	100	36%	0.4
Sparsely populated	25	57%	–

The network planning policy of European operators concentrates on serving most of the population and not on the amount of area covered [17]. That is, 2G area coverage is almost 100%, while 3G coverage is below 40%. This reflects that sparsely populated areas are served by the minimum service level as defined by national telecommunication authorities, i.e. voice (2G) and low-speed data connection (GPRS). Following this trend, we assume that LTE is deployed in dense urban, urban, suburban, and rural areas only. For instance, German regulation forces operators to serve "only" 90% of the population with broadband access [18], which practically allows them to skip scarcely populated areas. The average European population density excluding sparsely populated areas amounts to

$$\bar{p} = \frac{\sum_d a_d p_d}{\sum_d a_d} = 250 \quad \text{in [citizens/km}^2\text{]}. \tag{16.3}$$

The population density relative to the average, $p_d/\bar{p}$, is shown in the rightmost column of Table 16.3.

16.4.2 Long-term large-scale traffic models

The objective for the long-term large-scale traffic models is to determine the average served traffic at a certain time of day for a given deployment scenario. Abstracting the models from current European cell-planning maps, the following methodology allows us to deduce the daily traffic variations in terms of the actual traffic demand per unit area:

1. define the average traffic demand per active subscriber for different terminal types;
2. by virtue of a daily traffic profile the traffic volumes per subscriber are obtained;
3. define relevant scenarios of different terminal/subscriber mixes;
4. determine the number of active users per unit area for the considered terminal/subscriber mixes;
5. given the population densities for the respective deployments, the scenario specific network traffic per unit area in [Mbps/km²] can be derived;
6. the total network traffic of an average European country is obtained by weighted summing of the scenario-specific network traffic.

16.4 Traffic model

Table 16.4. Anticipated traffic demand ranges of terminal types in Europe in 2015

Terminal type k	Average rate r_k in [Mbps]	Daily[†] [MB]	Monthly[†] [GB]
PC	0.25 → 2	256 → 2048	8 → 64
Tablet	0.125 → 1	128 → 1024	4 → 32
Smartphone	0.03125 → 0.25	32 → 256	1 → 8
Reference PC (2010)	0.03125 → 0.125	32 → 128	1 → 4

[†] For notational convenience the exact values are rounded to the closest power of 2.

Traffic demand per active subscriber

The user-generated data volume is tightly connected to operator policies and data subscriptions plans. While the amount of traffic varies from country to country, studies within the EARTH project revealed that the average rates per active subscriber are independent of the deployment scenario d. For the envisaged terminal types k we therefore propose to define a range of prospective traffic demands per subscriber r_k (in a representative Western European country), which apply to all considered deployments. We furthermore distinguish between heavy and ordinary subscribers. While the actual data rates of heavy and ordinary subscribers strongly vary due to regional differences and operator policies, it is expected that an ordinary subscriber demands half the data rate of a heavy subscriber. The range of data rates that comprises both ordinary and heavy users are as follows:

- *PC users* demand for bandwidth equivalent of providing, e.g. SDTV or even HDTV for all active users, which translates to a data range of $r_k = [0.25, 2]$ Mbps/user [19]. This range is the equivalent of average and high-end DSL demands of 2010.
- *Tablet users* demand for half the bandwidth of PC users, $r_k = [0.125, 1]$ Mbps/user.
- *Smartphone users* demand for a quarter of the bandwidth of a tablet user [19], $r_k = [31.25, 250]$ kbps/user.
- As a benchmark *reference PC users* in 2010 demand for half the traffic volume of smart phone users in 2015, $r_k = [31.25, 125]$ kbps/user. This range is equivalent to average and high-end 3G traffic in 2010 [21]–[23].

Note that these traffic demands are substantially higher than that of current 3G networks, but should be considered as expectations for well-established mobile broadband markets beyond 2015. The data rate requirements per subscriber for the terminal types considered are listed in Table 16.4. We emphasize that the figures listed in Table 16.4 represent average traffic demands; typically strong temporal and geographical deviations with respect to these average values are experienced, e.g. one or two so-called heavy users may fully utilize a cell even for extended time periods.

Daily traffic variations

Clearly, not all subscribers are always active; rather the number of active subscribers changes between busy and off-peak hours. In today's networks 10–30% of the data subscribers are active in the busy/peak hours; as a European average we assume that

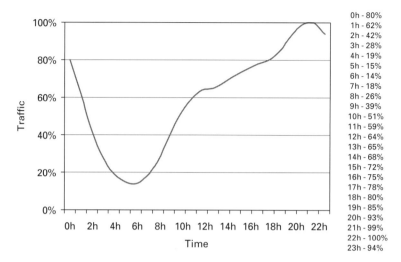

Figure 16.5 Normalized average daily data traffic profile $\tilde{\alpha}(t) = \alpha(t)/\widehat{\alpha}$, taken as a reference for an European country.

at peak hours a mobile broadband subscriber is active with probability $\widehat{\alpha}=0.16$. Thus, on average the generated traffic volume of terminal type k at peak hours of duration T yields $\widehat{v}_k(T) = \widehat{\alpha} r_k T$.

Let $\alpha(t)$, with $0 \leq \alpha(t) \leq \widehat{\alpha}$, define the daily variation of active users that captures the variations in average network traffic over a day. Based on internal surveys on operator traffic data within the EARTH project and the Sandvine report [20], the average traffic demand follows the daily variation $\alpha(t)$, illustrated in Figure 16.5. The EARTH project found that the daily traffic variations $\alpha(t)$ hardly depend of the deployment d; hence the daily traffic profile in Figure 16.5 is valid for all deployments. Aggregating the traffic demands over time gives the traffic volume per subscriber over duration T

$$v_k(t, T) = r_k \int_t^{t+T} \alpha(\tau) \, d\tau. \quad (16.4)$$

The average traffic volume over a whole day yields $v_k(24\text{h}) = r_k \bar{\alpha}$, with $\bar{\alpha} = \int_0^{24\text{h}} \alpha(\tau) \, d\tau = 0.095$. The obtained traffic volumes of the considered terminal types are summarized in Table 16.4.

Scenarios of terminal and subscriber mixes

According to the expectations of wireless Internet services, the fraction of broadband data subscribers of the whole population will increase from year to year and in the most mature European markets may reach 25% by 2015; however, the European average might be somewhat lower. Moreover, tablets, smart phones, and other mobile equipment, that are becoming increasingly popular already today, are expected to stimulate additional traffic demand in Europe; and even more so in North America, due to much lower mobile PC traffic compared to Europe. Let s_k denote the fraction of broadband data subscribers of the whole population for terminal type k, different scenarios can be constructed that reflect the expected share of mobile broadband subscribers in 2015:

- *Scenario #1:* $s_{PC}=20\%$ of the population are heavy PC users, requesting an average data rate of $r_{PC}=2$ Mbps/user.
- *Scenario #2:* $s_{PC}=20\%$ of the population are PC users, $s_{tab}=5\%$ of the population are tablet users, and $s_{fon}=50\%$ of the population are smartphone users, all of which are classfied as heavy users requesting an average data rate of $r_{PC}=2$ Mbps, $r_{tab}=0.5$ Mbps and $r_{fon}=0.25$ Mbps. This scenario serves as an upper bound on the envisaged traffic demand in 2015.
- *Scenario #3:* $s_{PC}=20\%$ of the population are PC users, $s_{tab}=5\%$ of the population are tablet users, and $s_{fon}=50\%$ of the population are smartphone users, of which one half is classified as heavy users with rates $r_{PC}=2$ Mbps, $r_{tab}=0.5$ Mbps and $r_{fon}=0.25$ Mbps, whereas the other half are classified as ordinary users with 50% lower rates, which is $r_{PC}=1$ Mbps, $r_{tab}=0.25$ Mbps and $r_{fon}=0.125$ Mbps. Based on [19], we consider this as the most relevant European scenario for 2015.
- *Scenario #4:* serves as a reference scenario for the contemporary traffic demand, where $s_{rPC}=10\%$ of the population are reference PC users in 2010, requesting a rate of $r_{rPC}=0.125$ Mbps.

Active subscribers

Given the geographical population data, the daily traffic variations and the terminal and subscriber mixes established in Sections 16.4.1, 16.4.2, and 16.4.2, the average number of active subscribers of a European country at a given time of day can be quantified. Provided that the population is served by N_{op} operators, each of which has a $1/N_{op}$ share of the total traffic volume, the average number of active subscribers of deployment d for terminal type k at time t is given by

$$U_{k,d}(t) = \alpha(t) p_d \frac{s_k}{N_{op}} \quad \text{in [subscribers/km}^2\text{]}. \qquad (16.5)$$

The number of active users $U_{k,d}(t)$ is scaled with the population density p_d. That is, taking the average population density $\bar{p}$ as a reference (see Section 16.4.1), the number of active users in dense urban, urban, suburban, and rural deployments are $\frac{p_d}{\bar{p}}=12, 4, 2$ and 0.4 times the average number of active users of the whole country, respectively.

Aggregated traffic demand

As the data volume per subscriber does not depend on the deployment scenario, the generated network traffic is proportional to the population density. Given the terminal specific data rates r_k, the number of active users in a given deployment $U_{k,d}(t)$, and the mix of terminal types, the generated area throughput of the network is determined by

$$\frac{R_d(t)}{A} = \sum_k r_k U_{k,d}(t) = \sum_k r_k \alpha(t) p_d \frac{s_k}{N_{op}} \quad \text{in [bit/s/km}^2\text{]}. \qquad (16.6)$$

Table 16.5. Estimated area throughput and number of active users at busy hours served by $N_{op} = 3$ operators in a dense urban cell in Europe of 2015

Scenarios and mixes		Peak area throughput [Mbps/km^2]	Active users per cell
#1	Heavy PCs	192	2–3
#2	Heavy PCs & tablets and smartphones	276	8–9
#3	Mix of heavy & ordinary PCs, tablets and smartphones	207	8–9
#4	Reference PC from 2010	6	1–2

Table 16.5 shows the peak area throughput $\widehat{R}_d$ and number of active users per cell at peak traffic served by $N_{op} = 3$ operators for a dense urban environment with $p_d=3000$ citizen/km^2 and 500 m inter-site distance between base stations [9].

Aggregation over time yields the total traffic volume per unit area served during a duration T, that is

$$\frac{V_d}{A} = \frac{1}{8} \sum_k v_k(t,T) \frac{p_d s_k}{N_{op}} \quad \text{in [Byte/km}^2\text{]}, \tag{16.7}$$

where the terminal specific data volume $v_k(t,T)$ is given by (16.4). Given the geographical distribution of the respective deployments a_d, the total traffic of a countrywide network is obtained by weighted summing

$$\frac{V}{A} = \frac{\sum_d a_d V_d}{A \sum_d a_d} = \frac{1}{8} \sum_k v_k(t,T) \frac{\bar{p} s_k}{N_{op}} \quad \text{in [Byte/km}^2\text{]}, \tag{16.8}$$

where the average population density $\bar{p}$ is defined in (16.3).

16.4.3 Statistical short-term traffic models

In order to model the fluctuation of the traffic in a short timescale, the packet distribution generated by the different types of application is modeled statistically. Since the same short-term traffic models per active user should be applied in all deployment areas, the traffic demands in different deployments are derived from the differences in the corresponding user density figures. A detailed description of the traffic models can be found in [9].

16.5 Green metrics

What metric we use to capture efficiency determines how we think and act. Therefore, it is important that we use metrics that guide us in the right direction. When we discuss energy efficiency we may refer to how much energy it takes to achieve a certain amount of work, or we may refer to how much work we can get achieved by using a certain amount of energy. The difference is subtle, but important, as discussed in Section 16.5.1.

In many areas where efficiency is important, such as transportation, the definition of "work" is straightforward; e.g. a vehicle, a person, or a unit of weight is moved a certain distance. In cellular networks it is not as easy to define what exactly one unit of "work" constitutes. The network provides connectivity over a certain area and it transports bits to mobile users. Users pay not only for the served number of bits but also for the possibility of using the network everywhere and anytime. Hence the area coverage provided by the network is important, even when no user is transferring any bits. In order to capture both of these aspects two different metrics will be introduced in Section 16.5.2.

16.5.1 Efficiency metrics vs. consumption metrics

A metric of energy usage can be expressed either as a *consumption index* or as an *efficiency index*. In the automobile industry both the "miles per gallon" or MPG metric and the "liters per 100 km" metric are commonly used. The MPG metric is an efficiency index, i.e. a car that consumes less fuel will have a higher metric compared to another car with higher fuel consumption. The "liters per 100 km" on the other hand is a consumption metric, i.e. a car that consumes less fuel will have a smaller metric. In essence these two indexes both contain the same information, and hence it is straightforward to convert any efficiency metric to the equivalent consumption metric and vice versa. However, there are subtle differences that may lure an observer to misleading conclusions.

The main benefit with an efficiency index, where the useful work is in the nominator and the consumed energy is in the denominator (work/energy), is that a larger metric means better performance, which many people find intuitive. The drawback of an efficiency index is that it is difficult to relate an efficiency increase with the achieved energy savings. As an example, 20% better MPG translates to only 16.7% less fuel (20% is 1.2 times larger distance correspond to $1/1.2 = 0.83$ of the original fuel consumption). The interpretation of an efficiency metric becomes yet more difficult if two different upgrade options are to be compared. This is visualized in Figure 16.6, drawing the efficiency metric (left) and the consumption metric (right) over the consumed energy E (for a fixed provided work). Figure 16.6(a) illustrates that for a given reduction in energy consumption $\Delta E = E_1 - E_2$, the lower the absolute energy-consumption E_1, the larger the increase of the efficiency metric $f(E) \propto 1/E$. This implies that improving a more inefficient system, with larger absolute energy-consumption, will result in a relatively small improvement of the efficiency metric. This may lead to an underestimation of the improvements that actually provide the highest energy savings ΔE. In contrast, a consumption metric where the energy use is in the numerator, $f(E) \propto E$, will always show the same improvement for a given ΔE. Consumption metrics are therefore preferred, since they guide people to take rational decisions.

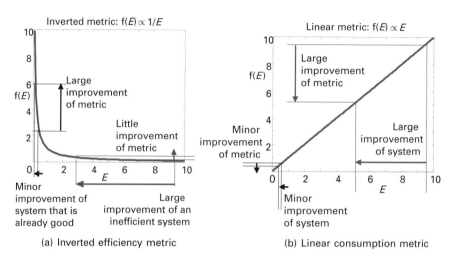

Figure 16.6 A linear consumption metric is easier to interpret than an inverted efficiency metric.

16.5.2 Energy-consumption metrics in cellular networks

To capture the energy-consumption perspective in the analysis, we employ the two following energy-consumption indices:

- power per unit area, measured in [W/m^2];
- energy per bit, measured in [J/bit].

The reason for arguing for two different metrics instead of only one is that they both are relevant and they provide complementary information about how efficient the energy usage in a network is.

The power per unit area metric is defined as the network average power usage divided by the coverage area of the network, P/A, and is measured by the unit [W/m^2]. The metric is a measure for the total energy-consumption and is closely related to the CO_2 emissions and the associated carbon footprint. Power per area unit is particularly relevant at low traffic loads, as in this case the network is coverage limited rather than capacity limited. Moreover, since the coverage area A, for which the system is to be evaluated, is typically a predefined constant the metric avoids a quotient of variables. This prevents misleading conclusions since when forming the quotient of variables it is impossible to understand whether an increase of the metric is due to the increase of the numerator, and/or the decrease of the denominator.

In order to be meaningful, the power per area metric needs to be complemented by a quality metric. When using the [W/m^2] metric it is important not to directly compare the power per area values of two systems, if they deliver vastly different performance. For instance, a dense network may consume more power per area unit than a sparse network, but the dense network may provide significantly higher user performance. However, given a minimum performance requirement, e.g. the 10 th percentile user throughput shall be at least 1 Mbps, the [W/m^2] metric provides a meaningful comparison of any two systems that both fulfill these requirement(s). This metric is also proposed in [3];

however, in ETSI this quantity is expressed as an efficiency metric, A/P, instead of the consumption metric adopted here.

The metric [W/user] is also commonly used and may be derived from the [W/m^2] metric by applying the inverse average user density in [m^2/user] as a conversion factor. If the focus is on evaluating the energy consumption by using a radio network system simulator, which is the case here, then the [W/m^2] metric is more convenient, since the system load is typically adjusted by varying the number of active users. If the purpose instead is to express the performance of a real network based on real measurements, then the W/user metric might be the better choice (since the area covered by a real network is difficult to estimate accurately).

The second metric, which measures the energy per bit in [J/bit], is defined as the network energy-consumption E during the observation period T, divided by the total number of bits RT that were correctly delivered in the network. Since the number of successfully transmitted bits is the average rate R, multiplied by the observation period T, this metric could equivalently be described as the average network power $P=E/T$ in relation to the average data rate R, expressed in [W/bps]. The energy-consumption metric [J/bit] focuses on the amount of energy spent per delivered bit and is hence an indicator of network bit delivery efficiency, which may be important especially in scenarios where the traffic load is high. The [J/bit] metric is commonly used in the literature, especially for theoretical studies and single link evaluations. However, in the literature the metric is often used as an efficiency metric in [bit/J], instead as the [J/bit] metric proposed here.

Since both the numerator (the network energy-consumption) and the denominator (the number of delivered bits) are typically variable, the metric is affected both by changes in the energy consumption and by changes in the number of delivered bits. Some caution is therefore required when using the [J/bit] metric. The first thing to note is that this metric approaches infinity as the traffic load goes to zero. It is also interesting to note that the metric improves when the amount of delivered bits, RT, grows faster than the required energy use E. In this respect the metric may be criticized as being self-optimizing. If we consider a newly deployed system built for area coverage that has only few users, then the energy used per bit will initially be extremely high. As time passes more users will join the network and the bits they generate will drive the metric down, since the extra users do not require any new base stations. Hence an operator may claim that the energy efficiency in terms of [J/bit] is improving at an impressive rate, while in fact the total energy-consumption has increased and nothing has been done to reduce the total energy-consumption of the network. Basing decisions only on this metric is certain to result in the conclusion that high data volumes are the best way to improve efficiency. Small cells will be considered more efficient than large cells, 4G more efficient than 3G and 2G, etc., simply because of the higher bit-rates provided.

16.6 Case study: energy efficiency of LTE

16.6.1 Assessment methodology

In this section the EARTH energy-efficiency evaluation framework (E^{3}F) is applied to evaluate the energy efficiency of a wireless network that covers an area A. Consider a large network with several deployment areas served by several types of base station (BSs), e.g. macro and pico BSs. Given the BS supply power (16.2) of $BS_{\ell,i}$, denoted by $P_{\ell,i,d}(t)$, where ℓ denotes the BS type and i accounts for the BS index, the consumed energy of deployment area d is determined by

$$E_d = \sum_i \sum_\ell \int_0^T P_{\ell,i,d}(t)\, dt. \quad (16.9)$$

The energy-consumption metrics devised in Section 16.5.2 may be expressed as $\Phi_d^{(W/m^2)} = \frac{E_d}{AT}$ and $\Phi_d^{(J/bit)} = \frac{E_d}{V_d}$, where V_d accounts for the traffic demand of deployment area d given by (16.6).

Following the methodology that quantifies the traffic volume over a large area established in Section 16.4.2, the power efficiency of the large-scale network may be assessed by weighted summing of the deployment-specific scenario snapshots:

$$E = \frac{\sum_d a_d E_d}{\sum_d a_d}. \quad (16.10)$$

The areal power-consumption becomes accordingly $\Phi^{(W/m^2)} = \frac{E}{AT}$. The consumed energy per transmitted bit is related to the areal power-consumption by the large-area average traffic density V/AT

$$\Phi^{(J/bit)} = \frac{E}{V} = \Phi^{(W/m^2)} \frac{AT}{V}. \quad (16.11)$$

We note that the energy E and the data volume V must be taken over the same area A.

16.6.2 Small-scale short-term evaluations

For short-term, small-scale evaluations a macro-cellular 3GPP LTE Release 8 network with regular hexagonal cell layout is implemented. Nineteen macro BS sites, each with three sectors, 10 MHz bandwidth operating at 2.1 GHz carrier frequency is assumed. Moreover, a 3 dB feeder loss between the BS site and the transmit antennas and 2×2 MIMO transmission with adaptive rank adaption is assumed. The inter-site distance (ISD) for the dense urban and urban environments is set to 500 m, whereas the ISD for suburban and rural areas is set to 1732 m. The users are uniformly distributed, with population densities corresponding to the respective deployment scenarios. The simulation parameters are taken from 3GPP specifications [9]. Two cases are simulated: i) base stations without sleep mode, and ii) base stations with micro sleep during idle transmit intervals, where neither data nor control signals are transmitted.

The power per area unit P/A, expressed in [kW/km²], is depicted in Figure 16.7. As can be seen, the power consumption increases with the served traffic in the network. In an urban scenario (see Figure 16.7(a)), with an ISD of 500 m corresponding to a coverage area of $A = 0.2165$ km² per site, the power per area unit is around $P/A = 4.15$ kW/km² at low loads, whereas it approaches 5.1 kW/km² at high loads. For the network with micro sleep-capable base stations, the corresponding figures are $P/A = 4.15$ kW/km² at low loads and above 5 kW/km² at high loads. For comparison, in an empty network when only control channels are transmitted, but no user data, the power consumption equals $P = 885\,\text{W}$ per site, which corresponds to a power per area unit of $4.1\,\text{kW/km}^2$. In the (hypothetical) extreme case, when nothing at all is transmitted (i.e. no data and no control channels) so that the RF output power is 0 W, we obtain $P/A = 3.3\,\text{kW/km}^2$. The power consumption per area unit for suburban and rural areas, shown in Figure 16.7(b), is substantially lower, which is due to the increased ISD of 1732 m, which corresponds to a coverage area of $A = 2.6\,\text{km}^2$. However, the system throughput per area unit decreases accordingly, due to the increased site coverage area.

Figure 16.8 contains the energy consumption per delivered bit, $E/(RT)$ in [kJ/Mbit], over the served data rate R in [Mbps]. Even though the total energy-consumption increases with the traffic load, the energy consumption per bit decreases with traffic. That is, the number of delivered bits increases faster than the network energy-consumption. The dominating reason for this is the fact that the power model (16.2) is associated with a fixed cost at $P_{\text{out}} = 0\,\text{W}$ RF output power and when the traffic increases, this fixed cost is shared over a larger number of bits, which results in the energy per bit decrease.

16.6.3 Large-scale long-term evaluations

In order to assess the expected performance of a countrywide area over a day, short-term, small-scale evaluations are combined with the long-term traffic models and the geographical distribution presented in Section 16.4. It is assumed that no coverage is provided in the sparsely populated and wilderness areas and hence, these areas are not included in the analysis. Moreover, the global traffic model in Section 16.4 is for the entire user population, whereas the short-term, small-scale evaluations are performed for a single carrier, which is served by a single operator. It is here assumed that the market share of the studied operator is 30% and that this operator also carries 30% of the total data traffic.

The outcome of the aggregation indicates that with the models and assumptions used in this evaluation, the average power per area unit is about $0.6\,\text{kW/km}^2$ without sleep modes, and slightly above $0.5\,\text{kW/km}^2$ for the network with micro sleep-capable base stations, corresponding to an energy saving of 15–20%. These values are almost independent of whether high, medium, or low traffic density is assumed, i.e. the power consumption is mostly insensitive to the traffic load. One reason for this is that given the traffic and deployment models used here, the network typically operates in the very low load regime, where transmissions of control signals dominate over data. These evaluations highlight the fact that on average cellular networks are primarily providing coverage, and therefore mainly operate at low traffic loads.

16.7 LTE technology potential in real deployments

In this section, we compare energy-efficiency figures obtained from system-level simulations as presented in the previous section, with global averages obtained from statistical data and corresponding trend analysis.

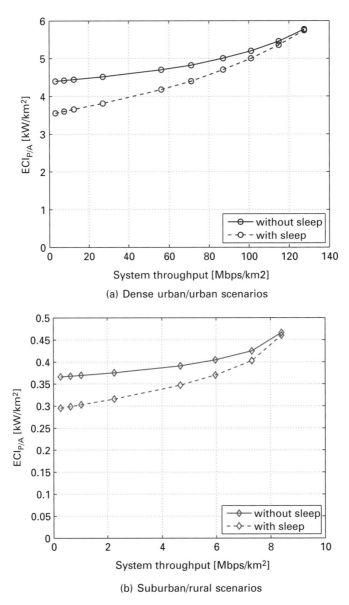

Figure 16.7 Power per area unit, P/A, of the downlink of an LTE radio access network as a function of the system load.

16.7.1 Global radio access networks

Results presented in [2] suggest that global radio access networks might increase from 3.3 million BS sites in 2007 to more than 11.2 million BS sites in 2020 and the total power-consumption is expected to grow from 49 TWh in 2007 to 98 TWh in 2020 in a business-as-usual scenario. During the same period the average traffic per site is predicted to increase from 62 kbit/s to somewhere between 11 to 18 Mbit/s. These numbers imply

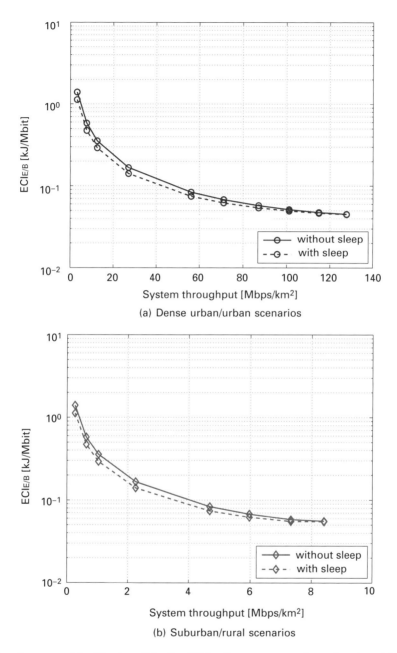

Figure 16.8 Energy per bit of the downlink of an LTE radio access network as a function of the system load.

that by 2020 the radio access network will have to improve its energy efficiency from about 28 J/kbit to 0.1–0.06 J/kbit. Taking into account current predictions on the number of mobile subscriptions, the average power per subscription is expected to decrease from about 16.6 kWh/year to 13 kWh/year.

16.7.2 LTE system evaluation

Considering the simulation results for a 3-fold sectorized 10 MHz LTE system as presented in Section 16.6, we obtain for the dense urban case data rates of up to 26 Mbit/s/site with acceptable cell-edge user performance. Besides average data rates per site, we may also conclude that LTE technology is able to provide much higher peak rates in 2020 for the following reasons:

- A large fraction of future traffic demand will be localized indoors and may be transported through small-capacity cells or offloaded to WiFi, which we have not considered here.
- Upgrading from 3-fold to 6-fold sectorized cells as well as 20 MHz bandwidth will almost quadruple the capacity per site.

Application of the E^3F to a countrywide LTE network with coverage provided by three operators yields energy efficiency of about 4.6 kJ/Mbit to 0.2 kJ/Mbit, and 49 kWh/year/sub to 50 kWh/year/sub for low and high traffic scenarios, respectively.

16.7.3 Evolution of LTE energy-efficiency over time

In order to compare the performance of an LTE network model with statistical data in a meaningful manner, it is essential to consider all energy-consumption figures at equivalent traffic conditions. In this regard, traffic demand per subscription for the low, medium, and high traffic profiles corresponds to the traffic scenarios as projected for the years 2013, 2017, and 2020 in [2]. Figures 16.9(a) and 16.9(b) show that the energy efficiency measured in both kJ/Mbit as well as kWh/year/sub is much higher than the projected development, at least for the case where an operator only carries 30% of the traffic. In a boundary scenario, where all traffic is carried by a single operator, all sites operate in higher load conditions; however, there are many fewer sites deployed and in this case the efficiency of the LTE system is comparable to the global average. Note here that the efficiency of the LTE system degrades especially in the low traffic scenario since currently the LTE system is optimized the for high performance rather than for low load efficiency. Figure 16.9 shows that a large saving potential exists in these cases.

It must be emphasized that the simulation-based analysis as presented in Section 16.6 assumes a pure LTE network and a rather specific mix of propagation conditions and traffic scenarios. In contrast, current and future cellular networks feature a mix of different cell sizes and standards. Despite these obvious differences a comparison of simulation results with real-world data and projections provides valuable insight into the principal effects of LTE deployment on cellular network efficiency. Here, we conclude that without additional improvements in scalability of the base station equipment, the rollout of LTE with full countrywide coverage would disrupt the current improvement trend of network energy-efficiency. Even though LTE BSs consume less power than legacy GSM equipment they tend to increase the power per subscriber when they are installed on top of existing networks. Especially in situations where the additional capacity exceeds the traffic demand (low traffic scenario), the kJ/Mbit efficiency of LTEs is insufficient for

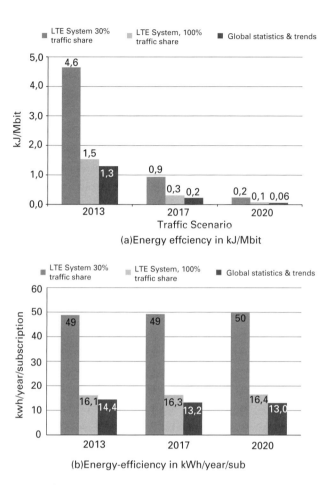

Figure 16.9 Impact of LTE energy-efficiency over time.

sustaining the current market trends. In this regard, technologies and improvements as put forward by the EARTH project must be seen as key enablers for the commercial viability of a large-scale LTE roll-out.

16.8 Fundamental challenges and future potential

Traditionally, radio access research both in academia and industry has focused on the challenge of achieving as high data rates as possible for a given maximum transmission power. By analyzing the challenges and the possibilities for the research on network energy-efficiency it is found that there is a second challenge, which is still not widely addressed or even accepted in the community. This second challenge addresses the power consumption of the system when it is not transmitting any data.

The first challenge is thoroughly addressed in research and specifications to allow for high peak data rates and capacity, for instance in LTE and HSPA. This very important work, which has been a main driver for the success of mobile telecommunication, as well as rapidly and steadily decreasing energy consumption per bit for the 3GPP technologies, will need to be continued although it is clear that it will be increasingly difficult to maintain the rate of improvement with sinking energy consumption per bit. The second challenge, exploiting system operation whilst not transmitting, either by improving the energy efficiency during idle operation, or by finding efficient solutions to eliminate empty resource block transmissions, has so far been a neglected research area. Consequently, this is where the big unexplored potential lies. It is important to understand that this second challenge does not only address empty cells and no-load scenarios. In fact the potential of the non-transmitting scenario depends strongly on the considered timescale. Considering a traditional O&M timescale of 15 minutes there may not be many periods, if any, without any transmissions at all. However, LTE scheduling decisions are made per ms, i.e. per every LTE subframe; when addressing this timescale instead, the possibility of idle subframes becomes considerable, even in fairly loaded cells.

EARTH-reference scenario simulations of a network covering the dense urban, the urban, and the sub-urban areas of a country or region, and assuming the medium traffic profile, yield that more than 98% of the subframes do not contain any transmitted data: this number can be seen as a form of theoretical limit for how large the potential is for the second challenge. As in this 98% of time about 97% of energy is consumed by the system, mechanisms addressing individual subframes without data transmissions could deliver up to 97% power savings. This is in addition to the power reduction achievable by features addressing subframes that are used for data transmissions.

Some examples of future potentials in energy saving are given by improvements at component level, e.g. introducing new power-management concepts that can adapt to varying traffic load. In fact introducing scalability into hardware components, and supporting them by dynamic power management, enables the adaptation of energy consumption to actual performance requirements. Further power savings are facilitated by the deactivation of components in time periods of no operation. These hardware characteristics could be exploited at the packet-scheduling level, to efficiently allocate radio resources, in order to properly manage empty subframes and drive components deactivation.

Finally, it should be made clear that different solutions can be envisaged by considering different time perspectives: starting from what energy saving can be achieved on short timescales that range from ms to μs scale, up to the possibility of savings when taking down cells or switching off whole BSs, something which will require actions on a timescale in the order of seconds or even minutes. As a matter of fact, all solutions on different timescales constitute part of the road toward the future, ultimately energy-efficient radio-access networks.

16.9 Conclusion

In order to identify the key drivers for energy savings the power consumption of mobile communication systems needs to be quantified. This includes sophisticated power models that map the radiated RF power to the supply power of a BS site, as well as traffic and deployment models that extend short-term small-scale evaluations to the countrywide power-consumption of a network over a whole day or week. Applying the EARTH E^3F to contemporary LTE Release-8 networks, numerical results reveal that for current network design and operation, the power consumption is mostly independent of the traffic load. This highlights the vast potential for energy savings by improving the energy efficiency of BSs at low load.

Acknowledgment

This work has received funding from the European Community's 7th Framework Programme FP7/2007–2013 under grant agreement n° 247733 project EARTH.

The authors gratefully acknowledge the invaluable insights and visions received from partners of the EARTH consortium.

References

[1] International Telecommunication Union (ITU), "Worldwide mobile cellular subscribers to reach 4 billion mark late 2008," Press release, 2008.

[2] A. Fehske et al., "The global carbon footprint of mobile communications - the ecological and economic perspective," *IEEE Communications Magazine*, vol. 49, no. 8, pp. 55–62, August 2011.

[3] ETSI TS 102 706 v1.1.1 (2009-08), "Environmental engineering (EE); energy efficiency of wireless access network equipment," Aug. 2009.

[4] ETSI TR 102 530 V1.1.1 (2008-06),"The reduction of energy consumption in telecommunications equipment and related infrastructure," June 2008.

[5] European Commission, "Code of conduct on energy consumption of broadband equipment," Version 4, Feb. 2011.

[6] ITU-T L.1400, "Overview and general principles of methodologies for assessing the environmental impact of ICT," Feb. 2011.

[7] GSMA, "Mobile energy efficiency, an energy efficiency benchmarking service for mobile network operators," Feb. 2011. [Online]. Available: www.gsmworld.com/documents/mee_meth_feb_11.pdf

[8] L. M. Correia et al., "Challenges and enabling technologies for energy aware mobile radio networks," *IEEE Communications Magazine*, Special issue on green radio, pp. 66–72, Nov. 2010.

[9] 3GPP TR 36.814 v9.0.0, "Further advancements for E-UTRA. Physical layer aspects (Release 9)," 3GPP, Technical Specification Group Radio Access Network, Mar. 2010.

[10] WINNER II, "Deliverable D6.13.7: Test Scenarios and Calibration Cases Issues 2", IST-4-027756 WINNER, Dec. 2006.

[11] International Telecommunication Union, Report ITU-R M.2134, "Requirements related to technical performance for IMT-Advanced radio interface(s)," 2008. [Online]. Available: www.itu.int/dms_pub/itu-r/opb/rep/R-REP-M.2134-2008-PDF-E.pdf

[12] S. C. Cripps, "RF power amplifiers for wireless communications," 2nd ed., Artech House Microwave Library, 2006.

[13] J. Xu, "Practical digital pre-distortion techniques for PA linearization in 3GPP LTE," Agilent Technologies, 2010.

[14] W. Kim *et al.*, "Digital pre-distortion linearizes wireless power amplifiers," *IEEE Microwave Magazine*, Sept. 2005.

[15] B. Debaillie *et al.*, "Opportunities for energy savings in pico/femto-cell base-stations," in *Proc. of Future Networks and Mobile Summit*, Warsaw, Poland, 2011.

[16] S. Borkar, "Design challenges of technology scaling," *IEEE Micro*, vol. 19, no. 4, pp. 23–29, July 1999.

[17] National Media and Infocommunications Authority Hungary, "Report on mobile internet coverage," 2010. [Online]. Available: www.nmhh.hu/dokumentum.php?cid=25074

[18] "Entscheidung der Präsidentekammer der Bundesnetzagentur fur Elektrizität, Gas, Telekommunikation, Post und Eisenbahnen" (in German), October 2009. [Online]. Available: www.bundesnetzagentur.de/cae/servlet/contentblob/138464/publicationFile/2807/Praes KammerEntschg_Id17404pdf.pdf

[19] UMTS Forum, "Mobile Traffic Forecasts 2010-2020," Report no. 44, May 2011. [Online]. Available: www.umts-forum.org/component/option,com_docman/task,cat_view/gid,485/ Itemid,213/

[20] Sandvine, "Mobile Internet Phenomena Report", 2010. [Online]. Available: www.sandvine.com/downloads/documents/2010GlobalInternetPhenomenaReport.pdf

[21] Akamai, "State of the Internet" report, 2010. [Online]. Available: http://wwwfp.akamai.com/dl/whitepapers/akamai_state_of_the_internet_q1_2010.pdf

[22] National Media and Infocommunications Authority of Hungary, "Report on mobile data usage". [Online]. Available: 2010. www.nhh.hu/dokumentum.php?cid=24769

[23] The Swedish Post and Telecom Agency, "Telecommunication Markets in the Nordic Countries", 2010. [Online]. Available: http://statistik.pts.se/pts1h2010e/download/PTS_ER_2010_26_Svensk_Telemarknad_2010_1h_en.pdf

17 Standardization, fora, and joint industrial projects on green radio networks

Alberto Conte, Hakon Helmers, and Philippe Sehier

17.1 Introduction

The urgent need to reduce the energy consumption of wireless ICT networks directly impacts the work of all actors involved in this industry, and in particular the groups and fora in charge of imaging, defining, and standardizing the evolution of telecommunication networks. The introduction of energy-saving mechanisms to current solutions and the definition of new techniques are issues addressed today by several experts, working in partnerships, all over the world.

Standardization groups represent, in the context of energy-saving techniques, as well as in other domains of telecommunication, a link between research activities, industrial product development activities, mobile network operators, and regulators. The standardization focus will depend on the role of each standardization body. Such a focus may be linked to performance evaluation and comparison, e.g. energy-efficiency benchmarking of entire mobile networks, or development of energy-efficiency metrics for individual radio base stations. The outcome of such work provides:

- guidance for vendors with respect to expected equipment characteristics;
- possibility for the telecommunication operator to compare offers from different vendors;
- guidance for mobile network operators related to deployment choices;
- possibility for the regulator to issue requirements or develop energy-related classification systems.

Another focus of standardization fora is related to the definition of functionality and open interfaces between equipments. The outcome of this work represents a common technological platform defined at a sufficiently detailed level to permit interoperability between vendors and hence the emergence of an ecosystem of researchers / suppliers / vendors on one side and mobile network operators on the other side. Different kinds of energy-saving functionalities may be addressed in this type of approach:

- functionalities favoring spectral efficiency which implicitly lead to energy efficiency;
- energy saving related to signaling reduction on the radio interface;
- energy saving related to support of efficient deployment schemes for radio base stations (e.g. HetNet);
- energy saving by switching off some equipment during low-load periods.

Given the importance of the energy-efficiency problem, a large number of additional initiatives have been launched, with the intention of driving, supporting and even preceding the standardization efforts. Some initiatives try to influence the overall ICT industry and the related standardization bodies. The Next Generation Mobile Networks alliance (NGMN), for example, groups a large number of mobile operators seeking to coordinate their needs to drive the evolution of the wireless technologies. NGMN has conducted several studies and delivered a number of recommendations, including energy-efficiency considerations, usually pushed toward 3G partnership project (3GPP). Other consortiums have been created that target the definition of technical solutions allowing the reduction of the power consumption of current wireless networks. They can focus on specific aspects (such as hardware and component-level improvements) or have an holistic view of the full system for cross-layer solutions. These groups (e.g. FP7 EARTH project) usually target incremental evolutions of existing networks in order to ensure an acceptable level of compatibility with equipment already installed. Finally, some initiatives look for game-changing solutions allowing for drastic gains in network efficiency. In this case backward compatibility is not a primary target and clean-state systems can be designed. GreenTouch is an example of such initiatives.

This chapter describes the activities conducted in the most important standardization bodies and fora as well as by the most relevant industrial and academic joint projects and consortiums.

17.2 Standardization fora

A large number of standardization bodies are involved in energy-efficiency-related activities, and an exhaustive description is not feasible. This section focuses on a selection of the most important fora around the world, with particular attention given to the groups working on mobile networks, such as

- *European Telecommunications Standards Institute (ETSI)*: involved in several activities related to measurements and efficiency improvement in support of several European commission mandates;
- *3GPP*: defining solutions for energy saving in 2G, 3G, and 4G networks;
- *Telecommunications Industry Association (TIA) and 3GPP2*: developing Green Energy Environment (GREEN) initiatives;
- *Alliance for Telecommunications Industry Solution (ATIS)*: which defines wireless access network energy-efficiency metrics and measurement methods;
- *Internet Engineering Task Force (IETF)*: with the Energy MANagement (EMAN) working group which specifies energy-management properties and control functions that will allow networks and devices to become energy aware;
- *China Communication Standard Association (CCSA)*: targeting energy-efficiency improvement particularly adapted to the mobile networks deployed in China.

GSMA (GSM Alliance) can also be mentioned, as it is conducting a study with operators with the objective of characterizing where power is consumed in operators and networks, and determining improvement directions accordingly.

17.2.1 ETSI

The ETSI performs energy-efficiency-related work in support of European Commission Mandates.

The current Mandates in this domain are:

- Mandate 439: Standardization in the field of standby and off modes power-consumption measurement for energy-using products;
- Mandate 450: Standardization in the field of measurements of no-load condition electric power consumption and average active efficiency of external power supplies;
- Mandate 451: Standardization in the field of power-consumption measurement of simple set-top boxes in active and standby modes;
- Mandate 462: ICT to enable efficient energy use in fixed and mobile information and communication networks.

Partnership activities

Some aspects of ETSI's energy-efficiency work is done in partnership with other organizations, including ITU-T Study Group 5, the Broadband Forum, the Home Gateway Initiative, CENELEC (for Mandates 439 and 451), and the Global eSustainability Initiative forum (GeSI). Links have also been established with the EC 7th Framework Programme (FP7) EARTH Project (Energy-Aware Radio and neTwork tecHnologies)

ICT has three inter-related focus areas as follows:

- *Common methodology*: ICT can contribute to setting the common methodology for the measurement of overall environmental impacts;
- *Enhancing energy efficiency in ICT*: various technical solutions to enhance energy efficiency within the ICT domain
- *Using ICT as an enabler in other relevant sectors*: ICT can be used as an enabler to reduce environmental impacts of other relevant sectors such as intelligent transportation system (ITS) and machine-to-machine (M2M) communications.

Several of ETSI's Technical Committees (TCs) are actively involved in specifying technologies to improve energy efficiency. For example, the Access, Terminals, Transmission and Multiplexing Technical Committee (TC ATTM) deals with energy efficiency for broadband with close collaboration with the ETSI Environmental Engineering Technical Committee (TC EE), which defines the energy-efficiency indicators and measurement methods. Working Group 5 of ETSI's Telecommunications and Internet converged Services and Protocols for Advanced Networking Technical Committee (TC TISPAN) is also currently working on how to use Next Generation Networks (NGN) to monitor and control power levels in Customer Premises Networks.

Common methodology for measurement

Work in this area is led by TC EE and includes:

- Life-cycle assessment (LCA), a system analytical method and model for assessing the environmental aspects associated with a product over its life cycle, i.e. material

extraction, manufacturing, transport, use, and scrapping processes. The current work items are DTS/EE-00014: "General definition and common requirements" and DTR/EE-00008: "Environmental Impact Assessment of ICT including the Positive Impact by using ICT Services";
- Requirements and measurement methods on power consumption of specific telecommunication products to enhance energy efficiency. TC EE's current work items in this domain are concerned with specifying:
  - Measurement methods for energy consumption of customer premises equipment: DTS/EE-00018 and DEN/EE-00021;
  - Measurement methods and limits for energy consumption in broadband telecommunications equipment: ETSI Standard ES 203 215;
  - Energy efficiency of wireless access network equipment: Technical Specification TS 102 706 [2];
  - Measurement methods for power consumption in transport telecommunication networks equipment: DES/EE-00023;
  - Measurement methods for energy consumption of router and switch networks equipment: DES/EE-00024;

Focus on energy efficiency in TS 102 706

TS 102 706 [2], produced by the ETSI technical committee Environmental Engineering (EE) defines harmonized methods to evaluate the energy efficiency of wireless access networks and their related base stations. The document covers: GSM/EDGE, WCDMA/HSPA, WiMAX, and LTE.

The metrics used are :

- Average power-consumption of RBS equipment: the RBS average power-consumption is based on measured RBS power consumption under reference configuration, reference environment, and under reference load levels.
- Average power-consumption of RBS site: the RBS site-level power-consumption is calculated based on RBS equipment power-consumption for reference RBS site configuration using correction factors for different power supply, cooling, and site solutions.
- Performance indicators for network-level energy efficiency: the network-level performance indicators are calculated based on RBS site-level reference power-consumption as well as on RBS coverage area for rural areas and RBS capacity for urban areas.

Note that TS 102 706 defines static and dynamic measurement methods.

Enhancing energy efficiency in ICT

Technical solutions to improve energy efficiency in networks and devices have been identified by TC EE, TC ATTM, and the 3rd Generation Partnership Project (3GPP):

- Use of alternative energy solutions in telecommunication installations (TC EE): Technical Report TR 102 532;

- Reverse power feeding (TC EE and TC ATTM TM6): Technical Reports TR 102 614 and TR 102 629;
- Digital Subscriber Line (DSL) power optimization (TC ATTM TM6);
- Energy control and monitoring in home networks (TISPAN WG5): Feasibility study in Special Report SR 085 014;
- Network energy saving for E-UTRAN (Enhanced UMTS Terrestrial Radio Access Network): Feasibility study underway in 3GPP RAN (Radio Access Networks) [4];
- Energy-saving management (3GPP services and system aspects (SA) WG5) [5].

The automated energy-savings management TC EE is also actively involved in standardization activities for best practices for deployment of telecommunications equipment, and for power supplies, bonding, and related topics.

17.2.2 3GPP

3GPP (Third Generation Partnership Project) is a standardization forum whose work is based on contributions from mobile network operators and equipment vendors. 3GPP Technical Specifications and Technical Reports are publicly available and can be downloaded from 3GPP's website; they are published by the partnering Standards Development Organizations (SDO), which are ARIB and TTC (Japan), ATIS (USA), CCSA (China), ETSI (Europe), and TTA (Korea). We will here describe activities performed in 3GPP's Technical Specification Groups (TSG SA, TSG CT, TSG RAN, and TSG GERAN), whose work is distributed on specialized Working Groups (WG). Standardization activities in 3GPP's Technical Specification Groups are organized in:

- "Work items" in which normative work is done, resulting in new or updated technical specifications (TS);
- "Study items" in which the obtained results are informative and published in technical reports (TR). A study item may prepare a subsequent work item on the same topic.

Until now, the work on energy saving has focused on the proposal and analysis of energy-saving solutions in the radio access network with associated telecommunication management (OA&M) and core network aspects. Energy-saving related topics have so far been handled in 3GPP Release 9 and Release 10, and activities continue in Release 11 with an extended number of contributing working groups (TSG SA WG2, TSG GERAN WG1 and 2).

3GPP Release 9: Basic signaling related to cell switch-off was standardized for UMTS (on the Iub interface between the NodeB and the RNC) and for LTE (on the X2 interface between eNodeBs). This work was conducted by the 3GPP working group TSG RAN WG3, and led to updates in the Technical Specifications TS 25.433, TS 36.300, and TS 36.423.

3GPP Release 10: Four working groups within 3GPP have completed study and work items, and published results, on energy-saving related topics during this release: TSG SA WG5, TSG CT WG1, TSG RAN WG2, and TSG RAN WG3. The study items resulted in the following Technical Reports:

- 3GPP TR 32.826: "Telecommunication management; Study on Energy Savings Management (ESM)" (TSG SA WG5) [4];
- 3GPP TR 36.927: "Potential solutions for energy saving for E-UTRAN" (TSG RAN WG3) [5];
- 3GPP TR 24.826: "Study on impacts on signaling between UE and core network from energy saving" (TSG CT WG1) [6].

Some analysis on physical layer aspects was also done by TSG RAN WG1 for LTE and UMTS. For UMTS the result of this work was captured in an internal technical report (TR 25.927). The LTE part was handed over to TSG RAN WG2 and TSG RAN WG3.

TSG SA WG5 also conducted a Rel-10 work item, resulting in a Technical Specification providing concept definitions and requirements for the telecommunication management part.

- 3GPP TS 32.551: "Telecommunication management; energy saving management (ESM); concepts and requirements" (TSG SA WG5)

In the context of this work item, energy-saving management functionality was also documented in functional Technical Specifications (TS 32.522 and TS 32.762). Other updated specifications were TS 32.425, TS 32.526, TS 32.626, and TS 32.766.

Note that the contents of 3GPP Release 11 is under discussion at the time of writing this chapter. The authors expect that 3GPP will decide to undertake further normative work related to energy saving in this release.

Overview of the 3GPP work for energy saving in E-UTRAN (Rel-9, Rel-10)

Backward compatibility and maintained user accessibility are two central requirements for the 3GPP energy-saving work, explicitly expressed in TR 36.927. TS 32.551 however provides some freedom linked to service impact: "The acceptable impact on services shall be determined based on operator's policy."

The need to be able to serve legacy mobiles (i.e. mobile equipment compliant to former 3GPP releases) significantly limits the possibility of introducing new energy-saving functions at the physical layer, e.g. in terms of reduction of power consumed in sending reference signals. However, TR 36.927 [4] identifies the possible usage of particular sub-frames for the purpose of energy saving. In this way, a particular subframe initially designed for multi-media broadcast over a single frequency network (MBSFN) for multi-casted multi-media services may be used in both FDD and TDD networks. This particular subframe contains fewer common reference signals (CRS) than normal subframes. Configuring as many as possible MBSFN subframes allows reduced eNB transmission time and hence reduced power-consumption by the power amplifier, which is responsible for a large part of the power consumption of any radio base station.

In TDD networks it is also possible to obtain energy saving by some configuration of subframe 1 and 6. These solutions were identified by TSG RAN WG2 and are already supported by the current specification.

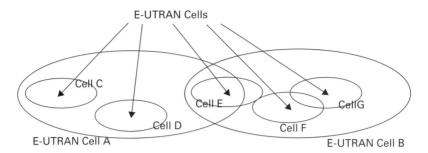

Figure 17.1 Radio base station overlaid scenario for intra-RAT (E-UTRAN) case. Cells A and B provide basic coverage, while cells C through G may be switched off to save energy when network load is low (from [4]). © 2011. 3GPP™ TSs and TRs are the property of ARIB, ATIS, CCSA, ETSI, TTA and TTC who jointly own the copyright in them. They are subject to further modifications and are therefore provided to you "as is" for information purposes only. Further use is strictly prohibited.

Due to limited possibilities of network energy saving by adaptation of the physical layer/link layer, the main part of the 3GPP work is linked to the following two scenarios defined in TS 32.551 and TR 36.927:

- Radio base station overlaid scenario: this scenario is in TR 36.927 derived into the inter-RAT scenario and the inter-eNB scenario.
- Scenarios based on coverage compensation (capacity-limited network): this scenario is in TR 36.927 described for LTE only.

Radio base station overlaid scenario

In this scenario, a network layer provides basic coverage and other network layers (intra-RAT or inter-RAT) provide capacity enhancements. Handling of such a scenario improves the possibility of the operator deploying, e.g. small cells for capacity enhancements without increased waste of energy during some periods of the day/night when such enhanced capacity is not used.

Cell switch off/switch on within the radio access network may easily be implemented in the UTRAN due to the master–slave architecture (each NodeB is controlled by a single RNC). In the E-UTRAN, which is based on a peer-to-peer distributed architecture, particular care needs to be taken. For this reason the solution that was standardized on the X2 interface in Release 9 does not contain any switch-off request: the decision is expected to be taken autonomously by the concerned eNB based on available information, which may include traffic load information reported by neighboring eNBs.

For switch-on, an explicit request on X2 was considered to be required. However, an efficient solution, from an energy-saving point of view, may require selective switch-on of cells when the network load increases, and hence extended support in the standard compared to the Release 9 solution. TR 36.927 [4] describes and compares different solutions for this purpose in intra-LTE and inter-RAT environments:

- switching on based on predefined low-load periods;
- switching on based on Interference over Thermal (IoT) measurement;

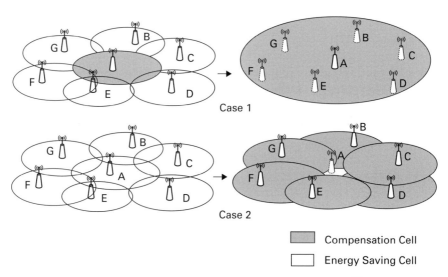

Figure 17.2 Two basic scenarios for coverage compensation (from [4]). © 2011. 3GPP™ TS$_s$ and TRs are the property of ARIB, ATIS, CCSA, ETSI, TTA and TTC who jointly own the copyright in them. They are subject to further modifications and are therefore provided to you "as is" for information purposes only. Further use is strictly prohibited.

- switching on based on UE measurement (network probing state);
- switching on based on position information.

Coverage compensation scenarios (capacity-limited network)
In this scenario, the network will adapt its configuration and cell sizes according to the load distribution within the network. In this way a single radio base station may extend its coverage in order to allow other base stations to switch off, or some surrounding base stations may jointly take over coverage from a central base station.

Impacts of cell switch-off for energy saving on management system and core network
From a management system point of view, cell switch-off functionality for energy saving creates new information and new network states that need to be handled by the management system. It is also important that a cell or eNB that is switched off for energy-saving purposes is not considered as being in outage due to some error or other abnormal condition. These impacts have been documented by TSG SA WG5 in the documents cited above.

TSG CT WG1 has, in a study item, identified a series of core network-related key issues mainly linked to cell switch-on / switch-off, but also linked to the coverage compensation scenario described above. These are analyzed in TR 24.826, and potential solutions are also identified and discussed. One of these key issues is related to mass signaling effects for idle mode UEs for the cases where the energy-saving cell is situated on a Tracking Area boundary (intra-LTE energy saving) or in the case of inter-RAT energy saving. It is considered that UEs in connected mode can be handled by the network via timely hand-overs to cells remaining permanently active. A broad area of access technologies is considered.

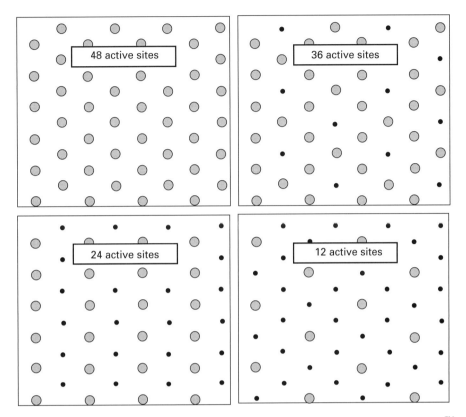

Figure 17.3 Coverage compensation scenarios analyzed using a grid approach (from [5]). © 2011. 3GPP™ TS$_S$ and TRs are the property of ARIB, ATIS, CCSA, ETSI, TTA and TTC who jointly own the copyright in them. They are subject to further modifications and are therefore provided to you "as is" for information purposes only. Further use is strictly prohibited.

The technical report does not foresee any need for major updates of technical specifications linked to the core network in the context of the present analysis.

Contents of the 3GPP work for energy saving in GERAN

Energy saving in GERAN is particularly important due to the very high number of GSM radio base stations deployed worldwide, and also because in some regions a significant number of these use diesel as the primary or as a backup power source. There has not been a large amount of focus on energy saving in GSM/EDGE networks so far in 3GPP, although some solutions have been agreed in previous Releases. In particular the Multi-Carrier Base Transmitter Station permits the replacement of analog circuitry (per carrier) by digital components handling entire sectors, suppressing the need for an analog combiner and hence increasing energy efficiency. Increased spectral-efficiency in the GERAN (voice services over adaptive multi-user channels on one slot–VAMOS) will also contribute to improved energy efficiency in terms of network capacity per consumed unit of energy.

A Release 11 study item has been opened by GERAN, which will result in a technical report "Solutions for GSM/EDGE BTS Energy Saving."

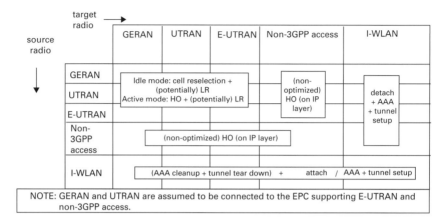

Figure 17.4 Combinations of source and target radio for energy saving (for UE attached for PS services) and possible procedures after switch-off of source radio. The main diagonal of the matrix describes intra-RAT energy saving, and the non-diagonal elements constitute inter-RAT energy saving (from [6]). © 2011. 3GPP™ TSs and TRs are the property of ARIB, ATIS, CCSA, ETSI, TTA and TTC who jointly own the copyright in them. They are subject to further modifications and are therefore provided to you "as is" for information purposes only. Further use is strictly prohibited.

17.2.3 TIA and 3GPP2

The Third Generation Partnership Project 2 (3GPP2) is a collaborative third-generation (3G) telecommunications specifications-setting project born out of the International Telecommunication Union's (ITU) International Mobile Telecommunications IMT-2000 initiative.

TIA (Telecommunications Industry Association) is one of the organizational partners of 3GPP2.

TIA Engineering Committee TR-45 (via its Subcommittees) has been working jointly with 3GPP2 (via its Technical Specification Groups (TSGs)) on preliminary work on green energy-efficient environment (GREEN) initiatives.

Preliminary GREEN study work includes the following:

- 3GPP2 vision for 2010 and beyond, published in 3GPP2 in June 2010 (as SC.R5003-0 v2.0). This document included a section on terminal battery and power-consumption improvements that addressed the issue of "developing and refining standards such that they are conducive to green technologies".
- **G**reen **R**evised **E**nergy **E**fficient **N**etwork (GREEN) technology initiatives study for cdma2000 wireless networks, published in 3GPP2 in December 2010 (as S.R0140-0 v1.0) approved for publication in TIA in February 2011 (as TSB-473). The intent of this document is to investigate protocol enhancements, which can potentially reduce power consumption, extend battery life, and better match the available power in both network and terminal equipment.

Current GREEN standards development work includes the following:

- Dynamic RAN power management of radio carriers and BTSs–work item (WI), work item (3GPP2-00285) initiated in January 2011 to support the ability to manage power for BTSs and radio carriers dynamically for energy savings.

- Dynamic RAN power management–systems requirements document (SRD). A group of member companies (China Telecom, Huawei, Alcatel-Lucent, ZTE, and Qualcomm) proposed a draft SRD in January 2011. The intent is to further study and address possible issues in the radio access network (RAN) (i.e. A-Interface) as part of this work.
  - TSG-S WG-1 initiated work on GREEN solutions and RAN power management–stage 1 (S.P0148) targeted for publication 4Q2011.
  - Joint 3GPP2 TSG development (being led by TSG-S WG-1) underway in May 2011 with dynamic RAN management experts in TSG-A to progress the work on S.P0148 based on numerous member company contributions on energy-savings mode architecture, enhancements to A-interface, power management notification, and more.

Future GREEN standards development work is anticipated to include the following:

- TSG-A (TR-45.4) RAN: Protocol support for GREEN BTS;
- TSG-C (TR-45.5) radio interface: Protocol support for new energy-efficient protocols;
- TSG-X (TR-45.8) packet data network: Protocol support for new energy-efficient protocols;
- TSG-S (TR-45.8) OAM&P: GREEN operations field guide to applying green BTS and other techniques and protocols to network deployments.

17.2.4 ATIS

The alliance for telecommunications industry solutions (ATIS) specification for the measurement of the telecommunication efficiency ratio (TEER) for radio base station (RBS) [1]. The specification covers LTE, CDMA (including CDMA2000, EV-DO, UMTS, GSM, and WiMAX.

The specification defines a methodology to be used by vendors, third parties, and test laboratories in the determination of RBS data and voice TEER. The measurement method enables fair evaluations of TEER related to equipment from different vendors, for the same technologies.

A static measurement method is used, where RBS power consumption is measured at various traffic loads. The TEER is defined as the ratio between the data and voice processed by the RBS and the power consumption. Weighting factors are applied on the measurements to be representative of real RBS load distribution over the day.

17.2.5 IETF/EMAN

IETF has approved the energy management (EMAN) [3] project in view of developing:

- requirements for energy management to specify energy-management properties that will allow networks and devices to become energy aware;
- energy-management framework to describe extensions to current management framework, required for energy management;

- energy-aware networks and devices MIB document;
- power and energy-monitoring MIB document;
- battery MIB document;
- applicability statement describing the variety of applications that can use the energy framework and associated MIB modules.

The project final deliveries are planned for the end of 2011 with the publication of several internet drafts on energy-aware networks and devices MIB, power and energy-monitoring MIB, battery MIB, and energy-management applicability.

17.2.6 CCSA

The China Communication Standard Association (CCSA) TC5, WG9 has completed the national standard on RBS EE static test method and the study item on dynamic testing is underway [3]. Static test methods are also considered for the RNC and Core Network.

The technologies covered are GSM and UMTS (including TD-SCDMA).

There are a lot of similarities between the test methods defined by ETSI and ATIS.

17.3 Consortium and joint projects

To assist, influence, and even precede the standardization evolution, several groups have been created in the recent past. Also, industrial or academic fora that already existed have started specific working groups or research actions targeting energy efficiency. An exhaustive description of all the groups and fora is not possible; however, it is possible to classify them regarding the approach and goals, and to provide a significant example for each type:

- groups aiming to assist and influence the standardization evolution (e.g. 3GPP)–an example is the NGMN alliance;
- groups targeting the technical improvement of existing technologies (e.g. LTE)–an example is the FP7 EARTH project;
- groups targeting a longer step in exploring new solutions and defining clean-slate systems able to offer drastic improvements in terms of energy efficiency – a major example is the GreenTouch consortium.

17.3.1 NGMN alliance

The next-generation mobile networks (NGMN) alliance was founded in 2006 by some of the major mobile network operators from Asia, Europe, and North America. It is an open forum with the goal of ensuring that the evolution of the standards for next-generation mobile terminals, networks, and services meets the needs and expectations of operators and end users. The NGMN Alliance focuses particularly on LTE technology and its evolutions (LTE-Advance), with 3GPP as the main standardization interlocutor. In 2011 the NGMN Alliance joined the 3GPP as a Market Representation Partner. All the results and recommendations are publicly available from the NGMN alliance website [7].

The NGMN alliance works one step ahead of standardization bodies and industry fora to provide a coherent view of what the operator community requires for successful deployment of mobile broadband solutions, in terms of cost, time-to-market, and end-user experience. The scope of interests of the NGMN alliance includes:

- deployment aspects like antenna design and configuration, FDD/TDD convergence, and field trials;
- operation efficiency, including self-optimized networks (SON) use cases and energy efficiency;
- wireless devices including certification;
- network architecture topics, like centralized RAN, backhaul, and heterogeneous networks;
- application enablers and related APIs standardization;
- spectrum, bandwidth aggregation, and global harmonization.

Since energy cost is a main part of the operator's operational expenditure (Opex), the energy-efficiency issue has been addressed within the operational efficiency project. The focus has been on the use of SON algorithms to enable and assist the dynamic switch-on/off of network equipments, and in particular base stations, as a function of traffic load. The project results (recommendations and proposed solutions) have been captured in [8].

The proposed first solution uses the redundancy of different radio access technology (RAT) layers, resulting from the deployment of collocated LTE + 3G/2G cells. In this context, LTE cells can be switched off when there is no load in the LTE cell, or when the remaining traffic can be safely moved to the 3G/2G layer. The dormant LTE cell is then woken up by a specific indication coming from the 3G/2G layer.

The main recommendations on the network elements (NEs) and the interface between base stations (X2 in LTE technology) are

- The network element shall provide an energy-saving mode with minimum power-consumption allowing a restart of the network element in less than 5 minutes triggered via the O&M or X2 interface.
- The network elements shall be informed about the status of neighbor sites.
- Energy-saving features shall be integrated in other SON use cases, like load balancing, cell/service outage detection and compensation, etc.
- Non-availability of sites due to energy-saving mode should be considered as normal and not be alarmed.

The main recommendations on network management are

- The network management system shall have the actual status of the network elements at all times. Moreover, it shall be able to remotely control the energy-saving functionalities (e.g. activate/deactivate an energy-saving mode on a single or group of network elements).

- The system supports automatic detection of low-load periods as a basis for operator or automatic decisions on the definition of time frames when the feature shall be active.
- It shall be possible to configure thresholds and rules of conditions to switch-on/off a site automatically.
- A failed restart of a network element shall be alarmed.

The NGMN alliance will continue to work and provide recommendations on energy-efficiency aspects in the coming years.

17.3.2 FP7 EARTH project

EARTH (Energy-Aware Radio and neTwork tecHnologies, [9, 10]) is a European project belonging to the EC 7th Framework Programme for Research and Technological Development (FP7). EARTH stands as the most significant FP7 project with respect to the energy efficiency of radio access and network technologies in the period 2010–2012.

EARTH is committed to the development of a new generation of energy-efficient equipment, components, deployment strategies, and energy-aware network management solutions to improve the overall mobile systems. The project is industry driven with strong contributions by academia. It has mobilized a large European consortium composed of: Alcatel-Lucent, BME, CEA, DOCOMO, Ericsson, ETSI, IMEC, Instituto Superio Tecnico, NXP, Technische Universität Dresden, Telecom Italia, University of Oulu, TTI Norte, and University of Surrey.

The target of EARTH is to enhance the energy consumption of mobile systems by a factor of at least 50%. The project is primarily focused on LTE technology, its evolution LTE-Advance and systems beyond, where a potential impact on standardization is envisaged, but it also considers 3G (UMTS/HSPA) technology for immediate impact.

Technical approach

The EARTH project addresses the whole mobile system and proposes technical solutions to improve its power efficiency at component level, link level, and network level. These three technical tracks can make use of a specific tool to evaluate the effective gain of the proposed solutions. The tool has been defined within the project under the name of the energy-efficiency evaluation framework (E^3F), and has been recently adopted by the GreenTouch consortium. The E^3F framework provides the means to analyze the energy efficiency of current solutions and to estimate the improvements brought by the envisaged techniques, in real networks.

At the radio component level, the main effort is on improving the power amplifier (PA) efficiency, known to be responsible for a significant portion of the energy consumed by a macro-cellular base station ([11]). Signal conditioning algorithms, like crest factor reduction for decreasing the PAPR and digital pre-distortion for increasing the PA linearity, are considered to enable the PA operation closer to saturation, corresponding to the maximum efficiency working point. Furthermore, PAs based on special architectures are investigated to achieve high efficiency values. For example, Doherty power amplifiers ([13]), which contain one main amplifier that is always active and an auxiliary one that is active only when signal peaks occur, show an efficiency profile matching well with the PAPR values of interest.

In addition to PA efficiency-improvement techniques, the EARTH project also explores other concepts on a component level. Solutions to adapt or deactivate transceiver components of a macro base station by following the signal level have been presented for a small RF-transceiver, a power supply unit, and the digital transceiver part. Furthermore, during low load periods, rapid component deactivation/reactivation allows power savings in time slots of no transmission (cell DTX). The benefit of cell DTX is highly dependent on how fast deactivation and reactivation can be supported by the functional unit. For example, LTE technology requires timescales in the order of tenths of microseconds, to suppress the power consumption at symbol-level speed.

At link level, the EARTH project addresses how radio interfaces are designed. This starts from solving the problems raised by the increasing number of transmitted reference signals (also known as pilots) necessary to achieve accurate channel estimation in multi-antenna systems. In fact, the amount of reference symbols grows strongly with the number of transmitting antennas. At full load, the energy consumed to transmit this overhead is overcompensated by the enhanced spectral-efficiency: in theory, up to 50 percent of reference signal overhead is acceptable to still see capacity gains by adding transmit and receive antennas at fixed total power [12]. However, the amount of the auxiliary information remains the same in low-load situations, where the radio interface carries only a fraction of its maximum capacity, which translates into a loss of energy efficiency.

Base station sleep and standby modes are also part of the link-level topics. Assuming adapted components (e.g. PA), various kinds of base station sleep modes should be possible in future standards:

- micro sleep modes, where base stations suspend transmission in the order of milliseconds, as in, e.g. cell DTX;
- deep sleep modes, where base station transmitters are shut down for extended periods of time.

While in micro sleep mode base stations are required to wake up almost immediately, in the deep sleep some transmit circuits are completely switched off, which implies that wake-up times are substantially longer. In this case new protocol and operating modes are needed allowing the suspension of the pilots emission while still supporting handover and initial network entry procedures.

On the network level, the potential for reducing energy consumption comes from the layout of network deployments and in their management, to adapt the network configurations according to the variations of the traffic. During low traffic periods, in densely populated areas, changing sectorized cells to omni cells and switching whole sites off (even whole radio technology layers) are expected to be the most efficient solutions. In rural areas, where cell switch-off becomes more difficult due to the risks of creating coverage holes, short-time cell-DTX and bandwidth adaptation are expected to be the most efficient solutions. At the network level, the use of cellular heterogeneous structures, and cooperative transmission from several base stations to one mobile device, are further enablers considered by EARTH.

The results of the technical studies will be validated in the operator testbed of a project partner and in system-level simulations. They will be combined into an integrated solution and the achieved energy saving will be quantified and demonstrated.

Expected impact

The EARTH project targets increased economic efficiency (cost/bit) of wireless access infrastructures. The largest energy-efficiency gains from EARTH solutions are expected to occur for the low-loaded parts of the networks.

The scientific results of the innovative energy-efficiency solutions developed by EARTH will be widely disseminated. Large parts will require amendments to standards or new standards in order to ensure multi-vendor interoperability and wide applicability.

The exploitation is one of the key interests of the EARTH partners. One of the major goals is timely commercialization of results by the industry and SMEs to enable roll-out of competitive new products and tools for cost-efficient operator networks.

For the equipment vendors, the project results are expected to include solutions that will be applied to base stations (in particular LTE eNBs), base station components, mobile terminal platforms, network management systems, and managed services.

Operators will exploit the defined network deployment and operation principles of emerging as well as existing networks. These principles are expected to deliver significantly lower energy cost compared to currently existing solutions.

17.3.3 GreenTouch initiative

GreenTouch [14] is a non-profit research consortium founded in 2010 by experts from industry, academia, government, and research institutions around the world. The consortium is open to all sectors of the ICT industry and academia and the list of its members is constantly increasing (more than 50 in February 2011), showing the interest the initiative has gathered over time.

GreenTouch targets the definition of new, clean-slate technologies that will be at the heart of sustainable communication networks in the decades to come. The possibility of concentrating on completely new solutions, natively designed with energy efficiency as main target, represents the fundamental characteristic of the GreenTouch initiative. The solutions proposed by the consortium are not constrained by any requirement of backward compatibility with existing networks. This allows efforts to be focused on the power-consumption issue, and to propose radically new techniques and approaches.

The need for such a drastic technology shift has been demonstrated by several scientific studies in the past. For example, in 2010 [15] used a combination of micro-economics (i.e. analysis of a specific known service and extrapolation of its evolution from market adoption trends) and macro-economics methods to estimate the size and evolving contributions to network traffic over the next decade. The study concluded that to support the expected traffic expansion the overall network power-consumption will continue to

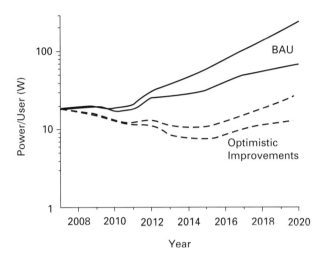

Figure 17.5 Total network power-per-user considering BAU trends and best-case (optimistic, dashed) efficiency improvements. High and low curves correspond to high- and low-traffic projections (from [GT1]).

grow if a business-as-usual approach (BAU) is taken. More troubling, even with an optimistic approach, ensuring a timely adoption of energy-efficiency improvements, is that the overall consumption will continue to grow, despite an initial reduction (see Figure 17.5). Without drastic modifications, ICT systems expansion will become unsustainable.

Starting from these conclusions the GreenTouch founders agreed on the need to fundamentally transform communications and data networks in order to significantly reduce the carbon footprint of devices, platforms, and networks. This translated into an aggressive goal: create the technologies needed to make communications networks 1000 times more energy efficient than they are today. An early goal for this initiative is to deliver, by 2015, a reference architecture, specifications, and demonstrations of key components needed to realize a fundamental re-design of networks that increase energy efficiency by 1000 times as compared to current levels.

Research activities and early results related to mobile networks

The GreenTouch research effort is structured on a classic access vs. core separation, further decomposed in wireline and wireless (mobile communications) working groups for the access network, and in optical and routing/switching working groups for the core network.

The mobile communications working group focuses on new, highly energy-efficient air interfaces, trade-offs between energy efficiency and mobile system design criteria like spectrum efficiency or service delay, power models of wireless access points, dynamic coverage and capacity management, and deployment and management strategies for energy-efficient mobile systems.

During its first year of work, the working group launched the three following collaborative research projects:

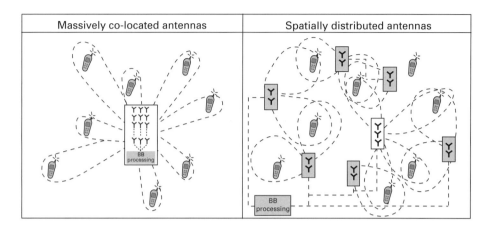

Figure 17.6 Massively co-located antennas vs. spatially distributed antennas.

- large-scale antenna systems (LSAS);
- green transmission technologies (GTT);
- beyond-cellular green generation (BCGG).

The LSAS project uses a very high number of active antennas to reduce the power requested for each radio link between the base station and the mobile terminals. The network architectures considered include the possibility to co-locate all the antennas in the same place to create extremely directive beams, or to distribute them in a given area and exploit spatial diversity with network-MIMO techniques (Figure 17.6). In both cases, the base-band processing is expected to be centralized to ease the synchronization and signal processing among all the radio heads.

In LSAS, the improvement of the wireless energy-efficiency comes through a combination of radiated power reduction and increased throughput (spectral efficiency). Beginning in 2011, an early stage proof-of-concept showcased that the total transmitted power over air can be reduced by a factor of two each time the number of antennas is doubled. The demonstrator had 16 co-located antennas, which enabled a reduction of transmitted RF power of 16 times. This technique is particular interesting for large (macro) base stations where the power consumption of the RF part is preponderant over the consumption of other parts (e.g. base-band processing).

The LSAS project addresses key research problems required to practically make large-scale antenna systems with up to a few hundred antennas. The main challenges for this project are:

- design the best architecture, as hardware and software partitioning between central and remote antenna units plays a crucial role in reducing radiated and processing energy expenditure;
- design the adapted algorithms needed, in particular for channel estimation and A/D conversion resolution;
- validate that theoretic gains can also be achieved in real-life deployments.

The GTT project focuses on the energy-efficient design of transmission schemes, radio resource management strategies, and signal processing algorithms based on the fundamental trade-offs in wireless communications. The improvement of energy efficiency and the reduction of energy consumption are usually associated with the cost in other operational metrics. This means the energy efficiency of the system has trade-offs, in particular between spectrum efficiency (SE) and energy efficiency (EE), and between service delay (DL) and power consumption (PW). The key issue is when, what, and how to trade off while maintaining satisfactory network service.

The announced stepped approach of GTT consists of (i) characterizing the identified trade-offs from Shannon's formula and then (ii) devising transmission schemes, resource management strategies, and baseband processing algorithms to realize the power saving promised by SE vs. EE and DL vs. PW trade-offs.

The third project, BCGG, aims to redefine the signaling and data channels in order to remove (or mitigate) their current limitations for dynamic energy management. In fact, today cellular technologies are all based on a continuous emission of broadcast signaling (e.g. system information, paging, synchronization, and pilots) to assist the terminals in the regular cellular procedures: initial network detection, attachment, mobility, reception of incoming call during idle mode, etc. Unfortunately, this continuous signaling emission sets hard limitations on the dynamic cell switch-on/off techniques. The principal idea of BCGG is thus to rethink these channels and to separate the signaling network from the data network. This allows the dynamic activation/deactivation of the data cells as a function of the network load, while maintaining the system availability by the always-on signaling layer.

Finally, it is worth mentioning that the mobile communications working group also established an architecture and metrics subgroup. It addresses general network energy-efficiency metrics suitable for various wireless architectures, and will create a roadmap for saving energy in future wireless access networks. The working group defined a framework for the calculation of energy-saving factors to be used by all projects in the working group. The framework is directly inherited from the EARTH project (E^3F).

17.4 Synthesis and classification of energy-saving solutions for wireless networks

By analyzing the different initiatives presented in the previous sections, it is possible to classify the most promising (and recurrent) techniques currently under investigation to improve the energy efficiency of wireless networks. In general, the considered solutions apply at several levels, from the basic technological building block to the system-level operation.

17.4.1 Technology and component level

Wireless networks should naturally benefit from the energy-efficient materials developed for the worldwide electronics industry.

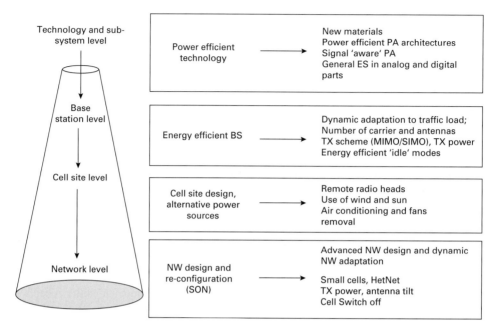

Figure 17.7 Representation of ES solution categories.

The power amplifier (PA) is responsible for a very significant part of the BS power consumption. The power efficiency of the PA has significantly increased during the 10 past years, thanks to digital pre-distortion techniques, and should further increase in the future, thanks to innovative PA architectures.

Signal-aware PAs should also play a significant role for further power-consumption reductions. The principle consists of ensuring that the PA always operates at its best efficiency point, even when transmitted RF power varies as a function of traffic load. These techniques enable significant savings in LTE, especially with a suitable MAC scheduler.

The last aspect concerns all digital parts of the base station such as baseband processing and control. Opportunities are offered to pass in idle mode the functions that are not used during periods of time. The saving potential is highly dependent on the digital architecture used, and may vary depending on the technological approach (DSP, ASIC or FPGA and combination).

Power-consumption reduction has a positive impact on the air-conditioning system whose power can be reduced, or even removed (passive cooling) in some cases.

17.4.2 Base station adaptation to traffic load

Ideally, the energetic cost per bit should be constant. The reality is still far from this objective; however several approaches enable some kind of proportionality between the power consumed and the traffic load. Several techniques are envisioned:

- use of a smart scheduler exploiting the PA capabilities (mentioned in the PA-aware section above); the scheduler should maximize the "no transmit" time when a PA switching capability is offered, or conversely, smooth the traffic when a PA adaptation capability is offered;
- adaptation of the number of transmit antennas depending on the traffic load; for example, modern base stations (BSs) use up to eight transmit antennas. When the traffic demand is low, it is possible to switch off some of the antennas/PAs and use a less spectrally efficient transmit mode;
- reduction of transmit band; for example, when several carriers are active in the same sector (e.g. a 3-carrier per sector WCDMA base station) a fall-back mode using only one carrier active allows sparing the transmission of all control overhead. A similar approach is possible with LTE, but instead of switching off carriers, narrower carriers are used (e.g. fall-back mode from 20 MHz to 5 MHz);
- power-efficient idle mode is also key when considering a low power BS that can be switched on and off, depending on the traffic demand.

17.4.3 Network architecture

Cloud-oriented architectures, where baseband processing is done centrally in baseband unit (BBU) pools, are currently enjoying a growing popularity. These architectures have significant energy-saving potentials. The main reasons are:

- Processing power is no longer dimensioned relative to the peak capacity of each BS, but on the basis of the capacity of the area covered by the cloud. This allows the benefit of statistical multiplexing gain: not all BSs in the given area are loaded at peak capacity at the same time. The gain in the amount of processing power is estimated at 60% for large clusters.
- Load balancing between BBUs is possible, allowing the activation of only those that are required. During low traffic periods, it is possible to switch off a large portion of the BBUs.
- Significant savings are obtained on thermal regulation. A single air-conditioning system is needed in the BBU pool. However, with massively distributed BB processing, the reduced need for processing at each site may allow passive cooling.
- Frequency of on-site inspections can be reduced resulting in less CO_2 emissions due to less frequent visits to sites.

17.4.4 Heterogeneous networks

It is a well-known theoretical result that the required transmitted RF power per cell depends on the cell size with a relation $P \approx R^4$, while the number of cells required increases as $1/R^2$. There is therefore a gain from an RF energy-consumption standpoint, when reducing the cell sizes.

In practice, cell sizes will not be reduced to zero. Practical trade-offs have to be found between costs (CAPEX/OPEX), throughput offered, and power requirements. A practical

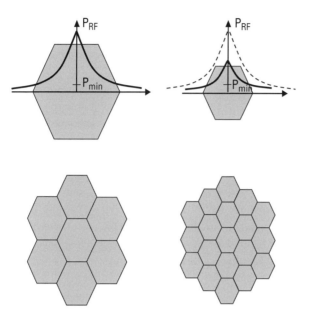

Figure 17.8   Relation between power and cell size.

compromise consists of mixing macro- and micro-cells, and locating micro-cells where the traffic demand is higher.

17.4.5   Air interface

To face the quickly growing traffic demand, in a situation where the spectrum resource is scarce, important efforts are made to increase the air interface spectral efficiency and decrease the energetic cost per bit. As a result, 4G will provide much higher throughput with a limited power-requirement increase in base stations and user equipment.

The key air-interface techniques aiming at spectral efficiency and coverage improvement being standardized in LTE are:

- multiple antenna transmission and reception ports, including multiple-input multiple-output (MIMO), for spectral efficiency and coverage improvements;
- coordinated multi-point (CoMP) or network MIMO, which can be seen as an extension of MIMO where the multiple antennas participating to the transmission with UEs are located on several BS sites. The key improvement brought by CoMP is a more uniform coverage. From an energy standpoint, CoMP allows a reduction of UE and BS transmit power, as well as the need to create additional sites to solve coverage holes issues.

It is also relevant to cite techniques that were introduced from 3G, aiming at improving the link budget, such as turbo codes, as well as all technological improvement, that make possible the implementation of more optimal algorithms for channel estimation and demodulation

17.4.6 Dynamic NW adaptation to traffic load

4G should mark the advent of massive deployment of small cells and home base stations. Obviously, there is a high variability in the number of users connected to a small cell, and there are large portions of time (e.g. at night) where no user is connected. Small cells are generally located under a macro-cell umbrella, and their role is to absorb the traffic in hotspots when the traffic demand is high. Being able to switch on and off the small cells base stations is a key aspect for energy saving in 3G and 4G networks. This power management has to be done without degradation of user experience, and without creating coverage holes. The 3GPP has extensively studied the solutions to enable efficient power management, from statically preprogrammed to fully dynamic. Centralized RAN architecture, that should be deployed in the coming years, have high power saving potentials, resulting from the combined gains of:

– Reduction of the number of baseband processing power, thanks to pooling effect
– Adaptation to the traffic load: in low load periods some processors can be switched off
– Power saving on temperature regulation, due to the fact that all baseband equipment are co-located.

17.5 Conclusion

Given the importance of energy consumed in wireless networks, and the potential improvements identified, a large number of initiatives have been launched in standards and fora, and several consortiums and projects have been formed.

A first aspect of the activity concerns the definition of energy-consumption measurement methods. All major standardization bodies have launched activities in their organization, and already produced specification documents.

Another aspect concerns the definition of features that enable energy saving in networks. Due to the backward compatibility requirements, improvements on the air interfaces are limited. The main activities concern the activation and deactivation of base stations depending on traffic load. The main challenge is to make these imperceptible to users.

Several projects and consortiums are preparing for the longer term with much higher energy-efficiency improvement targets. The most promising techniques are the use of many antennas, small cells, and heterogeneous networks. However, the development of these concepts will be heavily dependent on the technology availability at an affordable price.

References

[1] STEP-TEE-2011-040R2 (ATIS) Energy efficiency for telecommunication equipment: methodology for measurement and reporting of radio base station metrics.
[2] ETSI TS 102 706 v1.3.1 (2010-12) Environmental engineering (EE) energy efficiency for wireless network equipment.

[3] CCSA, ICS 33.060 Energy efficiency metrics and measurement method for mobile communication equipment base station.
[4] 3GPP TR 36.927 V10.0.0 (2010-03) Technical specification group radio access networks; Evolved Universal Terrestrial Radio Access (E-UTRA); potential solutions for energy saving for E-UTRAN (Release 10). [Online]. Available: www.3gpp.org/ftp/Specs/html-info/36927.htm
[5] 3GPP TR 32.826 V10.0.0 (2010-03) Technical specification group services and system aspects; telecommunication management; study on Energy Savings Management (ESM) (Release 10). [Online]. Available: www.3gpp.org/ftp/Specs/html-info/32826.htm
[6] 3GPP TR 24.826 V2.0.0 (2011-05) Technical specification group core network and terminals study on impacts on signaling between UE and core network from energy saving (Release 11). [Online]. Available: www.3gpp.org/ftp/Specs/html-info/24826.htm
[7] www.ngmn.org
[8] NGMN Alliance, "NGMN top OPE recommendations," September 2010. [Online]. Available: www.ngmn.org
[9] www.ict-earth.eu, European Community's Seventh Framework Programme FP7/2007-2013, grant agreement n. 24773
[10] L. Sanchez et al., "EARTH: paving the way for future energy efficient broadband wireless networks," *MONAMI 2010,* Santander (Spain), Sept. 2010.
[11] L. M. Correia et al., "Challenges and enabling technologies for energy aware mobile radio networks," *IEEE Communication Magazine,* Nov. 2010.
[12] B. Hassibi and B. M. Hochwald, "How much training is needed in multiple-antenna wireless links?" *IEEE Trans. Info. Theory,* vol. 49, no. 4, pp. 951–963, Apr. 2003.
[13] W.H. Doherty, "A new high efficiency power amplifier for modulated waves," *Proceedings of the Institute of Radio Engineers,* vol. 24, no. 9, pp. 1163–1182, Sept. 1936.
[14] www.greentouch.org
[15] D. C. Kilper et al., "Power trends in communication networks," *IEEE Journal of Selected Topics in Quantum Electronics,* vol. 17, no. 2, 2011.

Index

3rd Generation Partnership Project (3GPP), 190, 386, 389
   3GPP TR 32.826, 390
   3GPP TR 24.826, 390
   3GPP TR 36.927, 390
   3GPP TS 32.551, 390
   Release 10, 389
   Release 9, 389
3rd Generation Partnership Project 2 (3GPP2), 386, 394

adaptive modulation and coding (AMC), 292
Alamouti code, 154
amplify-and-forward (AF), 68, 130, 134, 150, 291
antenna interface, 363
antenna tilting, 215
ATIS (Alliance for Telecommunications Industry Solutions), 395
automatic repeat request (ARQ), 100, 152

bandwidth efficiency, 103
base station power control, 212
baseband interface, 364
Bernoulli injection model, 54

capital expenditure (CapEx), 10
CCSA (China Communication Standard Association), 396
cell breathing, 215
cell splitting, 236, 238
channel state information, 54, 55, 61
coding gain, 120
cognitive pilot channel (CPC), 171
cognitive radio, 237, 289
constant available power, 63
constant bit rate, 62
constant-power slot, 61, 62
constrained Markov decision process (CMDP), 275
cooperative communications, 67
cooperative network (CoopNet), 20
cooperative relaying, 214, 215, 289
cooperative routing, 255
coordinated beamforming (CBF), 248

coordinated multi-point (CoMP), 19, 143, 287, 406
cross-layer adaptation, 266

decode-and-forward (DF), 130, 151, 291
decoupling approximation, 71
delay-tolerant network (DTN), 78, 79
demand-response, 222
demand-side management (DSM), 211
deployment efficiency, 10
discrete time Markov chain, 56
distributed antenna array (DAA), 240
dynamic time slot allocation (DTSA), 313

EMAN (Energy MANagement) project, 395
energy-constrained relay, 70
energy efficiency, 6, 190, 289
energy efficiency evaluation framework (E^3F), 360
energy harvesting, 53
energy harvesting profile, 54
energy neutrality, 54, 63
energy saving, 288
energy unconstrained, 56
energy-unconstrained relay, 70
European Telecommunications Standards Institute (ETSI), 192, 386, 387

finite-state Markov channel (FSMC), 268, 292
fixed relay station (FRS), 310
fixed time slot allocation (FTSA), 313
FP7 EARTH project, 4, 359, 386, 398
frequency-shift keying (FSK), 100

green computing, 212
green energy-efficient environment (GREEN) initiative, 394
green energy environment (GREEN) initiative, 386
green modulation/coding (GMC), 100
GreenTouch, 400
GSM alliance (GSMA), 386

heterogeneous network (HetNet), 20
hybrid ARQ, 132, 153, 293

Index

IEEE 802.15.4, 108
IETF, 386
incremental amplify-and-forward (IAF), 135
incremental decode-and-forward (IDF), 132
intercell interference, 237

leaky bucket model, 54
Long-Term Evolution (LTE), 190
LP relaxation algorithm, 296

M-ary FSK, 103
M-ary QAM, 106
M-ary quadrature amplitude modulation (MQAM), 103
macrodiversity, 143
Markov decision process (MDP), 273
Markov Modulated Poisson Process (MMPP), 269
max-min fairness, 213
mechanical relaying, 81
metric-to-timer mapping, 69
mobile Internet, 83
Mobile VCE Core 5, 175
mode switching, 214
MPSK, 71
multicell beamforming (MBF), 243
multicell processing (MCP), 236
multi-hop cellular network (MCN), 309
multiple access selection algorithms, 69

network coding, 151
    random network coding (RNC), 151
network MIMO, 406
NGMN (Next Generation Mobile Networks) alliance, 396

offset-QPSK, 108
operational expenditure (OpEx), 10
outage probability, 57, 59
outer stage, 64
outer stage power management algorithm, 61

partially observable Markov decision process (POMDP), 276
power amplifier, 216
power amplifier (PA), 190, 363, 364
power-aware cooperative routing (PACR), 255
power-consumption model, 218
power control, 61

power efficiency, 365
power management, 61, 210
power-saving routing (PSR), 256

quality-of-experience (QoE), 84

radio access network (RAN), 395
radio access technology (RAT), 168
random relay station (RRS), 311
receiver based auto-rate (RBAR) technique, 293
relay selection, 291
relay-to-destination (R2D) channel, 294
remote radio head (RRH), 190, 363
resource allocation, 336
restless bandit problem, 298
restless bandit system, 292
retransmissions, 55

selective decode-and-forward (SDF), 131
semi-definite programming (SDP), 248
signal-to-noise and distortion ratio (SiNAD), 364
smart base station, 214
smart grid, 210, 220, 222
source-to-relay (S2R) channel, 294
spatial diversity, 68
spectrum efficiency, 6
    (SE), 403
splitting-based selection algorithm, 69
statistical energy neutrality, 63
stochastic programming, 224
storage buffer, 54
store-and-forward (SF) protocol, 152
symbol error rate, 71

TDD-CDMA, 312
telecommunication efficiency ratio (TEER), 395
TIA (Telecommunications Industry Association), 394
timer algorithm, 69
transition probability matrix, 58

variable harvested power, 64
video-on-demand, 85

wireless multicast, 155

ZigBee, 108